Integrated Circuit Design for Radiation Environments

Integrated Circuit Design for Radiation Environments

Stephen J. Gaul
Renesas Electronics Americas, Inc.
USA

Nicolaas van Vonno
Renesas Electronics Americas, Inc.
USA

Steven H. Voldman
Consultant and IEEE Fellow
USA

Wesley H. Morris
Silicon-X Corporation
USA

This edition first published 2020
© 2020 John Wiley & Sons Ltd

The right of Stephen J. Gaul, Nicolaas van Vonno, Steven H. Voldman, and Wesley H. Morris to be identified as the authors of this work has been asserted in accordance with law.

Registered Offices
John Wiley & Sons, Inc., 111 River Street, Hoboken, NJ 07030, USA
John Wiley & Sons Ltd, The Atrium, Southern Gate, Chichester, West Sussex, PO19 8SQ, UK

Editorial Office
The Atrium, Southern Gate, Chichester, West Sussex, PO19 8SQ, UK

For details of our global editorial offices, customer services, and more information about Wiley products visit us at www.wiley.com.

Wiley also publishes its books in a variety of electronic formats and by print-on-demand. Some content that appears in standard print versions of this book may not be available in other formats.

Library of Congress Cataloging-in-Publication Data Applied for

ISBN: 9781119966340

Cover Design: Wiley
Cover Image: © Petrovich9/Getty Images

Set in 10/12pt WarnockPro by SPi Global, Chennai, India
Printed and bound in Singapore by Markono Print Media Pte Ltd

10 9 8 7 6 5 4 3 2 1

Contents

About the Authors

Stephen J. Gaul received his B.S. and M.S. in Electrical Engineering Science from Iowa State University in 1980 and 1982, respectively. His undergraduate study focused on fields and waves, analog circuit design, and physics. His research during graduate study was supported by a grant from 3M (Minneapolis, Minnesota).

In 1983 Gaul joined Harris Semiconductor in Palm Bay, Florida. Harris Semiconductor was part of Harris Corporation, with many divisions, and Steve worked sequentially through several. These included the CMOS division as a process engineer, Bipolar division developing device simulation models, and the Custom Integrated Circuit Division working as a device engineer on a variety of programs to develop radiation hardened and other analog solutions. Several of these programs included process as well as circuit developments.

In the late 1980s, Gaul was doing research on bonded-wafer silicon on insulator (SOI) as a replacement for dielectric isolation (DI) wafer material processing. He fabricated the first bonded wafer bipolar devices at Harris Semiconductor in 1989 and also developed bonded-wafer material process flows and trench isolation for the first such analog components fabricated at Harris Semiconductor in the early 1990s. By the mid 1990s, Gaul had worked on a number of projects, including improving defect density in bonded/trenched material processing and solving an electro-chemical mismatch issue on circuits using thin film resistors. Additionally, he worked on yield enhancement and migration of DI telecom circuits to bonded wafer substrates.

Gaul joined Unitrode in Merrimack, NH, in 1995 to develop a bonded-wafer trench isolation process as a cost improvement for existing bulk isolated processes. He improved the yield of this process from 0% to greater than 90% in less than six months. Gaul also investigated safe operating area (SOA) issues in output devices in a 20 V BiCMOS process fabricated at an external foundry.

In 1997 Gaul returned to Harris Semiconductor, working on yield and process enhancement on submicron BCDMOS processes. He completed in-house college-level classes in RF circuit design and DSP in 1998, and his work at this time increasingly brought him into contact with circuit design tools and systems. By 1999, Gaul had taken a position as a process design kit (PDK) developer with Harris's design automation team.

Harris Semiconductor became a standalone company in late 1999 called Intersil (based on the Santa-Clara company that was part of the 1988 Harris Semiconductor acquisition of GE Solid State), and Gaul turned to managing PDK and Design Systems for all circuit design teams at Intersil. He pioneered the use and repurposing of design system code for devices (pcells) providing rapid development of PDKs to keep pace with the corresponding rapid development of newer submicron processes. His process and device background was leveraged to improve electrostatic discharge (ESD) performance in existing processes and to cost-improve an in-house 0.6-μm semiconductor process using a software shrink of 10%; this shrink was instrumental in bridging the design community to a smaller geometry BCDMOS that was in development, with a large cost saving to Intersil.

Gaul currently manages the engineering design infrastructure at Renesas Electronics Americas Inc. (REA), which purchased Intersil in 2017. His interests include radiation effects, circuit hardening, IP security, export law, and small-town government. Gaul has written and presented 17 papers at refereed conferences, including the SOI Technology Conference, RHET, IEEE Nuclear and Space Radiation Effects Conference (NSREC), and the Journal of Radiation Effects Research and Engineering. He holds 40 US issued patents in a variety of areas, from semiconductor process flows, device design, ESD devices, radiation hardened devices, methods of forming SOI materials, improvements in trench isolation, 3D circuits, silicon on diamond (SOD), engineering design automation (EDA) software, and circuit solutions.

Nicolaas van Vonno has a strong background in the design, development, and production of high-performance semiconductor integrated circuits, predominantly for radiation-hardened space and defense applications. Specific areas of expertise include analog and mixed-signal technology, radiation effects and semiconductor device radiation hardening.

Van Vonno received his Bachelor of Science degree in electrical engineering from the University of Florida in 1966. He then joined Harris Semiconductor (then Radiation, Incorporated, which was subsequently acquired by Harris Corporation) as a reliability engineer, with an initial assignment in reliability assurance for the US Navy Polaris and Poseidon SLBM programs, gaining experience in radiation effects, hardening technology, packaging, device fabrication, and quality assurance. Later, he transferred to a device engineering group, where he was responsible for device design, process development, and design/layout support. During this assignment, he participated in a broad range of projects reflecting Harris's support of many diverse technologies, including early metal-gate CMOS and silicon-on-sapphire (SOS) processes. As a device engineering supervisor, he led a group of device designers, with representative projects including MOS hardness issues and radiation testing in the total ionizing dose, neutron and transient gamma environments.

In 1982, Van Vonno was promoted to director, Analog Product Development, with responsibilities in the design, development, and transfer to production of analog circuits for tactical, strategic, and space applications. Projects included the development of a broad range of analog parts for the Trident II SLBM program and development of a series of hardened analog signal processing circuits for the MX (Peacekeeper) program, as well as the development of mixed-signal chip sets for tactical applications and the development of mixed-signal design and verification methodologies, with Cadence as a technical partner. Other projects included the development of a radiation-hardened cryogenic analog CMOS process for hardened 77 K and 40 K applications and the design of hardened readout multiplexer chips for use in directly hybridized HgCdTe IR focal plane array imagers.

Van Vonno was promoted to senior scientist in 1994, with responsibilities in radiation effects in electronics, technology development, and advanced projects. He then led the development of an innovative integrated fingerprint sensor using RF field technology. This project culminated in the successful spinoff of Authentec, Incorporated from Harris Semiconductor, and the eventual sale of Authentec, Inc. to Apple. The RF fingerprint sensor is currently a key feature of the Apple iPhone.

Van Vonno is currently a principal engineer with Intersil, a Renesas Company, Palm Bay, Florida. In this capacity, he is responsible for customer support and radiation effects research for the space product line, with added assignments in product development, legal support, and export control issues. He is an active researcher in enhanced low-dose rate ionizing radiation sensitivity (ELDRS) phenomena in complementary bipolar processes. He has been an active participant in the JEDEC community, including the JEDEC JC13.4 Radiation Effects Assurance committee. He is an active participant in business development and customer support activities in Europe and is fluent in several languages.

Currently, Van Vonno holds 16 US patents in the areas of silicon device processing, optics, packaging, and biometric sensing. He has over 60 refereed publications and conference presentations and is a Life Senior Member of IEEE. He has participated in a wide range of professional activities, including IEEE and international conference assignments. He has been an active participant in the IEEE Nuclear and Space Radiation Effects Conference (NSREC), with assignments including poster session chairman, guest editor, session chairman, awards committee chairman, short course instructor, and short course chairman. Other IEEE activities include two three-year terms as a member at large of the IEEE Radiation Effects Steering Group (RESG) and three years as vice-chairman, publications, of RESG. He served as chairman of the 2001 IEEE Radiation Effects Data Workshop and was chairman of the local arrangements of the 2006 Nuclear and Space Radiation Effects Conference. Van Vonno received the IEEE Radiation Effects Award in 2009.

Dr. Steven H. Voldman received his BS in Engineering Science from the University of Buffalo (1979); a first M.S. electrical engineering (EE) (1981) from Massachusetts Institute of Technology (MIT); a second M.S. EE degree from MIT; an MS in Engineering Physics (1986), and a PhD in EE (1991) from the University of Vermont under IBM's resident study fellow program.

In 1977 and 1978, Dr. Voldman interned at the Robert E. Ginna nuclear plant as support for the nuclear engineer. He was a member of the MIT Magnetic Mirror Fusion experimental research group with a specialization in plasma physics from 1979 to 1981. In 1982, he was the first person in the IBM corporation to address alpha particle effects in bipolar static ram (SRAM) applications. Dr. Voldman worked with G. Sai-Halasz on the development of the first SER alpha particle Monte Carlo simulation for the bipolar SRAM products. In 1984, he was assigned to IBM laboratory staff to develop a six-device cell CMOS SRAM SER Monte Carlo alpha particle simulation. In 1985, he was a member of the proton and neutron cosmic ray testing effort of DRAM and SRAM IBM products to evaluate the effect of cosmic rays on terrestrial applications. In 1984, Dr. Voldman also worked on CMOS latchup for the IBM technologies. He has been involved in latchup technology development for 30 years. He worked on both technology and product development in Bipolar SRAM, CMOS DRAM, CMOS logic, SOI, BiCMOS, Silicon Germanium (SiGe), RF CMOS, RF SOI, smart power, and image processing technologies.

In 2008, Dr. Voldman was a member of the Qimonda DRAM development team, working on 70, 58, and 48 nm CMOS technology. That year, he also initiated a limited liability corporation (LLC), and worked at headquarters in Hsinchu, Taiwan, for Taiwan Semiconductor Manufacturing Corporation (TSMC) as part of the 45 nm ESD and latchup development team. From 2009 to 2011, he was a senior principal engineer working for the Intersil Corporation on ESD and latchup development working on analog and power applications. His work at Intersil involved semiconductor development of analog and digital mixed signals, SiGe, bipolar-CMOS-DMOS (BCD) power, and power SOI technologies. In 2013, Dr. Voldman was a member of the Samsung Electronics development team working on latchup development in 14 nm technology.

Dr. Voldman is the first IEEE Fellow in the field of ESD for "Contributions in ESD protection in CMOS, Silicon On Insulator and Silicon Germanium Technology." He was chairman of the SEMATECH ESD Working Group, from 1995 to 2000. In his SEMATECH Working Group, the effort focused on ESD technology benchmarking, the first transmission line pulse (TLP) standard development team, strategic planning, and JEDEC-ESD Association standards harmonization of the human body model (HBM) standard. From 2000 to 2010, as chairman of the ESD Association Work Group on TLP and very-fast TLP (VF-TLP), his team was responsible for initiating the first standard practice and standards for TLP and VF-TLP. He has been a member of the ESD Association board of directors and education committee. He initiated the "ESD on Campus" program, which was established to bring ESD lectures and interaction to university faculty and students internationally. The ESD on Campus program has reached over 40 universities in the United States, Singapore, Taiwan, Malaysia, The Philippines, Thailand, India, China, and Africa.

Dr. Voldman has taught short courses and tutorials on ESD, latchup, and invention in the United States, China, Singapore, Malaysia, and Israel. He holds more than 260 US patents, in the area of ESD and CMOS latchup. He has served as an expert witness in patent litigation cases associated with ESD and latchup.

Dr. Voldman has also written articles for *Scientific American* and is an author of a 10-book series on ESD, electrical overstress (EOS), and latchup, including: *ESD: Physics and Devices, ESD: Circuits and Devices, ESD: Radio Frequency (RF) Technology and Circuits, ESD: Failure Mechanisms and Models, and ESD: Design and Synthesis, Latchup, EOS, ESD: Analog Design and Synthesis, ESD Basics: From Semiconductor Manufacturing to Product Use,* and *ESD Testing: Components and Systems,* as well as contributing to the book *Silicon Germanium: Technology, Modeling and Design,* and international Chinese editions of the ESD books. He is also a chapter contributor to the text *Nanoelectronics: Nanowires, Molecular Electronics, and Nano-Devices.*

Wesley H. Morris received his BS in Physics and Chemistry (dual majors) from Florida State University in 1977.

In 1981, Morris began his career at RCA Solid State as a semiconductor device engineer and worked there until 1986. His first job involved transferring a special Rad Hard manufacturing process developed at RCA Princeton labs from the Somerville Technology center to RCA's wafer manufacturing lab in Florida to begin manufacturing Rad Hard standard chip products. Morris successfully integrated and implemented the Rad Hard process and production line in Florida to begin manufacturing RHSOS products using silicon on sapphire (SOS) wafers. RCA began designing and producing SOS Rad Hard SRAMS, microprocessors (1802, 1804) and ASICs using the Rad Hard SOS wafer manufacturing process. Several RH SOS devices were selected and flown on JPL's Galileo and Jupiter missions and other NASA/DOD satellite systems, including the GPS satellite (MilStar) and first DBS (Direct TV) satellites included with RH SOS components in the satellite electronic payloads. In 1986, RCA awarded Morris the RCA solid-state division award for technology excellence for his development work with the RH SOS project. The RH SOS development team were also nominated for RCA's Sarnoff award.

In 1986, Mr. Morris joined Harris Semiconductor as a principal engineer in Palm Bay, Florida, and was assigned to the military and aviation (MAD) division supporting the development of (novel) SOI wafers to develop a radiation hard manufacturing process using the silicon on insulator (SOI) wafer. At the time, RH CMOS SOI products were produced using SIMOX SOI technology that use ion implantation to form the buried oxide layer.

In 1988, after successfully developing and demonstrating fully functional 0.7 µm RH SOI SRAM memories, he moved to Austin, Texas, to represent the company at SEMATECH, which was a USA semiconductor industry research consortium formed by the top 10 US semiconductor companies to focus on improving silicon manufacturing tooling. His first assignment was to transfer IBM's 4Mb DRAM back end of line (BEOL) process from IBM to SEMATECH in Austin. After that, he was put in charge of TCAD modeling for SEMATECH process developments supporting phase 2 and 3 manufacturing processes, as well as working with SEMATECH'S wafer Fab and the electrical device test lab to set up integration and tool selections SEMEATECH would use for all the manufacturing and device test operations.

In 1992, Morris proposed and initiated the Vertically Modulated Well project, which was to develop a new approach to wafer doping and wafer manufacturing architecture to simplify the CMOS manufacturing process and reduce cycle time, wafer manufacturing costs, and advance CMOS device isolation scaling. The problem was, legacy methods used for bulk silicon isolation needed to prevent latchup as CMOS scaling evolved to increase latchup resistance. Wafer production costs were also barriers. In 1995, the VMW project was successfully integrated into two SEMATECH member companies, AMD and NCR (Lucent), and robust latchup mitigation method and lower wafer costs were proven out. It was a good start at the right time.

In 1993, Morris founded Silicon Engineering Inc., which provided CMOS engineering (consulting) services to both SEMATECH and the SEMI industry for TCAD modeling, process, and chip developments. He began working with Eaton Corporation in Beverly, Massachusetts, in 1995, supporting EATON's global launch of the high-energy implant tools and the HDBL concept, which was developed at SEMATECH. He provided a technical interface to Eaton's customers and promoted adoption of the high-energy implant technology and sales (EATON/Axcelis) of implant tools at foundries in SE Asia (Taiwan, South Korea, Singapore), and, the United States and Europe.

In 2003, Morris was invited to publish new work at the IRPS conference and presented data for the first time that by implementing a novel boron doping structure called the buried guard ring; any bulk silicon CMOS device could be made latchup immune. Up to this time, this was considered impossible for bulk wafers and only possible with SOS or SOI thin-film technology. BGR was a better structure than the HDBL.

In 2004, Morris founded Silicon Space Technology, serving as CEO, CTO, and inventor of the company's RH/HT process and device IP. During this time, he led the company developments of a RH/HT technology platform based on modifications to the manufacturing process and IC design architecture to manufacture RH/HT standard products using commercial foundries. The technology developed enabled hardening of advanced (bulk) CMOS devices for operation in both radiation and high-temperature environments. Highlights of the work were published at the HiTEN conferences in 2011, 2014, and 2015. SST proved that bulk CMOS devices could be hardened and manufactured in commercial foundries to operate reliably in both high-radiation and high-temperature environments.

In 2014, SST provided multiple RH/HT IC components and also a company test board design to develop an experimental flight controller (board) co-developed with COSMAIC for the RHEME project. Using only RH/HT devices (SRAMS and Microcontrollers, and ASIC), the flight boards would be launched in three different Earth orbits aboard USAF/NASA experimental satellites. RHEME 1 was launched in March 2017 and mounted outside the International Space Station directly in the space environment. By June 2019, RHEME 1 had completed 806 days in space without any component or system failures. A paper summarizing results for the RHEME 1 flight was published at the 2018 NSREC conference. RHEME 2 was launched in December 2018 aboard an AF Satellite into an MEO/Polar orbit, again mounted outside the satellite (directly in space). RHEME 3 is scheduled for launch in December 2019. His work at SST proved that bulk silicon wafers could be designed and manufactured at modern wafer fabs (IDM or foundry) using IP he invented to modify the processes to withstand extremely high levels of radiation and high temperatures > 250C, and findings were published at IMAP'S HiTEN conferences in 2011, 2014, and 2015.

In 2017, Wes Morris founded Silicon-X Corporation and is currently active as technology consultant.

Preface

The world has certainly changed since the first satellite was launched into orbit (*Sputnik*, October 4, 1957). Now, many nations as well as commercial companies have launch capabilities, and there is an ever-increasing number of objects in Earth's orbit. The Union of Concerned Scientists (UCS) Satellite Database provides a good estimate of the current satellite count based on open-source information, indicating over 1900 operational satellites currently in orbit around the Earth (1957 as of November 30, 2018). While there are satellites that serve military purposes, the majority serve commercial purposes – purposes like GPS, communications, and weather monitoring. All of these orbiting objects rely on radiation-hardened electronics in order to stay operational in the space environment.

Designing for the space environment or any number of other radiation environments requires skills and knowledge that design engineers are not likely to encounter or learn while designing for terrestrial purposes. *Integrated Circuit Design for Radiation Environments* fills the knowledge gap, providing a front-to-back guide for designing integrated circuits intended for radiation environments. The text is divided roughly into three sections. The first of these includes Chapters 1 through 4 and provides the fundamental background history, the sources of radiation, essential nuclear physics, and basic semiconductor effects knowledge. Chapters 5 and 6 are aimed at two of the practical aspects of designing for radiation environments, radiation testing, and the development of device models. The remaining chapters are devoted to radiation-hardening topics such as semiconductor processing and layout, SEU reduction, latchup prevention, and a look at some advanced topics.

Chapters 1 through 4 of *Integrated Circuit Design for Radiation Environments* cover the historical aspects of radiation as well as the fundamentals needed by the practitioner. Chapter 1 provides the reader with a narrative of the role of radiation, since its discovery highlighting its impact on the evolution of integrated circuits and their use in radiation environments. The various radiation environments are presented in Chapter 2, starting with an overview of the relevant atomic and nuclear physics. The aim of this chapter is fulfilled as the various radiation environments are then discussed in turn. Chapters 3 and 4 cover the cumulative and transient effects of radiation, respectively, on semiconductor materials and structures. These are two very important chapters, as they cover essential fundamentals for anyone designing or evaluating radiation-hardened integrated circuits. Chapter 3 introduces the idea that radiation loses energy as it traverses the solid state. The various energy loss mechanisms for charged particles, electrons, neutrons, and photons traversing matter are discussed in the first half of the chapter. The remaining half of Chapter 3 discusses charge trapping and transport in silicon dioxide, including bulk and interfacial effects, two very important concepts for understanding the effects of total ionizing dose (TID) in integrated circuits. Transient effects caused by cosmic rays and other sources of radiation are a major concern in the space environment, and these are presented and discussed in Chapter 4. There are many types of transient effects besides the ambiguous yet ubiquitous single-event upset (SEU) description, including multiple-bit upset (MBU), single-event transient (SET), single-event functional interrupts (SEFIs), single event disturb (SED), single-event snapback (SESB), single-event latchup (SEL), single-event burnout (SEB), single-event gate rupture (SEGR), and single-event hard error (SHE). All of these are discussed in Chapter 4.

Simulation is the main subject discussed in Chapters 5 and 6, although from two completely different aspects. The ability to simulate radiation environments for testing and qualification purposes is covered in Chapter 5. The types of simulated environment needed for testing depends on the actual radiation environment, but can include TID, single-event effects (SEEs), neutron or proton irradiation, and transient gamma testing. The simulated radiation environments as well as the equipment and facilities available for testing are discussed. A review of circuit simulators, device models, and post-radiation device characteristics is provided in Chapter 6. The radiation effects on various devices are discussed after an introductory section covering circuit simulators, intrinsic models, and device modeling.

Chapters 7 through 9 take a closer look at radiation hardening from the viewpoints of semiconductor process and layout techniques, SEU reduction, and latchup suppression. Chapter 10 discusses radiation effects in advanced technologies. A good review of semiconductor process and layout-hardening methods is provided in Chapter 7. Bulk CMOS radiation hardening can include solutions in the substrate, epitaxial region, wells, isolation, and buried layers. Processes that passively isolate devices from the bulk substrate include silicon on insulator (SOI), silicon on sapphire (SOS), and silicon on diamond (SOD). Chapter 8 reviews circuit solutions for radiation hardening in digital architectures, primarily memories. The chapter reviews bipolar SRAMs, CMOS DRAMs, CMOS SRAMs, CMOS DRAM, arrays with CMOS SRAM error correction code (ECC), and CMOS SRAM cells with redundancy. Chapter 9 discusses circuit solutions to address CMOS latchup, including input circuit solutions, latchup prevention circuits, latchup detection networks, and power supply current limitation techniques. Finally, Chapter 10 looks at radiation effects in emerging technologies, including terrestrial as well as space-based systems.

The study of radiation effects and radiation effects hardening in electronic circuits is indeed a deep dive, considering the efforts of the many researchers who have contributed to the current state of the art. *Integrated Circuit Design for Radiation Environments* is intended to be a starting point for study aimed at novice as well as experienced circuit designers. While it is impossible to cover all of the prior research in one text, the authors hope that this text provides the essential knowledge about radiation environments, effects, and hardening needed by the circuit design community.

Enjoy the text,

Stephen J. Gaul
Nicolaas van Vonno
Steven H. Voldman
Wesley H. Morris

Acknowledgments

The authors are thankful for the years of experience and support from companies such as Radiation Inc., Harris Corporation, RCA, Intersil, and IBM, that were early pioneers in the production of radiation-hardened integrated circuits.

The authors are also thankful for the institutions that support and promote the radiation effects community by publishing journals, hosting conferences, and promoting learning through lecture series, invited talks, and tutorials. Conferences and symposia hosted by the IEEE, including the Nuclear Space and Radiation Effects Conference (NSREC), International Reliability Physics Symposium (IRPS), and the International Physical and Failure Analysis (IPFA) Symposium, as well as the EOS/ESD Association's Electrical Overstress/Electrostatic Discharge (EOS/ESD) Symposium provided both inspiration and motivation for the development of this text.

The authors acknowledge the gift of radiation effects knowledge provided by the many researchers and scientists that came before us as well as those that are our colleagues. We hope that we properly cited and recognized your works.

And finally, the authors would like to thank the publisher and staff of John Wiley and Sons Ltd. for their support for this text.

Stephen J. Gaul
Nicolaas van Vonno
Steven H. Voldman
Wesley H. Morris

Glossary of Terms

A

alpha particle An alpha particle is a helium nucleus (α) consisting of two protons and two neutrons bound into a particle identical to a helium nucleus. The alpha particle is generally produced during the process of alpha decay of radioactive material, however, alpha particles can also be produced by a silicon nucleus spallation event that is an inelastic collision of a natural cosmic particle (neutron) and the silicon nucleus. Alpha particles emitted from spallation events are energetic and can cause soft errors in silicon devices. As a proton carries one unit of charge, the alpha particle is doubly ionized. In CMOS silicon devices the figure of merit relating natural neutron flux (sensitivity to failure) is the FIT (failure in time). FIT reference is used to quantify the number of single bit failures of a silicon device located at other altitudes to a reference altitude (sea level at New York City in time). FIT units are the number of bit failures per 1 billion neutron particles/cm^2 passing through the silicon device.

analog design The design of continuous time circuitry in which data values are expressed as voltage or current levels. This is in contrast with digital circuitry, in which data values are expressed in binary format.

application specific integrated circuit (ASIC) An integrated circuit that is designed and tested for a specific, narrowly defined application in a subsystem or system. An ASIC is typically not readily suited for other applications.

atmospheric neutron A neutron generated by the interaction of high energy ions ("cosmic rays") with the upper layers of the atmosphere. These nuclear reactions generate fission fragments including neutrons, which cause further nuclear interaction and result in a downward shower of neutrons and other particles. These interactions and resulting showers are the only practical way to detect and characterize high-energy ions of solar or galactic origin.

avalanche breakdown A reverse-biased p-n junction breakdown phenomenon in which a current multiplication process leads to very large reverse currents. The multiplication process occurs when carriers in the depletion region are accelerated by the applied electric field to energies sufficient to create electron–hole pairs via collisions with bound electrons.

avalanche multiplication The current multiplication process outlined in *avalanche breakdown*.

B

ballast resistor   A small-value resistor used in multicell power transistors to provide individual current limiting and balancing at the cell level through debiasing.

Bethe equation   Equations developed by German-American nuclear physicist Hans Bethe in 1930 (non-relativistic) and 1932 (relativistic) that describe the stopping power of materials when traversed by charged particles.

bipolar junction transistor (BJT)   A transistor structure that uses both electron and hole charge carriers. In contrast, unipolar transistors such as field-effect transistors, use only one type of charge carrier. The BJT structure, uses two junctions between semiconductor types of opposing polarity and is, hence, available as an npn or a pnp device.

bremsstrahlung   Literally, *braking radiation*. Electromagnetic radiation produced when a charged particle changes its velocity. A typical example is an electron that is deflected by another charged particle or an atomic nucleus.

bulk damage   Damage caused by particles traversing a crystalline material. The damage takes the form of vacancies (an atom missing from a lattice position) and interstitials (extra atoms not at a lattice position).

C

capacitance-voltage (C-V)   A method or type of plot commonly used to characterize MOS capacitors. Measurements are usually taken quasi-statically by slowly sweeping the voltage on the capacitor while measuring the capacitance using a small high frequency signal superimposed on the capacitor voltage.

complementary metal-oxide semiconductor (CMOS)   A semiconductor that contains both n- and p-channel metal-oxide semiconductor field effect transistors (MOSFETs), which enable it to consistently improve performance while decreasing power consumption. PMOS is a p-channel MOS and NMOS is an n-channel MOS.

Compton scattering   The scattering of a photon by a charged particle. The scattered photon will have a longer wavelength (lower energy) after transferring a portion of its energy to the charged particle.

continuous-time random-walk (CTRW)   A generalization of the random walk where both the time and distance/direction variables are random. The CTRW formalism was introduced by Elliott Montroll and George Weiss to describe anomalous diffusion (diffusion processes with nonlinear relationship to time).

Compact Model Coalition (CMC)   A working group in the EDA (Electronic Design Automation) industry formed to select, promote and maintain the use of standard semiconductor device intrinsic models.

cosmic rays   Ionized, energetic high atomic number atoms or molecules, usually of solar or galactic origin. This is an inexact term as ions or molecules are clearly not "rays," but it has persisted, especially in nontechnical literature. Cosmic rays originating from the sun are anisotropic, while the galactic cosmic ray flux is isotropic.

critical charge The minimum amount of collected charge from a heavy ion or proton strike that will result in a single-event effect of any type. Critical charge is typically expressed in picocoulombs (pC).

cross section The number of observed single events per unit fluence. Fluence is expressed in ions/cm^2 and hence the unit of cross section is simply square centimeters (cm^2).

D

Deal convention A naming convention proposed by B. E. Deal in 1980 that standardized the terminology for oxide charges associated with MOS structures. It provided for four general types of charges; fixed oxide charge, mobile ionic charge, interface trapped charge, and oxide trapped charge.

destructive SEE A SEE causing permanent damage to the device. See *SEE*.

diode A p-n semiconductor structure that displays conductance in one direction only, enabling uses such as rectification of AC power.

diode–transistor logic (DTL) An early form of logic in which the gating function is performed by a diode network and the amplifying function is performed by a transistor.

dose As used for ionizing radiation such as gamma rays or X-rays, the energy absorbed in a given semiconductor volume. The commonly used dose unit used in electronics work is radiation absorbed dose (rad), defined as 1 rad = 0.01 J/kg. The equivalent CGS unit is the Gray (Gy), defined as 1 Gy = 100 ergs/g of matter, so that 1 Gy is 100 rad. Note that both units are specific to the material being irradiated due to differences in stopping power; in silicon devices they are expressed as rad (Si) and Gy (Si).

dynamic random access memory (DRAM) A digital memory device that stores digital information as charge on a low-value capacitor. As any charge stored in a capacitor will in time leak off, a periodic refresh cycle is required for continuous data retention.

E

E' center A well-known defect in crystalline silicon dioxide (quartz) as well as silicon dioxide layers used in semiconductor processes. The defect arises from neutral oxygen vacancies (Si–Si bonds) in silicon dioxide of poor crystallinity. The Si–Si bond is very weak and can be broken by the capture of a hole, which forms the E' center. The E' center is so named because the uncompensated spin of the electron in the dangling bond produces a signal in electron paramagnetic resonance (EPR) measurements.

Ebers-Moll The first bipolar transistor model introduced by J. J. Ebers and J. L. Moll in 1954 and used in the first SPICE simulators. It was a simple model with only 16 model parameters. A later model bipolar model, Spice Gummel-Poon (SGP – see *Gummel-Poon*), has largely replaced the Ebers-Moll model.

effective LET The linear energy transfer value modified to account for the change in total energy transferred from an incident ion when the path of the ion is not normal to the irradiated surface of that volume (q. v. LET). See *LET*.

electronic stopping The slowing down of a charged particle due to the inelastic collisions between bound electrons in a material and the particle.

enhanced low dose rate sensitivity (ELDRS) An effect in which a device exhibits enhanced total dose response at dose rates below 50 rad(Si)/s as compared to the response at dose rates in the standard 50–300 rad(Si)/s range. A dose rate of 0.01 rad(Si)/s is used as the standard low dose rate as it more accurately reflects the actual space environment.

electrical safe operation area (E-SOA) A set of operating conditions that ensures unconditionally reliable operation of a given device in a given radiation environment.

electron A subatomic particle (symbol e− or β−) with a negative electric charge. The electron is considered an elementary particle as it has no known substructure. The electron mass is approximately 1/1836 that of the proton.

electron–hole pair (e–h pair) The fundamental unit of carrier generation and carrier recombination processes when considering semiconducting materials. The generation of an electron-hole pair occurs when an electron in a material transitions from the valence band to the conduction band, leaving behind a hole in the valence band. Recombination is the reverse of this process, an electron in the conduction band recombines with a hole in the valence band.

epitaxy A semiconductor unit process in which additional semiconductor material layer is added to a surface of a semiconductor wafer. The process takes place at high temperature and the added material matches the crystal arrangement of the semiconductor wafer surface. A typical semiconductor wafer is about 500 μm thick, whereas, epitaxial thickness is generally in the 1 to 20 μm range.

F

flatband voltage The voltage that, when applied to the gate of a MOS capacitor, results in no net charge in the MOS capacitor oxide and, hence, no electric field in the MOS capacitor oxide. The value depends on the doping of the semiconductor bulk region and any charge located at the semiconductor-oxide surface or in the bulk oxide itself.

G

gamma rays Energetic electromagnetic radiation (photons) emitted from the nucleus of an atom by radioactive decay and having energies in a range from 1×10^4 to 1×10^7 eV. Gamma rays from, for example, ^{60}Co are widely used in electronic device testing to simulate space environments.

graphical user interface (GUI) User interface that allows users to interact with electronic devices through visual icons instead of with text only.

Gummel plot The combined plot of base and collector currents of a bipolar transistor versus the base-emitter voltage. When plotted on a semi-logarithmic scale, the Gummel plot is very useful in device characterization and modeling.

Gummel-Poon A type of bipolar transistor compact model developed by Hermann Gummel and H. C. Poon at Bell Labs in 1970 that made improvements over the Ebers-Moll model (see *Ebers-Moll*) by accounting for the effects of low and high levels of current injection.

H

hard errors An SEE-induced circuit error that is permanent and cannot be corrected by any means.

heavily doped buried layer (HDBL) A single charge boron (ion) implant that is done at high energy (typically > 1.5MeV) and at high implant dose (> 8e14 boron ions). Originally developed at SEMATECH as an alternative to P–/P+ epi to mitigate latchup in CMOS devices by increasing the trigger current. HDBL and variants have been demonstrated effective in increasing the PNPN SCR trigger current and also gettering of metals and oxygen contaminates in silicon wafers by a damaged layer formed by the HDBL implant at the end of range (Bragg peak).

heavy ions Ionized high atomic number atoms or molecules; contrast with, for example, protons or neutrons. Heavy ions in the space environment are also (and inaccurately) known as cosmic rays and can be of solar or galactic origin; those originating from the sun are anisotropic, while the galactic heavy ion flux is isotropic.

heavy ion fluence The total number of ions per unit area incident on a surface during a given exposure time.

heavy ion flux The time rate of flow of ions per unit area incident on a surface.

heavy ion sources A source of specific ion species in a focused and usually monoenergetic beam.

heavy ion testing The irradiation of semiconductor device samples using a heavy ion beam or other heavy ions source while monitoring the sample (s) for single-event effects.

high-voltage CMOS (HVCMOS) See *CMOS*.

I

inelastic scattering A scattering process where the kinetic energy of an incident particle is not conserved. Some of the incident particle energy is lost or increased when it interacts with other particles.

integrated circuit (IC) A group of resistors, transistors, and capacitors set on a flat piece (chip) of semiconductor material, usually silicon.

interface trap Silicon lattice dislocations formed at the boundary of SiO_2 in CMOS MOS devices that are created by oxidation grown in (wafer) silicon crystal. Oxidation of silicon to form silicon dioxide (SiO_2) creates defects at the silicon-silicon dioxide interface that are sensitive to radiation and form both electropositive and electronegative SiO atoms, which then form locally fixed inversion regions that conduct minority carriers. In CMOS devices, interface traps can be permanent and cause excessive leakage in NMOS transistors; observed as electrically threshold shift to depletion mode (lower VTn). In PMOS transistor, the effect is opposite, with V_t shift to enhancement (higher V_{Tp}).

ionizing radiation Radiation that is sufficiently energetic to strip electrons from atoms, creating an electrical charge difference of the ion. Ionizing radiation can lead to excess electron–hole generation in the dielectric layers of a semiconductor device, leading to volume charging and changes in device parameters.
See *radiation*.

J

junction field effect transistor (JFET) A majority carrier device similar to a MOS transistor that uses the depletion region of a p-n junction to modulate the flow of current between a drain and a source.

L

latchup A high-current state in an integrated circuit caused by the turn-on of intentional or parasitic four-layer structures in the substrate. Latchup can be destructive or nondestructive but can cause latent damage as well. A true latchup can only be terminated by interrupting the part's power supply. Latchup may be triggered by electrical conditions, electrostatic discharge (ESD), single-event effects or high temperature. Latchup event can be destructive or transient nondestructive. Destructive latchup result in permanent loss of the CMOS device if the holding current is lower than the supply voltage and shorts VDD to VSS causing high current destruction to the CMOS chip. Latchup in CMOS CORE regions are mostly transient upset events where the supply voltage (VDD) is lowest and the latch parasitic saturated holding voltage is higher than (VDD) voltage (nondestructive). Latchup events can be caused by radiation or high temperatures and are initiated by a diode polarity shift in blocking to a forward bias state locally at first, cascading the local potentials in neighboring devices regions (unstruck) and propagating the latchup event, mostly in CORE circuits. In IO circuits where the supply voltage is higher, the source potential (VDD2) is higher and can drive the SCR load more efficiently into lower saturation voltages and cause permanent destructive risk to the CMOS IO device region.
latchup testing Characterizing a part for latchup susceptibility through electrical stressing or heavy ion irradiation.
laterally diffused power MOSFET (LDMOS) transistor A power MOSFET device structure that uses a double-diffused drain structure to improve breakdown voltage.
See *MOSFET*.
linear energy transfer (LET) The amount of energy per unit length deposited by an ion traversing a material, often expressed in MeV.cm^2/mg. The milligram term accounts for the density of the material of interest. This unit is a source of frequent confusion, so we perform a simple derivation. The energy loss per unit length is dE/dx, which may be expressed in MeV/cm. The density of the material is $1/\rho$, which may be expressed in cm^3/mg; LET is $1/\rho \times dE/dx$, and the unit of LET then becomes (MeV/cm $\times$ cm^3/mg) or MeV.cm^2/mg.

M

manmade radiation environment A radiation environment caused by manmade events such as nuclear weapon detonations and nuclear power generation. See *radiation*.

McLean two-stage model A model first developed by F. B. McLean and co-workers to explain experimental results of hole transport through silicon dioxide following ionizing radiation. In the first stage, holes transporting through the oxide release hydrogen. In the second stage, a positive bias across the oxide causes the holes as well as the hydrogen to transport to the $Si–SiO_2$ interface. The hydrogen ions react with the Si–H bonds at the interface forming H_2 and dangling Si bonds.

mean excitation energy A property that is the measure of the mean energy required to ionize a bound electron in a material. It is a measure that is specific to the material and used when calculating the stopping power of materials for ionizing radiation.

memory An electronic circuit function that stores binary data through the use of bistable elements, capacitors, or other means.

metal-oxide semiconductor field effect transistor (MOSFET) A type of transistor with an insulated gate located above a channel region that connects a source and drain region. The channel conductivity can be changed with the amount of applied voltage. Its main advantage is that it requires very little input current to control the load current.

mixed signal design A combination of digital and analog design, resulting in a mixed signal device. Mixed signal functions can contain digital circuitry such as logic and memory as well as analog functions such as A/D converters, amplifiers, and power stages. Mixed signal design was first applied in analog to digital converters and in switched-capacitor circuits used for discrete-time signal processing. The performance of a switched-capacitor filter depends on capacitance ratios only, making them more suitable for use in integrated circuits.

multi-bit upset (MBU) A single event effect that induces upsets of multiple cells where two or more of the error bits occur in the same logical word (or frame for FPGAs).

multiple-cell upset (MCU) A single-event effect that induces several cells (e.g. memory cells or flip-flops) in an integrated circuit to change logic state at one time.

muon An elementary particle similar to the electron, with an electric charge equal to that of the electron. The muon is an unstable subatomic particle with a mean lifetime of $2.2\,\mu s$.

N

natural radiation environment A radiation environment caused by natural processes such as solar activity and charged particle trapping. See *radiation*.

neutron A subatomic particle (symbol n or n^0), which is electrically neutral and has a mass slightly larger than that of a proton. The nucleus of an atom consists of protons and neutrons.

noise In electronics, a random fluctuation in an electrical signal, an inherent characteristic of all electronic circuits. Electronic device noise is produced by several different effects and varies greatly depending on temperature and operating conditions. Noise effects include thermal noise, shot noise and $1/F$ noise.

nondestructive SEE A single-event effect (SEE) that causes a temporary change in device operation or signal levels, which will cause the CMOS devices to fail either temporally if the circuit is recovering in time independently, this type is called single-event transient (SET), or through external intervention such as a reboot. This type is referred to as nondestructive latchup (SEL). See *SEE*.

nuclear power Commercial electrical power generated by nuclear reactions such as uranium fission.

O

off-chip driver (OCD) A circuit that sends a signal "off-chip" a driver circuit, typically consisting of the output of an inverter circuit.

overshoot clamp network A circuit that absorbs an overshoot signal used to minimize overshoot reflections and latchup.

P

particle In the context of radiation effects in electronics and nuclear physics in general, a small localized object to which can be ascribed several physical or chemical properties such as volume, mass, and charge. Examples include subatomic particles such as the electron, neutron and proton, and ionized atoms known as heavy ions (natural heavy ions examples are any atom in the first four rows of the periodic chart that are atomic and have kinetic energy). A low LET heavy ions example is the doubly ionized helium nucleus, which are called alpha particles and is emitted by spallation in which a silicon atom is struck by neutrons, giving rise to a secondary energetic particle. See *solar particle*.

photoelectric effect A mechanism whereby photons absorbed by a material result in the emission of electrons. The photon interacts with an absorber atom in the material and completely disappears. An energetic photoelectron is then ejected by one of the bound shells of the atom. The energy of the photoelectron is dependent on the energy of the photon, not the intensity of the illumination as would be predicted by a more classical analysis. While the photoelectric effect was first observed by Heinrich Hertz in 1887, it was explained almost 20 years later by Albert Einstein in 1905. In CMOS devices, charge carriers (electrons and hole) are generated and mobilized when exposed to light (photoelectric). However, if a CMOS device is exposed to a nuclear burst that emits high levels of X-rays or gamma rays, then this dose effect generates secondary photo currents (electrons and holes) within the bulk silicon that are free to flow to the nearest cathode and anode for termination, the charge collection process occurring at the CMOS well, any diffusion, and, finally, the VDD or VSS metal grid, if they are not recombined during emission lifetimes. The CMOS device n- or p-well isolation regions vary widely in doping concentrations and depths, which affects "minority carrier lifetimes" and can be modified to engineer a decrease in lifetime (termination) as a way to mitigate (reduce) the photoelectric effect.

polaron A quasiparticle originally proposed by Soviet physicist Lev Landau in 1933 to describe an electron moving in a dielectric crystal. The idea has been broadened to include interactions between electrons or ions and the atoms in materials that result in a lower energy state than a non-interacting state. A region of polarization is formed around a charged particle moving through a material when the atoms or electrons in the material shift positions slightly in response to the charged particle. This region of polarization surrounding the particle along with the charged particle is called a polaron.

photon An elementary particle representing a quantum of the electromagnetic field, including electromagnetic radiation such as light, and the force carrier for the electromagnetic force. The photon has zero rest mass and always moves at the speed of light within a vacuum.

power MOSFET A large MOSFET device of either polarity designed to handle significant power and used for control and switching of power circuitry. A power MOSFET can have several device structures, including vertical and lateral architectures. Its advantages are high switching speed, ease of driving due to the isolated MOSFET gate and good efficiency at low voltages.

proton A subatomic particle (symbol p or p+) with a positive electric charge of one electronic charge and a mass slightly less than that of a neutron. Protons are the most abundant particle in the space environment. The number of protons in a nucleus determines the element and is referred to as the atomic number (represented by the symbol Z).

process design kit (PDK) A set of files and scripts used during the development of integrated circuits that provides context for the semiconductor process used during the integrated circuit fabrication. While the contents of the kit are somewhat dependent on the design software tools used during development, they usually include design and foundry process layer definitions, Design Rule Checking files (DRC rules), Layout Versus Schematic checking files (LVS rules), device schematic and layout examples or libraries, and device simulation models. The PDK provides an overlay to commonly available EDA software tools so that correctly designed files are provided to the semiconductor process foundry.

R

radiation The emission or transmission of energy in the form of waves or particles through space or through a material. Radiation can take the form of electromagnetic radiation, particle radiation, or acoustic radiation. The existence of gravitational radiation has been proven and has been the subject of intensive research.

radiation belts The dipole nature of Earth's magnetic field causes charged particles such as electrons and protons to oscillate along the magnetic field lines between the poles, effectively trapping the particles in a quasi-stable region for long periods of time. See *van Allen belts*.

radiation damage In the context of radiation effects in electronics, radiation damage can take two forms. Ionizing radiation strips electrons from atoms in dielectric layers, causing volume charging and shifts in device parameters. Nonionized particles such as neutrons cause direct damage to the semiconductor lattice, also causing shifts in device parameters.

radiation hardening The reduction of radiation sensitivity of semiconductor devices. Hardening can be accomplished by a number of methods, including hardening of the basic semiconductor process, hardening through circuit design, or a combination of these approaches.

radiation induced leakage current (RILC) Gate leakage current caused by radiation. The very thin gate oxides used in advanced processes are prone to leakage due to tunneling because the probability of tunneling increases exponentially with decreasing gate oxide thickness. While very thin oxides tend to be very hard to radiation, defects created in the gate oxide due to charge trapping during electrical overstress (stress induced leakage current, SILC) and/or irradiation provide a leakage path. RILC and SILC are significant factors inhibiting device miniaturization and scaling.

recombination and generation Semiconductor processes associated with electrons and hole pair (EHP) generation or recombination.

regenerative feedback A positive feedback process that occurs in MOSFETs and silicon- controlled rectifiers which leads to switching of the current and voltage state of a device.

retrograde well In CMOS processes, a device well that has a concentration profile in which the doping concentration increases with depth. Retrograde wells are used to improve the latchup (q. v.) characteristics of the device and are usually created through multiple ion implantations. See *well, triple well*.

S

safe operation area (SOA) A region of the current-voltage (I-V) characteristic in which device operation is unconditionally nondestructive.

saturated cross section The maximum observable cross section, manifesting as a saturation of the cross section vs. LET curve.

second breakdown A breakdown phenomenon that is associated with thermal runaway and failure of a semiconductor device. Second breakdown occurs at a higher current than electrical breakdown.

sensitive volume A region or multiple regions within CMOS devices that are sufficiently deep or lightly doped enough to give rise to charge generation along the ionized particle trajectory to cause enough charge to generate a bit upset (single-event upset). Thin film silicon devices like SOS and SOI form thin silicon layers in which less charge is generated by a rad particle (nuclear, atomic, or photo) and are more resistive to upsetting CMOS bit state as "less charge is generated since they are thin (typically < 0.5 µm)." Bulk CMOS devices are thick and are far more sensitive due to material thickness. Charges generated then are in proportion to the silicon material thickness. However, bulk CMOS devices can also be made less sensitive to rad particle effects using substrate engineering methods.

silicon controlled rectifier A power-switching device consisting of four semiconductor layers of alternating polarity, forming a closed loop npn/pnp BJT structure. The application of a small signal to one of the BJT bases will trigger the device into a low-resistance state through the feedback inherent in the BJT pair. This structure can be parasitic, causing latchup (q. v.).

silicon diode A structure composed of two regions of opposite polarity which displays rectifying characteristics.

single-event effect (SEE) Any measurable or observable change in state or performance of a microelectronic device, component, subsystem, or system caused by the passage of a single energetic particle.

single-event burnout An event in which a single energetic-particle strike induces a localized high-current state in the device, resulting in catastrophic device failure or in permanent degradation, and which is usually characterized by a significant increase in leakage current.

single-event gate rupture (SEGR) An event in which a single energetic-particle strike results in a breakdown and subsequent conducting path through the gate oxide of a MOSFET, MOS capacitor, or floating-gate memory. SEGR is a destructive effect in all cases.

single-event hard error (SHE) An irreversible change in operation that is typically associated with permanent damage to one or more elements of a circuit. Examples of radiation-induced hard errors are single-event gate rupture, single-event burnout and destructive latchup.

single-event latchup (SEL) An abnormal high-current state in a device caused by the turn-on of a real or parasitic four-layer structure by the passage of a single energetic particle through sensitive regions of the device structure, and resulting in the loss of device functionality. SEL can be destructive or nondestructive and can cause latent damage

single-event transient (SET) A momentary voltage excursion (voltage spike) at a node in an integrated circuit caused by the passage of a single energetic particle.

single-event upset (SEU) A logic state change of a bistable element caused by the passage of a single energetic particle, which then amplifies the CMOS parasitic network to spread the charge collections.

single-event functional interrupt (SEFI) A nondestructive single-event effect that causes the component to reset, hang, or enter a different operating condition or test mode.

silicon on diamond (SOD) A device structure consisting of a diamond substrate with a silicon device layer arranged thereupon, with the silicon layer used for active and passive device fabrication. This structure enables dielectric (passive) isolation that improves performance and radiation hardness.

silicon on insulator (SOI) A device structure consisting of a silicon substrate with a silicon dioxide insulating layer and a silicon device layer arranged thereupon, with the silicon layer used for active and passive device fabrication. This structure enables dielectric (passive) isolation which improves performance and radiation hardness.

silicon on sapphire (SOS) Invented by RCA in response to CMOS latchup. RCA believed CMOS was unstable if latchup was possible and developed the world's first CMOS dielectrical isolation transistor structure to make latchup-immune CMOS devices, beginning in the 1970s. A device structure consisting of a sapphire (single crystal Al_2O_3 that is pulled from a sapphire melt to form thin sapphire ribbons that are optically transparent). Then the sapphire substrate is used to deposit a silicon epitaxial layer via epi CVD process to form a silicon semiconductor film (about 0.5 µm or less) on the sapphire base material, which is then cut into circular wafers. The silicon layer is then used to form active area regions to fabricate CMOS transistors and open areas for isolation. SOS devices are nonplanar, and the active islands are independent and fully dielectrically isolated. This structure enables dielectric (passive) isolation, which makes the SOS device latchup immune and reduces the sensitive volume that then improves resistance to natural radiation (SEU, SEL, SET, and DOSE RATE UPSET (manmade)).

snapback An electrical phenomenon that occurs in semiconductor devices where a device switches from a high voltage/ low current state to a high current/low voltage state. This occurs in bipolar junction transistors, MOSFETs, and silicon controlled rectifiers (SCR).

soft error A circuit error that can be corrected by electrical means such as rebooting, rewriting, or resetting.

soft error rate (SER) The rate of soft errors per unit time for a given radiation environment.

solar flare A rapid and intense increase of the brightness of a limited area of the Sun's surface. Flares are caused by a sudden release of stored magnetic energy in the solar atmosphere and are an extremely high-energy event, with released energy as high as 1×10^{25} joules. X-rays and UV radiation emitted by flares affect Earth's ionosphere and disrupt long-range radio communications.

solar activity The emission of radiation and particles by the sun.

solar activity cycle The activity of the sun varies over time, with an 11-year period between solar maximum and solar minimum.

solar particle A particle of solar origin. The nuclear fusion processes in the Sun emit large amounts of radiation and particles, mostly protons, but with substantial amounts of helium atoms and high atomic number atoms. These charged particles can be further accelerated by solar flares or coronal mass ejections. Solar particles can get trapped by Earth's magnetic field and populate the van Allen belts. The abundance of energetic protons creates a significant radiation hazard to electronics and to manned spacecraft. See *particle*.

static random access memory (SRAM) In contrast to a dynamic random access memory (q. v.), a SRAM stores digital information in bistable elements such as flip-flops. This architecture will retain the digital information until the supply voltage is removed, eliminating the need for a refresh cycle.

stopping power The retarding force acting on particles due to their interaction with matter, resulting in a loss of particle energy.

surface potential model A physics-based MOS model that results in a formulation for the drain to source current that is continuous over the entire range of device operation. The model solves for the surface potential at the two ends of the channel in order to calculate the terminal charges. Surface potential models represent a fourth generation of MOS model and provide accuracy for device geometries of 100 nm and smaller.

T

thermal neutron A free neutron with a kinetic energy of about 0.025 eV, which is the most probable neutron energy at a temperature of 290 K.

thermal safe operating area (T-SOA) A set of operating conditions that ensures unconditional thermal stability of a given device in a given application environment.

threshold LET The minimum ion linear energy threshold (LET) required to cause a single-event effect (SEE), either destructive or nondestructive. See *LET*.

thrust-specific fuel consumption (TSFC) Fuel efficiency of an engine with respect to its thrust output.

Taiwan Semiconductor Manufacturing Company (TSMC) modeling interface (TMI) A C-based modeling application program interface designed to improve simulation times and reduce memory usage.

transient latchup A latchup-triggering event that does not reach the full forward biased latchup saturation stage at a voltage lower than the supply voltage. It is nondestructive and will recover in time without permanent failure from either a radiation or high-temperature transient event.

transient latchup upset (TLU) A latchup event caused by a transient event. The transient can be a single pulse, or any transient event.

transient safe operation area (T-SOA) The safe operating area associated with transient phenomena.

transient soft error rate The soft error rate during transient switching of circuits. In latch circuits (e.g. SRAM cells), the sensitivity is significantly higher during a switching process of the circuit.

transistor–transistor logic (TTL) An early form of logic in which the gating and amplification functions are performed by npn transistors. There are low-power and high-power versions of TTL.

trapped particle A charged particle trapped in the Earth's magnetic field; see also *Radiation belts*.

triple well A variation on the single-well bulk CMOS process that has wells of both polarities to accommodate the P- and N-channel devices used.

trench isolation a type of device isolation in a semiconductor process that uses a dry etched trench that is refilled with silicon dioxide either alone or combined with other semiconductor materials.

U

undershoot clamp networks A circuit that absorbs an undershoot signal to minimize reflection and latchup events

V

van Allen belts The trapping regions named after their discovery by James van Allen in 1958. These trapped particle regions represent the dominant ionizing radiation source for the near-Earth environment. See *radiation, radiation belts, solar particle*.

W

well A bulk CMOS process requires regions of opposite polarity to accommodate the P- and N-channel devices used. A well is a specific doping region (masked) of opposite doping (N or P dopants) formed in the silicon semiconductor wafer substrate to form the doping isolation regions in the CMOS silicon device. Wells and diffusion are formed by ion implants and thermal diffusion activation to form diffusion regions that are called p-wells, n-wells, deep n-wells, isolation wells, and triple wells. Triple wells are formed from a deep n-well in which an (iso) p-well is formed for biasing flash memory. In SOI wafers transistors formed using "depletion mode," the p- or n-wells are lightly doped.

X

X-ray A form of electromagnetic radiation with a wavelength ranging from 0.01 to 10 nm and energy ranging from 100 eV to 100 keV.

Z

Zener diode A type of diode that allows current to flow in both forward and reverse directions. The reverse current flow occurs only when the Zener voltage is reached. The current flow at a reverse bias voltage equal to the Zener voltage is caused by the Zener effect. This effect is a type of breakdown that is observed in Zener diodes with Zener voltage from about 6 to 8 volts. It is caused by a diode design that has a narrow depletion region. The reverse bias voltage creates an electric field that is intense enough to pull some of the valence electrons into the conduction band by breaking their covalent bonds.

1

Introduction and Historical Perspective

1.1   Introduction

This book provides a structured approach to the design of radiation-hardened integrated circuits (ICs), covering digital, and analog and mixed-signal applications. The
emphasis will be on space applications, but we will also briefly discuss other environments including nuclear reactors, weapons spectra, and the interesting and challenging
environment found in high-energy particle physics facilities. We will begin with a brief
historical introduction to radiation and its interactions with semiconductor materials.
An understanding of basic physics is vital to efficient design, as without this knowledge the design process is all too likely to devolve into a trial-and-error exercise. In
today's highly competitive semiconductor business, time to market is a vital and probably the most important consideration, with major development programs often canceled
because of schedule slips. Delays in the time to market may well result in a part that is
no longer competitive or even relevant. It should also be noted that an understanding of
the physics behind radiation effects on device structures enables the effective designer
to look at simulation and modeling tools with a critical eye, defining and correcting
inconsistencies that would otherwise lead to redesigns.

It may well be asked what this radiation effects in electronics fuss is all about, with
apologies to the late National Semiconductor analog circuit designer Bob Pease. Why
does a satellite cost hundreds of millions of dollars, and why do the components that
make up its systems sell for several orders of magnitude more than their commercial
equivalents? Why do we closely follow and try to predict the cyclical activity of the
Sun, and why indeed do we speak of "space weather" at all when anyone knows that
"weather" in a vacuum is an absurdity? The answer is simple: space is one of the harsh-
est and most difficult environments we know, and being a vacuum merely adds to the
difficulty. The range of space radiation types is broad and diverse, ranging from ultraviolet (UV) and X-ray radiation consisting of energetic photons to relativistic heavy ions
with tera-electron volts (TeV) energy, against which shielding is impractical, to solar
protons and electrons. The radiation type and intensity for a given application depends
on its position in space, both spatially and temporally, ranging from the various Earth
orbits seen by commercial satellites to the intense radiation belts surrounding Jupiter.
Understanding the design methodologies used to harden electronic components against
these environments requires at least a working understanding of those environments,
as each one has its own interactions with matter and its own resulting damage mechanisms. This damage affects all electronics at the component, subsystem and system

Integrated Circuit Design for Radiation Environments, First Edition.
Stephen J. Gaul, Nicolaas van Vonno, Steven H. Voldman and Wesley H. Morris.
© 2020 John Wiley & Sons Ltd. Published 2020 by John Wiley & Sons Ltd.

levels. Satellite systems commonly have 15-year lifetime requirements, with trends to higher numbers, and most orbits are out of reach of today's manned space transportation systems. Interplanetary exploration missions are, of course, well outside our reach, and these limitations result in a requirement for systems that can operate reliably for a long time in a harsh environment without any servicing. *Operation* here goes well beyond, for example, the cruising to Mars of an interplanetary probe; the entire landing sequence on these missions must be executed independently from Earth control, as the time delay for radio signals ranges from 4 to 24 minutes, depending on the Earth–Mars distance at the time.

1.2 Discovery of X-Rays, Radiation, and Subatomic Particles

As with most scientific disciplines, the physics of radiation and its effects on electronics builds on a long history of research, discovery, and expensive lessons. In radiation effects that history is relatively short, however. The atomic theory of matter, on the other hand, goes all the way back to Demokritos (460–370 BCE) in ancient Greece, who postulated that all matter was composed of invisibly small atoms, after the Greek *–a* (not) and *–tomos* (to divide), indicating his belief that these particles were not further divisible. In this theory, there was empty space between the atoms. This was an enormously advanced concept at a very early time, and it took until the nineteenth and twentieth centuries to experimentally identify subatomic particles such as protons, neutrons, and electrons.

Light was also believed to be composed of particles, but inconsistencies were identified beginning in the eighteenth century. In particular, the diffraction and interference of light through narrow slits was difficult to explain using a particle theory, and by the mid-nineteenth century, wave models began to be generally accepted. In 1865, James Clerk Maxwell developed the Maxwell wave theory [1], which demonstrates that light is an electromagnetic wave phenomenon. This wave interpretation accounted for many of the observed properties of light, but some could still be only explained by the particle theory. It took until the development of quantum mechanics in the twentieth century to fully understand the dual-wave/particle nature of photons, visible, or otherwise.

The systematic study of subatomic particles started with investigations by J. J. Thomson of electrical discharges in vacuum using the Crookes tube [2], which was an outgrowth of earlier gas-filled Geisler tubes known to produce light with colors dependent on the gas species. The Crookes cathode ray tube, shown in Figure 1.1, was a high-vacuum device enabled by the greatly improved vacuum pump technology of the time. It consists of a cold cathode and an anode in a glass envelope; a high-voltage between the two electrodes leads to high-field electron emission from the negative cathode. The electrons are then collected by the positive anode, but some continue toward a portion of the tube containing a target. Those electrons striking the glass envelope across from the cathode cause phosphorescence, forming a negative shadow in the shape of the target. It was soon determined that these "cathode rays" were, in fact, negatively charged particles, as they could be easily deflected by either an electric or magnetic field. With known magnetic and electric field strength, the speed and the charge to mass ratio (but not the absolute value of these last two parameters) of the then-still-unknown particles could be readily derived. Thomson found [3] that these particles had uniform mass and charge, regardless of the anode material or anode voltage. This work built on earlier

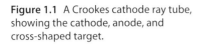

Figure 1.1 A Crookes cathode ray tube, showing the cathode, anode, and cross-shaped target.

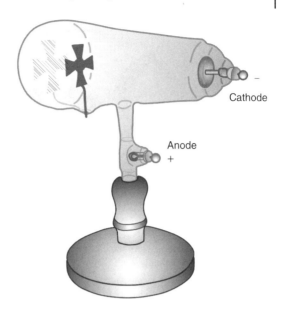

theoretical work by Lorentz and Zeeman and led directly to the discovery of the electron by Thomson in 1897, which has since resulted in an amazing number and range of applications and in fundamental changes in human society. The charge-to-mass ratio puzzle was solved in 1910 by the American Robert Millikan, who determined the charge of the electron using UV light irradiated oil drops suspended between positively and negatively charged plates. The Crookes tube went on to evolve into the cathode ray tube, which dominated television and computer displays before its replacement by various flat-screen technologies.

The end of the nineteenth century saw a period of intense radiation physics activity resulting in three breakthrough discoveries: X-rays by Wilhelm Röntgen, radioactivity by Henri Becquerel, and the isolation of radioactive elements by Marie and Pierre Curie. Using apparatus similar to that used by Thomson with voltages to the order of 20 kV and a thin metal "window" to allow the then-unknown "particles" to travel outside the tube, Röntgen somewhat accidentally discovered X-rays in 1895 [4]. While conducting experiments with remotely located phosphorescent screens, Röntgen unintentionally left a protective cardboard cover over the X-ray tube apparatus and found that the cover did not block the emitted radiation. He found these "unknown" ("X") rays to be able to penetrate many substances, including the human body. A well-known anecdote claims that this was the hand of Röntgen's wife initially, showing the bone structure on the screen to his amazement. This imaging capability was a powerful tool that led to immediate diagnostic applications in medicine.

An X-ray tube based on the earlier Crookes tube from the early 1900s is shown in Figure 1.2. Electrons emitted from the cold cathode at the right strike the angled target, producing X-rays by a combination of Bremsstrahlung and X-ray fluorescence processes. The target is composed of a high-density metal, with W or Pt common in this application. In practice, only a small percentage of the electrons bombarding the target produce X-rays; the majority lose their energy as heat in the anode, which is dissipated by the external air-cooling arrangement on the anode connection at the left of the tube.

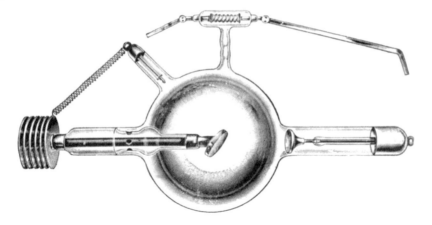

Figure 1.2 An X-ray tube. Note the cathode (right) and the anode structure, consisting of the angled high-Z metal target (usually W or Pt) and the external air cooling arrangement.

Operation of the tube slowly getters the residual gas in the tube so that higher voltage must be applied for continued operation, resulting in the production of higher energy X-rays as the tube ages. The X-ray tube shown in Figure 1.2 employs a "softener" electrode (at the top of the tube) so that gas pressure can be regulated in the tube. The "softener" contains a small amount of Pd; upon heating, hydrogen will outgas from the Pd sample and thus increase the pressure in the tube.

Formal radioactivity studies started at the end of the nineteenth century. Henri Becquerel discovered radioactivity in 1896 while investigating phosphorescence phenomena caused by crystalline potassium uranyl sulfate. He found that these emissions were quite penetrative and did not depend on an external power source, and that photographic film was a convenient detector of what eventually turned out to be alpha particles. Becquerel's work was followed by pioneering studies of radioactive elements and their extraction by Marie Sklodowska-Curie and her husband Pierre Curie at the Sorbonne in Paris, in which months of work and tons of pitchblende, a uranium ore, resulted in the isolation of a tenth of a gram of a highly radioactive substance that eventually was found to be polonium chloride. Significantly, the short- and long-term physiological effects of ionizing radiation were then not at all known or perhaps even suspected, and these materials were handled casually and refined without any safety precautions. Marie Sklodowska-Curie discovered radium shortly after the first isolation of polonium, but died in 1934 of anemia caused by long-term exposure to radioactive materials. Radium was used for therapeutic applications in treating skin cancer, and the writer (van Vonno) well remembers my dermatologist parents' small and barely portable lead vault containing a sample of *Union Minière du Haut-Katanga* radium chloride used for that purpose. The element was also used to provide self-luminescent paint on the dials of watches, an unfortunate early insight into the dangers of radioactive substances was gained through greatly shortened life expectancy for the young women doing this work. Several current units of radioactivity are named after these early pioneers. The Becquerel is a unit equal to one disintegration per second; the Curie is a much larger unit and is equal to the disintegration rate of a gram of radium, which is 3.7×10^{10} disintegrations per second.

The discovery of radium by Curie laid a solid foundation for the discovery of numerous other subatomic particles. Ernest Rutherford at Victoria University attempted to focus the radiation emitted by a radium sample, and found [5] that rather than being collimated, the resulting particle beam was split into three components by an external magnetic field, suggesting three distinct forms of radiation. These were termed alpha, beta, and gamma rays by Rutherford and Paul Villard, after the first three letters of the Greek alphabet. The gamma rays were found to be penetrative and not deflected by magnetic or electric fields and eventually turned out to be energetic photons. The beta rays were the most easily deflected and were found to be electrons, using methods developed by Thomson and Millikan. The alpha particle proved to be the most difficult to characterize; its deflection in the magnetic field was opposite to that of the beta particles, indicating a positively charged particle.

In 1907, Rutherford completed investigations that showed the alpha particle to have twice the charge and 7000 times the mass of the electron. Subsequent spectroscopic analysis of trapped alpha particles allowed Rutherford to show that this particle consisted of two protons and two neutrons, a structure identical to a helium nucleus. This work was important in several ways, including the fundamental concept that a sample of radium is transmuted into other elements as it undergoes radioactive decay. The intensity and duration of the radioactivity of a given sample was also found to vary widely, and the concepts of exponential decay and half-life were developed to quantitatively express these effects.

By the turn of the twentieth century and into the 1930s, many prior scientific observations and problems were finding explanations as the understanding of the atom progressed. First among these was black-body radiation. The current classical explanations failed to predict observation at either low or high frequencies until Max Planck solved the problem in 1900 using the idea that energy could be quantized. He introduced Planck's constant, h, as the smallest quantum of action and the proportionality constant between energy E and frequency v in the equation

$$E = hv \tag{1.1}$$

in which h is Planck's constant (4.13567×10^{-15} eV-s).

At the time, Planck viewed the quantization of light energy as applying only to the absorption and emission of radiation rather than a physical attribute, but the quantum properties of light would prove to be more fundamental. Einstein's explanation of the photoelectric effect in 1905 made further use of the idea that energy was quantized, building on the earlier discovery of the photovoltaic effect by Becquerel in 1839 and on observations by Heinrich Hertz in 1887. Einstein's solution resolved conflicts between the observation that the energy of emitted electrons increases with light frequency (made by Philipp Lenard several years earlier) and the prediction from Maxwell's wave theory of light that the energy should be proportional to light intensity. Einstein's photoelectric effect solution showed that light itself was quantized and has both particle and wave properties.

Even earlier problems would see solutions in the early twentieth century. Consider the observation of spectral lines in solar radiation or the spectral emission lines from a hydrogen flame. Almost a century earlier, German physicist Joseph von Fraunhofer carried out a systematic study of the absorption lines in the solar spectrum; his designation for the wavelengths with principal features labeled with the letters A through K

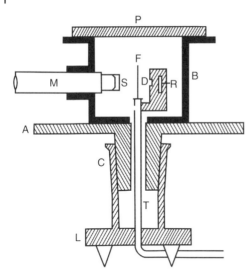

Figure 1.3 Geiger and Marsden apparatus (1913) designed to measure the angular scattering pattern of alpha particles hitting a gold foil.

is still used today. During the nineteenth century many contributors including Anders Ångstrom, Charles Wheatstone, Gustave Kirchhoff, Robert Bunsen, and William Crookes made advances to early spectroscopy. By the 1880s, Swedish mathematician Johann Balmer and the Swedish physicist Robert Rydberg had independently arrived at mathematical descriptions for the spectra of the hydrogen atom, but understanding the reason behind these empirical formulas would have to wait for explorations of atomic structure and new models of the atom developed in the twentieth century.

Work by Rutherford and colleagues at Cavendish in 1907 involved the use of alpha particles to search for the as-yet-unidentified positively charged component of the atom, which clearly had to be present in order to achieve overall charge neutrality. A collimated alpha particle beam was used to bombard Au foils, and the resulting interactions were imaged on a fluorescent ZnS screen. The density at the atomic level of even a heavy element such as gold was found to be very low, with the alpha particles penetrating the foil readily. Rutherford also observed scattering, in which alpha particles were deflected far from the beam axis, and the infrequent occurrence of scattering phenomena supported the understanding that the atom is composed of subatomic particles but is largely empty. Rutherford placed the positive charge of the atom in a central nucleus with electrons orbiting the nucleus. An apparatus (1913) used to measure the angular scattering pattern is shown in Figure 1.3. A collimated beam of alpha particles is produced by the radium source R hitting the gold foil F. The scintillation screen S and microscope M can be rotated to check for off-axis alpha particles.

The idea that the positive charge in the atom is located in a nucleus provided a new framework for how the atom could be arranged. Rutherford placed the electrons in orbits around the positive charge, but orbiting electrons posed a problem to physicists. An electron in an orbit is being accelerated, and accelerated charges emit electromagnetic waves and lose energy. This energy loss would make the Rutherford atom unstable. In 1913, Niels Bohr proposed an atomic model [6] that provided an explanation for the spectral emission lines of hydrogen. Bohr proposed electron orbits at certain discrete distances from a small, positively charged nucleus. These orbits were associated with a definite energy and could be occupied by an electron without loss of energy. Electrons could jump to a different orbit by absorbing or emitting a photon, depending on the energy difference between the two orbits:

$$\Delta E = E_2 - E_1 = h\nu \tag{1.2}$$

Bohr found that the energy levels resulting in the observed spectral emission line of hydrogen required the angular momentum L of the electron around the nucleus to be an integral multiple of Planck's constant h divided by 2π

$$L = mvr = n\frac{h}{2\pi} = n\hbar \tag{1.3}$$

in which m = electron mass, v = electron velocity, r = orbit radius, n = 1, 2, 3, ...

The Bohr model is significant because it applies a quantum rule (n in Eq. (1.3)) to a classical treatment of electron motion; however, this semi-classical approach was only a first-order approximation of the quantum behavior of atoms. The Bohr model was supplanted by more successful quantum mechanical descriptions of the atom by Heisenberg (matrix mechanics) and Schrödinger (wave mechanics) in the 1930s.

The discovery of the proton, reported in 1919, was the result of experiments by Rutherford in 1917 that proved the hydrogen nucleus was present in other nuclei. The neutron itself remained a mystery for some time, with its existence inferred but not proven due to the difficulty of detecting an uncharged particle. Positive proof of the neutron's existence was developed by Chadwick [7] in 1932, based on alpha particle investigations using a Wilson cloud chamber. Later work using the newly developed mass spectrograph characterized the atomic weight of the various elements, and it was noted that the atomic weight of a given element was an average of its constituent isotopes, explaining the long-term question of why atomic weights are fractions and not integral numbers as expected. The atomic model of a heavy nucleus composed of protons and neutrons and surrounded by electrons in discrete shells was now complete. In the process, the importance of beams of energetic particles in studying nuclear reactions and atomic structure had been conclusively proven, and it was equally clear that particle energies would have to be greatly enhanced in order to probe deeper. Early accelerator developments included the Cockroft-Walton voltage multiplier and van de Graaff high voltage electrostatic machines, and this work was later followed by cyclotrons and linear accelerators.

Transmutation of various elements through neutron irradiation was developed as a commercial process in the 1930s, facilitated by improved neutron sources and leading to fabrication of small quantities of isotopes such as ^{32}P and ^{33}P for medical applications. These capabilities led to research into previously unknown elements, culminating in the synthesis by Fermi of the first transuranic element, neptunium, through neutron irradiation of ^{238}U. Experiments by the German physicists Otto Hahn and Fritz Strassmann in 1938 continued the investigations into neutron irradiation and showed that one of the byproducts of ^{238}U fission was barium, which as element 56 is a long way from uranium at element 92 on the periodic table. This result clearly showed that nuclear reaction products were not necessarily limited to isotopes close to the original element in the periodic table. The exiled German physicist Lise Meitner then calculated the energy released by a single "split" ^{238}U nucleus to be 200 MeV, a very high energy yield. This "splitting the atom" concept was a practical application of the mass−energy relationship ($E = mc^2$) proposed by Albert Einstein in 1905 [8, 9] and introduced the concept of commercial energy generation through a neutron-mediated fission chain reaction. Much of this research was carried out in the Allied countries by refugees from the Axis regimes, and resulted in the first subcritical reactor (1942) at the University of Chicago, the well-known Manhattan Project in the USA and its culmination in the first nuclear weapons.

1.3 The Nuclear Age

The nuclear age was given tremendous new impetus by the wartime need for more powerful weapons and the fear of these weapons being developed by Germany. The Manhattan Project was a United States nuclear weapons development program with support from Canada and the United Kingdom. The project [10, 11] was initiated in 1939 and grew rapidly as World War II progressed, with a large part of its scientific staff derived from motivated German exiles. Research and production of fissionable material and completed weapons took place at secret sites in the participating countries. Two types of atomic bomb (Figure 1.4) were developed during the war, using ^{235}U and ^{239}Pu respectively as fissionable materials. The first type was a simple ^{235}U device using explosive charges in a gun barrel-like arrangement to drive two opposing subcritical masses together into a single supercritical mass. This approach required significant quantities of ^{235}U, which constitutes only 0.7% of naturally occurring uranium and is chemically identical to the dominant ^{238}U isotope; separation of these closely similar materials required breakthrough work in uranium separation and enrichment methods.

In contrast to the gun-type ^{235}U weapon, the second device was a much more complex plutonium implosion weapon using ^{239}Pu as the fissionable material. This technology uses precision-machined shaped explosive charges to compress a near-critical-mass plutonium core, bringing it to critical mass. This approach required the development of volume production technology, in which nuclear reactors are used to neutron-irradiate uranium producing plutonium by a transmutation process; it also required extreme precision in machining the shaped charges. The plutonium must then still be separated from the irradiated uranium, a difficult task as the two elements are chemically nearly identical as in the ^{235}U/^{238}U separation. The first nuclear device was actually an implosion-type ^{239}Pu bomb detonated at the Alamogordo test range in New Mexico on 16 July 1945.

Figure 1.4 Little Boy (left) and Fat Man (right) atomic bombs. Little Boy was a gun-type ^{235}U weapon whereas Fat Man utilized ^{239}Pu and a shaped charge. Source: US government DOD/DOE.

Subsequently, a gun-type ^{235}U device code named Little Boy and an implosion-type device code named Fat Man were used to bomb Hiroshima and Nagasaki, using Boeing B-29 aircraft as a delivery system and effectively terminating World War II.

After the war, the Manhattan Project went on to develop thermonuclear (fusion) weapons and conducted a lengthy series of tests of these devices in the Pacific. The project also laid the foundations for commercial nuclear power generation and developed compact reactors for marine propulsion systems. The Manhattan Project segued into the US Atomic Energy Commission (AEC) in 1946, transferring control over US nuclear weapons research, device production, and device storage to civilian authorities.

The first subcritical research nuclear reactor was built [12] by Enrico Fermi and Leo Szilard at the University of Chicago and was soon followed by a number of large reactors to meet the requirements for plutonium transmutation processing for fissionable material. After World War II, the development of commercial power generation reactors based on Manhattan Project technology proceeded quickly. This work remained under government control, as one of the byproducts of a uranium-based reactor is plutonium and it was well recognized that the proliferation of this material needed to be closely controlled. The first practical commercial reactor (1951) in the United Sates was a 100 kW facility, and subsequent declassification of reactor technology development led to widespread acceptance of commercial nuclear power. Nuclear power plants now generate nearly 6% of the world's energy from over 400 reactors in 31 countries, with France the leading user of nuclear energy; over 70% of all electrical power in France is derived from nuclear sources.

Nuclear power for marine propulsion systems was also investigated and was found to be particularly suitable for submarine applications, in which the technology provided nearly unlimited underwater endurance. A test reactor was developed by 1953 and the first nuclear submarine (USS *Nautilus*) was launched in 1955, laying the foundation for today's nuclear Navy. Nuclear propulsion was later extended to other marine applications such as aircraft carriers, but the early dreams of nuclear-powered aircraft were found to be impractical. Proposed civil engineering applications of nuclear explosives for such projects as canal excavation were similarly found to be highly impractical due to the large amounts of radioactive material released.

As late as World War II, science arguably still had an incomplete understanding of the interaction between radiation and living organisms, or indeed between radiation and the materials that would be used in yet to be discovered solid-state devices. Early nuclear weapons were believed to be blast-effect devices only, with little understanding of the long-term physiological and genetic damage involved. As systems became more complex and relatively radiation-hard vacuum tube technology gave way to solid-state devices, we were about to learn the radiation sensitivity of this new technology, in many cases the hard way.

1.4   The Space Age

Systematic radiation effects studies for the natural environment began in 1946 with US scientists using surplus WWII V-2 rockets to launch exploratory instruments into suborbital flights. The van Allen belts were discovered [13] as part of International Geophysical Year research in 1958 by James van Allen of Johns Hopkins University. As seen

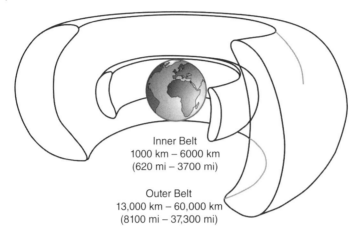

Figure 1.5 Cut-away view of the van Allen radiation belts, showing the inner and outer belts. Source: NASA.

in Figure 1.5, the van Allen belts are concentric toroidal volumes of charged particles trapped by Earth's magnetic field. The geomagnetic field lines trap low-energy charged particles including electrons, protons and some heavy ions. In the magnetic trapping mechanism the particles execute a helical motion around the magnetic field lines; they then reflect back at the poles and execute an oscillating motion between the poles. The particles also drift around Earth due to the Lorentz force; electrons drift east as they are negatively charged while protons drift west, positively charged. The combination of magnetic trapping and east/west drift result in three-dimensional regions of charged particles (Figure 1.6). The inner belt has a relatively stable composition containing mostly protons with a lesser amount of electrons and some charged atomic nuclei. Electrons in the inner belt have energies in the hundreds of KeV, whereas protons in

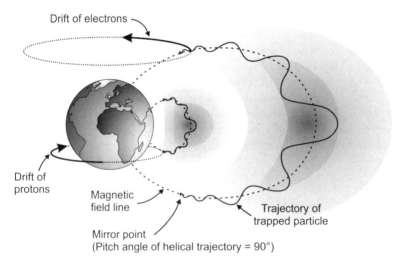

Figure 1.6 The motion of trapped charged particles in the Earth's magnetosphere. Note the helical motion of particles around magnetic field lines and the reflection near the poles.

that belt have energies in the hundreds of MeV. The outer radiation belt varies in size and shape and consists almost entirely of electrons with energies in the 0.1–10 MeV range. The outer belt exchanges particles with Earth's ionosphere but also traps solar wind electrons and protons. The belts are a highly dynamic region and the distribution, concentration, and energy of the trapped particles vary constantly. The system is not restricted to two belts; the Starfish Prime exoatmospheric thermonuclear burst (1962, see below) "pumped up" the belts and created a temporary third belt between the two existing ones. In 2013, a temporary third belt generated by a coronal mass ejection on the Sun was reported [14]; it formed outside the outer (electron) van Allen belt, existing for about a month before being absorbed by the outer belt.

To provide a sense of the complex structure of the van Allen belts, Figures 1.7 and 1.8 show simulated proton flux contours of the lower belt for proton energies of ≥ 0.1 and ≥ 400 MeV, using the AP8 simulation tool. AP8 performs proton flux modeling for the near-Earth environment while the complementary AE8 code performs the same function for electrons. Both these codes were updated in 2012 resulting in the AP9/AE9 tools, which revised AE8/AP8 to improve space system orbital radiation profiles based on more recent data. They also provide better coverage in energy, time, and location for trapped energetic particles and plasma, as well as estimates of instrument error and modeling of statistical fluctuations in the space weather.

The US Vanguard program was started in 1955 with the objectives of orbiting an artificial satellite using modified Redstone rockets, tracking the satellite, and conducting scientific observations and experiments. However, the first actual satellite, *Sputnik*, was

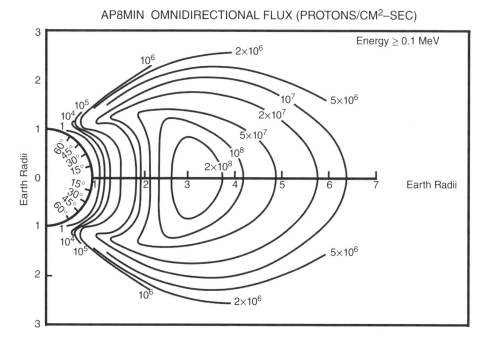

Figure 1.7 Example of a simulation code output: calculated constant intensity flux contours for ≥ 0.1 MeV protons as a function of altitude using the AP8 code. The unit of altitude is Earth radii (R_e). Source: US National Space Science Data Center [15], Figure 65, p. 87.

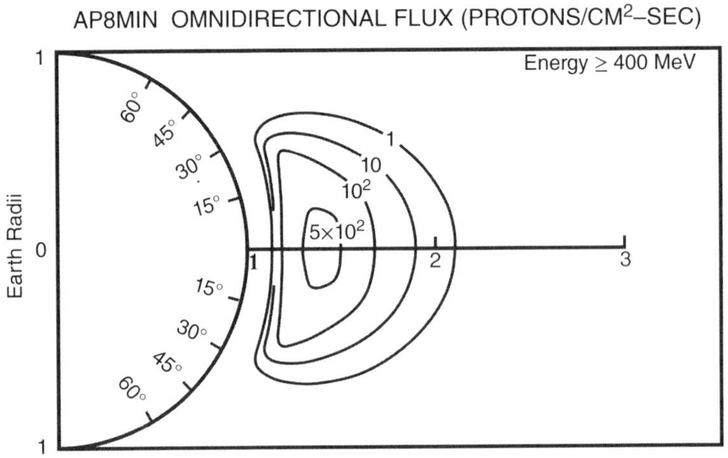

AP8MIN OMNIDIRECTIONAL FLUX (PROTONS/CM^2–SEC)

Figure 1.8 The corresponding constant intensity flux contours for >400 MeV protons as a function of altitude, again using the AP8 code. Note the different belt structure for these high-energy particles. The unit of altitude is Earth radii (R_e). Source: US National Space Science Data Center [15], Figure 70, p. 92.

put into orbit by the Soviet Union in 1957 from Kazakhstan at what is now the Baikonur cosmodrome. *Sputnik* was a small 83 kg vehicle that orbited Earth in 98 minutes, with a simple radio beacon transmitting temperature and pressure data. The airframe was pressurized and included an internal pressure monitor to detect the expected micro meteor strikes. The signature beeping of the *Sputnik* radio beacon abruptly introduced the Space Age in an atmosphere of mutual Cold War anxiety and propaganda and resulted in an intense space race between the United States and the Soviet Union that has, in some aspects, survived the end of the Cold War. After the launch of *Sputnik 2*, the United States overcame a number of booster failures and launched *Explorer 1* and *Vanguard 1*. The National Aeronautics and Space Administration (NASA) was formed in 1958 and took over the Vanguard program, putting space exploration under civilian control.

By 1962, the Cold War between the United States and the Soviet Union was at its most intense, and the 1992 Comprehensive Test Plan Treaty was still decades in the future; above-ground tests of both fission and fusion (thermonuclear) devices were carried out on a regular basis by both sides. The STARFISH PRIME test [16] (Figure 1.9) detonated a 1.4 megaton exoatmospheric thermonuclear device at 400 km altitude above Johnston Island in the Pacific. The effects of this test were noted in Hawaii, 1400 mi away, and included power grid disturbances and microwave communications disruption; this gave the United States its first practical experience with electromagnetic pulse (EMP) phenomena. Fully one third of the early telecommunications satellites then in orbit were permanently damaged or destroyed. These failures led to the first on-orbit repair of a satellite, correcting the gain degradation of a bipolar junction transistor (BJT) using a change in its operating point to thermally anneal out the radiation damage. The van Allen belts were greatly intensified ("pumped up") and a temporary "third belt" [14] was formed. It is interesting to note that the belts are just now settling into their original extent and intensity, some 50 years after the event.

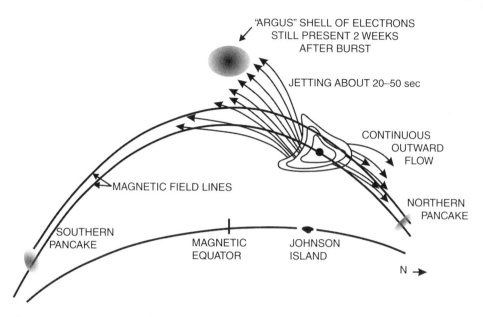

"ARGUS" SHELL OF ELECTRONS
STILL PRESENT 2 WEEKS
AFTER BURST

JETTING ABOUT 20–50 sec

CONTINUOUS
OUTWARD
FLOW

MAGNETIC FIELD LINES

NORTHERN
PANCAKE

SOUTHERN
PANCAKE

MAGNETIC
EQUATOR

JOHNSON
ISLAND

N →

Figure 1.9 No clear idea yet of the real consequences: the Starfish Prime high altitude nuclear event, 9 July 1962, Johnston Island. Source: US DOD [16], Figure 5, p. 20.

The Apollo moon landing and exploration program represented the culmination of years of NASA manned space flight experience and captured the world's imagination as well as providing a considerable propaganda advantage. The landing of *Apollo 11* was the first human venture to another celestial body and made history for the human race. The space race and its propaganda value lost much of their immediacy as the Soviet Union collapsed in 1992, and public interest in space exploration faded further as a consequence of the two space shuttle disasters, poorly defined objectives, and persistent funding cuts. Interest in manned space flight dropped yet further as the US space shuttle program wound down with no clear plans for a replacement capability, reducing the United States to contract its space station support flights using Russian vehicles. At the time of this writing (2018) a transition from government to commercial space transportation is proceeding in a spectacularly successful manner, and NASA is increasingly tightening it focus on robotic interplanetary science missions to Mars and beyond.

The leading science-fiction author Arthur C. Clarke is considered the inventor of the basic principles of the communications satellite, having proposed the use of artificial geosynchronous satellites for relaying radio signals in a paper published in *Wireless World* [17] in 1945. The relaying aspect of a communications satellite is clearly defined by the industry term *bent pipe*, as such a satellite supports a number of receive/transmit channels and is in that respect no different from a microwave relay station on Earth. There are three principal orbital regimes: the low-Earth, Molniya, and geosynchronous (or sometimes Clarke) orbits. A "geosync" satellite revolves around Earth in a circular equatorial orbit at the same angular velocity as the surface of the Earth itself and hence remains in a fixed position overhead [17] with respect to an observer on that surface.

A ground-based antenna can hence track a geosynchronous satellite with only minor pointing corrections, saving cost and complexity and justifying the extra cost of inserting a satellite into the high-altitude geosynchronous orbit. The main drawbacks of geosynchronous satellites are increasing crowding in the 37 000 km orbit and reduced coverage in the polar regions. The high altitude also requires increased transmitter power, higher signal-to-noise performance receivers, and better antenna gain and directivity for equivalent link quality, as well as adding a sometimes objectionable 0.25 second delay through the link.

The geosynchronous Earth orbit (GEO) is equatorial and will have poor polar coverage; this led to the development of Molniya orbits, with the name reflecting the not unexpectedly Russian origin of this approach. The first half-day orbit Molniya orbit satellite was launched by the USSR in the 1960s and was used to relay television signals from Moscow to the Russian Far East. The Molniya orbit is highly inclined and eccentric, with the satellite above the far northern latitudes for much of its orbit and with a nearly constant ground footprint. The half-day period means that at least three satellites are needed to provide 24-hour coverage, providing a compromise between single GEO satellites and the multiple satellite "constellations" needed for low Earth orbit (LEO) systems. The polar regions also have a more intense cosmic ray environment, which must be considered in the design of satellite electronics for this application.

The third and final communications satellite orbital regime is LEO, which is defined as a circular orbit typically some 400–500 km above the surface. Since the very high frequencies used for satellite communications are line of sight only, these satellites have a very limited ground coverage of perhaps 1000 km, resulting in a requirement for many satellites in a "constellation" such as the Iridium, Teledesic, and Globalstar voice telephone systems in order to provide continuous coverage. LEO satellites are less expensive to launch than geosynchronous or Molniya satellites and have reduced signal strength and noise performance requirements; recall that electromagnetic wave propagation follows a $1/r^2$ relationship, so the effect is dramatic. By some accounts the Iridium system was introduced well ahead of its time and with a poorly understood business model, but it currently provides truly global coverage by 66 operational satellites orbiting in a circular 780 km orbit. The Iridium system introduced many satellite design challenges due to the large number of satellites in the constellation, with the manufacturer building the "birds" on a mass-production line rather than the massively expensive (USD 250–500 million) one-off approach used for larger satellites. The cost constraints inevitably led to reliability tradeoffs, which, in turn, drove the need for a number of spares in the constellation. Figure 1.10 provides a perspective on the various orbits and a sample of their satellite occupants.

The descriptions of these three distinct orbital regimes suggest that satellite electronics and their radiation response have to be specifically designed for their intended orbit of application. For example a spacecraft in a highly eccentric Molniya orbit will fly though the van Allen belts four times a day and will incur a much greater amount of radiation damage by electrons and protons than a LEO "bird" flying well below the inner electron belt. Similarly, a geosynchronous satellite does not have the geomagnetic shielding advantage provided by Earth's magnetic field and will hence be exposed to a larger heavy ion flux than satellites in lower orbits.

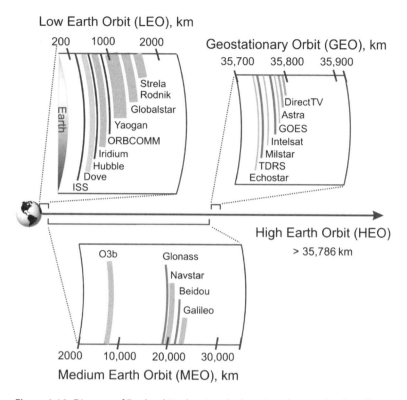

Figure 1.10 Diagram of Earth orbits showing the location of a sample of satellites.

1.5   Semiconductors – Revolution, Evolution, and Scaling

The semiconductor revolution started in 1948 with the invention [18, 19] of the transistor by J. Bardeen, W. B. Shockley, and W. H. Brattain at Bell Laboratories in the late 1940s. Figure 1.11 shows a portion of the patent drawings for a point contact transistor issued to Bardeen in 1950. The prototype had two gold foil traces separated by a plastic wedge and the entire combination pressed against a small (and unoxidized) bar of germanium. The Au–Ge contacts were then stabilized by running a current pulse through them, resulting in opposing P-N junctions. The base width of this lateral device is set by the Au trace spacing. This was the most radically different electronic invention since the advent of the vacuum tube in the first decade of the twentieth century and has in time driven profound changes to human society. Both the vacuum tube and the transistor are amplifying devices; the fundamental difference is the means by which current flow is supported and controlled. In a vacuum tube, current flow results from the motion of charge carriers (which in this case are electrons) between electrodes within a high vacuum envelope. In contrast, a transistor is a solid-state device in which current flow results from the motion of charge carriers (in this case, electrons and holes) in a crystalline solid. Electrons and holes are an integral part of the solid, while the electrons in

Oct. 3, 1950   J. BARDEEN ET AL   2,524,035
THREE-ELECTRODE CIRCUIT ELEMENT UTILIZING
SEMICONDUCTIVE MATERIALS
Filed June 17, 1948   3 Sheets–Sheet 1

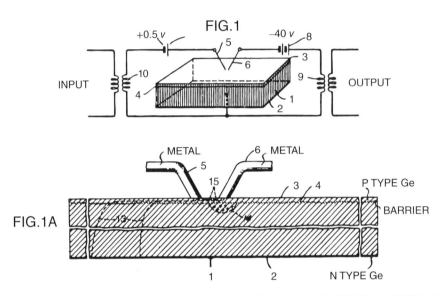

Figure 1.11 Point contact transistor patent issued to J. Bardeen et al. in 1950. Source: US Patent Office.

a vacuum tube must be generated by either thermionic emission from a hot cathode or by high field emission from a cold cathode. In addition, the vacuum tube requires much higher supply voltage and is a bulky and fragile device due to the need for a vacuum envelope. The elimination of the thermionic cathode provides an immediate reduction in system power, as filament power is to the order of a Watt even for small 6.3 V receiving tubes. The filament also represents an inescapable time-dependent failure mechanism, as its electron emissivity (or that of the cathode it heats, in indirectly heated devices) drops as a function of time and the filament itself has a limited lifetime. Properly designed and applied solid-state devices have no intrinsic wearout mechanisms, and will, in theory, operate indefinitely. With the introduction of flat-screen displays and the resulting elimination of vacuum cathode ray tubes, vacuum tube technology is now limited to traveling wave tube (TWT) microwave power amplifier and specialized high-end audio amplifier applications.

As indicated by the name, semiconductors are based on a given semiconducting material, which is defined as any of a class of solids (such as Ge or Si) characterized by electrical conductivity that is between that of a conductor and that of an insulator, with a pronounced and often negative temperature coefficient. These materials were first described by Faraday in 1833, who noted a negative thermal coefficient of resistance (TCR) in samples of silver sulfide (Ag_2S), as opposed to the positive TCR commonly found in metals. The first potentially useful phenomenon in semiconducting materials was the Hall effect, discovered [20] by E. H. Hall in 1879 in metallic conductors. The Hall effect turned out to be a useful diagnostic of carrier concentration and mobility and provided the first evidence of holes, which are electron vacancies and which effectively function as positive charge carriers.

Practical applications of semiconductors lagged as materials technology struggled to catch up with theoretical work, with the invention [21] of the basic metal-oxide semiconductor field effect transistor (MOSFET) structure by J. E. Lilienfeld as early as 1925 as a good example. It took until the 1960s for technology to progress to the point of realizing a practical implementation of this deceptively simple device, and the Lilienfeld patent was still a factor at the time of introduction of the BJT in the late 1940s. In 1928, A. J. W. Sommerfeld proposed the Fermi-Dirac-Sommerfeld statistical theory of electron conduction in solids, followed by the introduction by A. H. Wilson of the energy band theory of solids and the concept of an electron vacancy acting as a positive charge carrier. Rectifying *point junction devices* were already well known at this time, with simple galena (PbS) crystals contacted by a W wire in widespread use as detectors in radio receiving apparatus. Silicon-based rectifiers appeared at the onset of WWII, driven by microwave detector requirements for critical radar technology, and efforts to improve these devices drove advances in single-crystal silicon and germanium fabrication, purification and metrology. Crude semiconductor devices such as copper oxide or selenium rectifiers were even then beginning to displace vacuum rectifiers in low-power applications.

The advances in single-crystal materials processing resulted in greatly accelerated semiconductor technology following World War II. Shockley developed the theory [18] of p-n junctions and of BJTs in 1949, a year after the development of the first successful point contact "transfer resistance" or "transistor" [19] device (discussed above) by Bardeen and Brattain in 1947. In 1948, Bardeen and Brattain filed for a patent that was eventually issued to Bell Labs – US patent #2,524,035, "Three Electrode Circuit Element Utilizing Semiconductive Materials." Later, Ge alloy junction transistors became the first practical semiconductor devices, and diffusion as a means of introducing impurities into semiconductor materials was developed at General Electric. The advantages of solid-state devices in military applications were immediately obvious and led to greatly enhanced research activity under sponsorship of the US DoD. Bell Laboratories research resulted in the planar epitaxial process in 1960, which laid the foundation for practical integrated circuit technology. Shockley, Bardeen, and Brattain received the Nobel Prize in 1956 for their pioneering work.

The planar process introduced the notion of creating diffused regions by the use of successive oxidation/photoresist/exposure/etch/diffusion operations, exploiting the fact that a silicon dioxide (SiO_2) layer of the proper thickness is well suited to block impurity (dopant) diffusion and will hence enable selective diffusion. The planar process sequence also introduced the concept of simultaneously defining all of the die on a wafer in a single set of operations, which was a great reduction in die cost and which predictably drove increases in wafer size. Production single-crystal Si wafers were 0.5″ in diameter in 1961, grew to 1.0″ in 1964, and continued steady growth through 4.0″ in 1975 to the present 12.0″ and even 18.0″. It did not take long to realize that putting more than one transistor on the same die was a straightforward operation, with Jack Kilby at Texas Instruments [22] disclosing the initial concept in 1959, describing the device that would subsequently be called an *integrated circuit* as having "a plurality of electrical circuit components in a wafer of single-crystal semiconductor material … a plurality of junction transistors defined in the wafer …," with the objective of creating "a body of semiconductor material ... wherein all the components of the electronic circuit are completely integrated." In this case the "material" was still Ge, and the interconnecting was done with simple Au wires, which was a major scaling limitation, as it was still a die-by-die rather than a wafer-by-wafer operation.

The interconnect problem was quickly solved by Robert Noyce at Fairchild, who used a thin patterned layer of evaporated Al to interconnect the elements on the chip; other than many more layers, greatly reduced feature sizes and some exotic metallurgy, this general approach method to IC manufacture continues to this day. Kilby won the 2000 Nobel Prize in physics, somewhat belatedly, for his part of the invention of the integrated circuit. Later work by Noyce developed the same set of concepts in Si, which was a better material for higher-temperature applications and which also features a much more durable oxide that works well as a diffusion barrier and has a stable oxide/semiconductor interface. It was also realized that the "plurality of transistors" on one chip would need to be isolated from each other, and this was accomplished by Kurt Lehovec at Sprague, who invented a system of separate diffused regions that used back-biased junctions to isolate the devices. Later work at Radiation, Inc., the predecessor to several of the authors' company, developed a system of passive isolation using oxide layers instead of diffusions; both isolation systems are still in use, with the passive isolation approach yielding several advantages in harsh environment applications. Figure 1.12 shows a classical

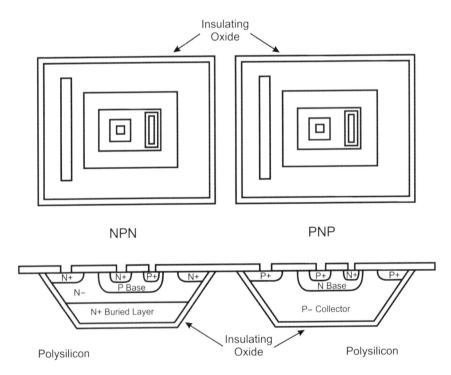

Figure 1.12 The "single-poly" dielectrically isolated (DI) structure. Small, single-crystal regions are insulated from each other and from the polycrystalline silicon substrate by a thick thermal oxide. Source: Adapted from Harris Semiconductor Custom Integrated Circuits Division Rad-Hard/Hi-Rel Data Book 1987, Figure 2, p. 1–5. Reproduced with permission of Renesas Electronics Americas, Inc.

"single-poly" dielectrically isolated (DI) structure, in which small, single-crystal regions are insulated from each other and from the polycrystalline silicon substrate by a thick thermal oxide.

The connections to the external world have been made mostly by thin Au or Al wires, with early work at Bell Labs in electroplated Au beam leads for difficult environments; flip chip and bump bonding are now preferred for large IO count (in the thousands) functions such as field-programmable gate arrays (FPGA) and logic devices. The die and wires are contained in a compact package of either hermetic ceramic or injection-molded plastic construction. In integrated circuits, the maximum number of transistors per unit area turned out to be a function of photoresist and photomask tolerances and minimum dimensions, and ICs grew quickly from a few transistors in the diode-transistor logic (DTL) of the early 1960s and 1 K static random-access memories (SRAM) of the next decade, to the first primitive microcontrollers of the 1980s and the multimillion gate FPGA devices used extensively in today's systems.

Early in the game, Gordon Moore of Intel proposed in a 1965 *Electronics* paper [23] appropriately titled "Cramming more components onto integrated circuits" what would eventually be known as Moore's law (which is not a law at all, in this writer's opinion, but rather a nicely accurate rule of thumb), predicting (Figure 1.13) that the transistor count can be expected to double each year, with later refinements claiming a doubling every 18 months or 2 years and trying to insert defect density terms.

What these pioneers did not predict was the sheer magnitude, the extreme competitive nature, and the cyclical characteristics of the industry that would result from this work, with average selling prices for a given part routinely falling 10% or 20% per year. No other commodity in today's society shows regular, predictable selling price erosion,

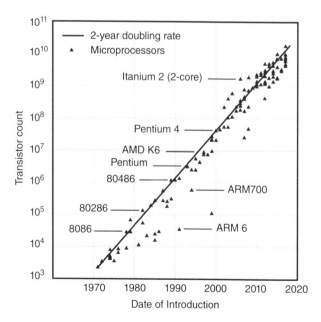

Figure 1.13 A log-linear plot of Moore's law (or rule of thumb), showing the level of integration as a function of time for microprocessors.

and the pressure at the manufacturing level can be extreme; if the selling price drops, the costs of building the part have to drop by that percentage or by more in order to retain a commercially viable business. The integrated circuit has revolutionized the electronics industry and indeed the whole of human society, yielding capabilities that could not have been envisioned just a few years ago. Moore's law itself is running into nonlinearity as critical dimensions drop below UV wavelengths and lithography runs into limitations, as well as localized variations in doping density; this is driving research into 3D structures and alternative materials.

After years of bipolar transistor dominance, the 1925 Lilienfeld patent for an electrostatically controlled transistor was revisited at Radio Corporation of America and at Fairchild in the late 1960s, and practical MOSFET devices were soon developed. It was soon realized that the cleanliness of the then-current fabrication processes was not at all adequate to produce a stable and high-yield MOSFET device, notably in the areas of mobile ionic contamination and processing chemical purity. Initial metal-oxide semiconductor (MOS) applications used P-channel MOS (PMOS)-only devices, necessitating negative logic and supplies, but improved processing soon enabled N-stable MOS (NMOS) parts. Work in 1963 by F. M. Wanlass at Fairchild [24] took advantage of the availability of MOSFET devices of both polarity to create complementary MOS (CMOS). In this logic format the PMOS and NMOS devices were stacked, with both gates common, resulting in a function with no DC power dissipation; either the top P-channel or the bottom N-channel are turned on at any time, but not both, and the output voltage swings very near the power supply rails. There is thus no DC current drawn by the circuit, and the only power dissipated is during switching and is given by $P = CV^2f$. This low-power capability contributed to higher levels of integration and enhanced system complexity, especially in portable battery-powered and medical applications, and enabled today's large-scale devices such as FPGAs and microprocessors. The continued decrease in minimum feature sizes also drove the move to CMOS-based products, as bipolar devices do not scale as well as CMOS.

1.6 Beginning of Ionizing Radiation Effects in Semiconductors

Initial radiation effects studies were triggered by nuclear weapon testing and, not surprisingly, concentrated on bulk displacement damage (DD) effects as caused by neutrons and protons. DD leads to lattice damage and to an increased density of recombination sites, which cause minority carrier lifetime reduction in the bulk semiconductor and gain degradation of the BJT. This was first assumed to be the dominant mechanism in semiconductor radiation effects and experimentation was carried out starting in the late 1940s using alpha particles, neutrons, and electrons as well as X-ray radiation to approximate the nuclear device spectrum. This concentration on bulk radiation effects was driven by the prevailing (the only, really, at the time) semiconductor technology, which consisted of bipolar junction devices and diodes; the MOSFET device has a much different set of mechanisms but a practical version of this device was at the time still a decade or more in the future.

We have earlier mentioned the extensive above-ground test program of both nuclear and thermonuclear devices carried out by both sides during the Cold War. The July 1962

STARFISH exoatmospheric high-altitude test [16] was a key event; it caused widespread disruption of electrical power grids and damaged several early communications satellites then in orbit. The *Telstar 1* satellite command system failed later in November 1962, and this failure was traced by ground measurements to degradation caused by surface-related effects in semiconductor components. The system was subsequently restored to normal operation by the first on-orbit correction of a radiation-induced problem. The Telstar failure and recovery confirmed that a repair could be performed by thermally annealing out the oxide damage, and recovery was indeed accomplished by changing the operating points of the failed transistors and letting the resulting heat anneal out the damage. These failures did not go unnoticed, and the Soviet Union indicated that they were aware of these effects and that the vulnerability of space systems would be considered as a potential way of destroying or incapacitating them.

The need for survivable space systems was soon recognized by the United States and an intensive program of radiation studies was initiated, building on the Telstar experience. The successful ground-initiated repair of the Telstar satellite advanced the radiation effects field into surface damage effects, in contrast to the BJT bulk damage mechanisms investigated in earlier research. The understanding of the basic mechanisms of charge-trapping dynamics in semiconductor oxide layers was first applied to the understanding of BJT behavior, as MOS devices were not yet in their current predominant position. The MOS device is a majority carrier surface device that is insensitive to minority carrier lifetime, and was initially believed to be inherently hard in the ionizing radiation environment. This was soon found to be entirely incorrect, and the MOS device was subsequently shown to be sensitive to ionizing radiation with much more complex failure mechanisms. Initial work advanced the use of composite or doped oxides or the use of composite or non-SiO_2 insulators such as nitrides or oxynitrides instead of simple thermally grown oxides. As in normal oxides, the radiation hardness of these alternative materials is driven by the relative charge trapping rates of holes and electrons; in thermal SiO_2 the electron mobility is many orders of magnitude greater than the hole mobility, leading to excessive positive volume and interface trapping in the film. Composite and non-SiO_2 materials were found to have a more reasonable electron–hole mobility ratio and were effectively used in a number of hardened processes. The need for these exotics eventually diminished as the industry realized the importance of process cleanliness in yield, stability and reliability, leading to the development of greatly improved fabrication tools. It was found that a combination of modified thermal processing and specific layout ground rules provided adequate hardness for most applications. Exotic materials did persist in such technologies as silicon-on-sapphire and the later silicon-on-insulator approach, in which the combination of a thin device layer with a highly disordered crystalline structure was found to yield parts that were not only hard to ionizing radiation but to single-event effects (SEE) from protons and heavy ions as well.

The intensive early research into surface effects soon produced the analytical tools needed for understanding these mechanisms. The capacitance-voltage (C-V) plot was recognized as a powerful adjunct to conventional current-voltage (I-V) characterization. Trapping in dielectric layers was found to involve a much more complex set of processes than had perhaps been expected, with the positive charge buildup depending critically on dose, dose rate, electric fields in the oxide layer, layer thickness, operating temperature and process variables, with even the crystallographic orientation of the silicon substrate playing a role. The trapped positive charge was found to be easy to anneal out using temperatures well below the relatively high processing temperatures, and it was also realized that the trapping rates and mechanisms varied throughout the layer, with the identification of bulk and interface traps of very different behavior and consequences. Any irradiation with energy in excess of the 9.3 eV bandgap of SiO_2 will drive electron–hole pair generation and will result in net positive trapped charge due to the carrier mobility differential. The much-higher electron mobility and the lower density of electron traps quickly depletes the oxide of these carriers. The holes, on the other hand, have both reduced mobility and a region of hole traps near the SiO_2/Si interface to overcome. A great deal of quantitative work has been done on mobility and trapping mechanisms and on the dynamics of oxide trap density (N_{ot}) and interface trap density (N_{it}). This topic will be discussed in greater detail in Chapter 3.

The packing density of semiconductor processes is characterized by their minimum ("critical") dimensions, which are the driver for Moore's law and which have been on a constant decline since the beginning of the semiconductor industry. Hardened processes follow this trend as well, although a successful hardened process may be several generations behind the current industrial leading edge. The total dose hardness of a given gate or field oxide has been shown to depend strongly on processing parameters, and these oxides were found to be dramatically harder if grown at lower temperatures (in the 850–1000 °C range as opposed to the commonly used 1100–1200 °C) and in dry oxygen as opposed to wet (steam) O_2. The metal gate electrodes used in early MOS processes led to exposure of the gate oxide during metal deposition, and energetic deposition methods such as radio frequency (RF) sputtering and e-beam deposition were found to damage the oxide; hardness required earlier methods such as filament or crucible evaporation. The thinner gate and field oxides found in higher-density processes were found to be much harder to ionizing radiation, which is a result of the reduced oxide volume available for charge trapping and the resulting reduction in trapped charge per unit area. The hardness trend continues to this day, with current deep submicron processes using gate oxides of a few nanometers showing total dose hardness to high levels. In these processes, the gate oxide is so thin that it is no longer the driving parameter [25] for hardness, with field and device leakage and the resulting increases in quiescent power dissipation turning into the key issues. In these thin oxides, the oxide trap density scales with thickness, but the interface trap density does not, and this last parameter then turns into the limiting factor. Figure 1.14 shows the influence of oxide thickness on MOSFET threshold and flatband voltage shift response under total dose irradiation [26].

1.7 Beginning of Single-Event Effects in Semiconductors

SEE are a broad range of phenomena generally caused by *single* energetic protons, heavy ions, or neutrons. A *single event* implies a phenomenon caused by one "single" particle, as opposed to cumulative radiation damage by a very large number of, for example, X-ray photons or protons. This definition also implies that SEE are a random phenomenon, in which a given event can happen tomorrow or not for another 100 years. Suspected SEE in digital electronics were first reported during above-ground nuclear tests over the 1954–1957 time frame, with many anomalies observed in monitoring systems; it is not clear if these effects were due to actual single events or to the very high levels of transient radiation generated by these devices. SEE was first predicted [27] in the context

Figure 1.14 Normalized MOS device threshold and flatband voltage shifts at 80 K as a function of oxide thickness. The dashed line represents the square of the oxide thickness while the circles are experimental results. Source: Benedetto et al. 1985, Figure 6, p. 3919. © 1985 IEEE. Reprinted, with permission, from [26].

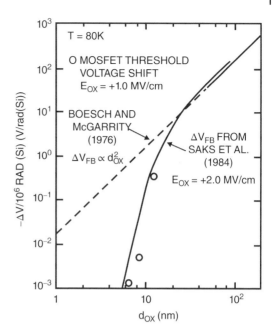

of feature size reduction trends in commercial integrated circuits. Actual single-event upsets (SEUs) were first reported [28] in 1975, with upsets observed in bipolar bistable elements (flip-flops) during a total mission duration of 17 years, and the definition of SEE was subsequently expanded [29] to include a broad range of nondestructive and destructive effects. As we will see in Chapter 2, high-energy charged particles exist as part of the natural background and are referred to as galactic cosmic rays (GCRs), somewhat inaccurately, as these are particles and not "rays" at all, but the name has persisted. The van Allen belts contain a high density of energetic solar protons; the proton has a low mass and is hence a low linear energy transfer particle (LET of 2–4 MeV cm²/mg, in which the LET is a measure of the energy lost [dE/dx] by the particular particle species per unit length for a given material density) at any but the most extreme energies, but protons can indeed cause [30] SEE issues in softer commercial parts. The energetic protons and heavy ions encountered in cosmic rays are highly penetrative, and no maximum energy has been documented. Protons make up the bulk of the low-energy end of the cosmic ray flux vs. energy spectrum (see Chapter 5) and can be shielded against, but the high energy end of the spectrum has a finite (but very low) flux out to the GeV range and shielding is ineffective, given satellite weight constraints. System hardening must in this case be accomplished by other means.

The first SEE phenomena observed were nondestructive "soft" errors in digital electronics, an effect known as SEU. This is strictly limited to digital systems and is defined as a soft error caused by an energetic (and therefore highly ionizing) ion or proton striking a sensitive node within a microelectronic device, resulting in a bit flip in a bistable digital element. Applying the SEU label to any other form of SEE is frequently done but is incorrect. In an SEU event, an incoming proton or heavy ion strikes the silicon chip, creating an ionized track of electron–hole pairs. The resulting charge is separated by the electrical fields in the device and then appears as a charge pulse (and hence a voltage pulse) at a circuit node. If the struck node is a sensitive one, such as an input of a flip-flop memory element, the flip-flop will change logic state and remain in that erroneous state until the system is reset or other means of error correction are carried out. SEU is a process-dependent effect and early integrated circuits using BJT technology were particularly sensitive, as was found in the early months of the Hubble Telescope mission; the use of bipolar random-access memories led to predictable operability issues every pass over the South Atlantic Anomaly.

Atmospheric neutrons have also been found to cause SEU [31] and other effects; while the neutron itself is neutral and cannot cause direct ionization, nuclear reactions in the material will generate cascades of charged secondary particles such as protons, some of which may have sufficiently high LET to cause SEE. These neutron effects are found in polar orbits or in high-altitude aircraft, where the atmospheric neutron flux is highest. A third source of SEU problems is alpha particle SEU; the alpha particle is a He nucleus emitted by many radioactive substances. Alpha particle SEU from residual radioactive materials in ceramic integrated circuit packages was first reported by May and Woods [32] of Intel in 1979 and led to improved control of radioactive isotopes in these materials. In 1979 researches from IBM and Yale reported a new mechanism whereby sea-level heavy ions could cause SEU in electronics; this was not initially a difficult problem, as the heavy ion flux at sea level is very low, but the advent of large "server farms" in the Internet age has brought this problem into considerable prominence. The issue, of course, is that these servers use commercial off-the-shelf (COTS) parts as a part of necessary cost control in a very competitive business; the use of any form of hardened or even just SEE-characterized parts is simply prohibitive for these applications.

SEU has broadened into multiple-bit upset (MBU), which is commonly found in large deep submicron memories. The memory cells in these devices have approached and in some cases are smaller than the diameter of the ionized column, and the resulting charge deposition can affect several bits at the same time. This makes error detection and correction (EDAC) algorithms of limited use.

Single-event transient (SET) effects are a later development in which the output of a continuous time analog function shows short-term transients. The length and width of these transients depends on the circuit design and application, and in many cases can be addressed by simply low-pass filtering the output of the device. SET has also arisen in deep submicron digital applications, in which a transient is not necessarily a problem until it is digitized by, for example, a comparator; this can result in setting an erroneous state in a bistable element, requiring error correction or a system reset. SET has also become a major issue in space power management systems, in which a voltage transient can cause immediate permanent damage to sensitive low-voltage, high-current loads such as mass memory or FPGA devices.

In addition to these nondestructive effects, there is a broad range of destructive single-event phenomena, implying permanent damage to the part. Single-event latchup (SEL), single-event gate rupture (SEGR), and single-event burnout (SEB) are commonly encountered.

SEL is similar to conventional electrically caused latchup, in which parasitic bipolar transistors in the substrate form four-layer devices. These are similar to the silicon-controlled rectifiers (SCRs) that have been used in power switching for decades; a four-layer device has two stable states, one OFF and one ON. The ON state results from a PNP–NPN combination being configured in a regenerative loop (i.e. p-type, n-type, p-type transistors combined with n-type, p-type, n-type transistors). The device's resistance in this state is extremely low as both transistors will saturate. In an integrated circuit, this low-resistance path results in a short-circuit between the supplies, and the resulting overcurrent will quickly destroy the part. SEGR applies to MOSFET devices, with power MOSFETs with large gate area most susceptible. In this mechanism, a heavy ion travels through the gate oxide, leaving a conductive track. This in itself would not necessarily be destructive, but these large devices have considerable gate-to-source and gate-to-drain capacitance, and this will discharge through the conductive path and lead to localized melting and interconnect failure. SEB is quite similar to SEL in that parasitic four-layer devices are triggered by a particle-induced voltage pulse, resulting in localized damage and a nonfunctional device. All three of these destructive effects are commonly mitigated by straightforward derating, in which a power MOSFET of considerable higher voltage rating is used. Hardening at the chip level is difficult, and in the case of SEL and SEB involves layout changes such as guard rings and isolation regions, epitaxial substrates, or passive isolation technologies such as silicon on insulator (SOI). SEGR is a simpler mechanism and is addressed primarily by the use of derating (which effectively implies a thicker gate oxide), which introduces performance compromises.

Evaluation of the SEE sensitivity of components and validation of hardening approaches are done by irradiating and monitoring samples while electrically active, using an energetic heavy ion or proton source. In most cases, these sources use either a van de Graaff high-voltage generator or a cyclotron to accelerate the particles, with a magnet system used to provide a monoenergetic beam. SEE testing is a demanding discipline that requires an in-depth knowledge of the part's design and process details, and sufficient experience to interpret what can often be a very unpredictable response. The fixturing and its signal processing electronics will be partially in the beam and will need to be shielded, while the signals need to be transmitted to the remote experiment room by the use of coaxial cables. These cables themselves have considerable capacitance and will require line drivers on the fixture end to retain signal integrity. The sample may be tilted at positions other than normal beam incidence in order to increase the effective LET of the particle being used; this simplifies beam tuning, which is an expensive and time-consuming exercise, but introduces nonlinearities, and the results do not always correlate with normal beam data. It should also be noted that SEE testing is a classical accelerated test, with particle fluxes for typical qualification testing ranging up to 10^4 ions/(cm^2 s). High flux testing has many similarities to conventional accelerated testing, in which the results correlate up to the point when the test enters a nonlinear regime and the results stop making sense. In-depth SEE testing and error rate modeling requires irradiation at a number of ion energies; this data is used to develop a cross-section vs. LET curve, which provides data for use in mission and orbit simulation. This SEE test methodology is very useful for predicting heavy ion error rates for space applications, and can also be used to determine terrestrial neutron error rates. In this case, a large number of parts must be evaluated using a neutron beam, with a number of neutron energies required to develop a cross section vs. LET curve as in heavy ion testing.

1.8 Summary and Closing Comments

In this Introduction we have provided a necessarily brief historical overview of the evolution of the physics of radiation effects. We then traced this sequence through the Nuclear Age, discussing the development of nuclear weapons and power generation, the evolution of space flight, and the attendant need for hardened parts for these diverse environments. Next, we provided a more detailed summary of the evolution of space systems, from early suborbital experiments to manned space flight. Brief discussions of nuclear weapons and commercial reactors and their much different environments follow; we then segued into a discussion of the historical development of semiconductor technology. The Introduction concludes with a more detailed discussion of ionizing radiation effects and SEE on semiconductor components.

Problems

1.1 Exponential Decay

If a process reduces a quantity of material at a rate that is proportional to the current amount of material, that process can be described by exponential decay

$$\frac{dN}{dt} = -\lambda N. \tag{1.4}$$

The solution to Eq. (1.4) is straightforward and has the form

$$N(t) = N_0 e^{-\lambda t} \tag{1.5}$$

where N_0 is the initial quantity of material at $t = 0$.
a) Use Eq. (1.5) to derive the mean lifetime $\tau = \langle t \rangle$.
b) Describe the significance of the mean lifetime dependence on N_0.
c) Use Eq. (1.5) to derive the half-life $t_{1/2}$ in terms of the mean lifetime τ.
d) Use the half-life values from Table 2.3 to calculate the time until a given quantity of the material reaches a "safe" level. For the purpose of this problem, use a safe level of 10 ppm.

1.2 The Rydberg Formula

In the 1880s, Swedish physicist Johannes Robert Rydberg (1854–1919) was working on a formula to describe the relation between the wavelengths in spectral lines of alkali metals. Rydberg found that his calculations were simplified if he used the wavenumber (the inverse of wavelength or the number of waves in a unit of length), equal to $1/\lambda$. He realized that Balmer's formula for hydrogen

$$\lambda = \frac{hm^2}{m^2 - n^2} \tag{1.6}$$

for $n = 2$, $h = 3.6456 \times 10^{-7}$ m, and $m = 3, 4, 5, \ldots$ was a special case for a more general formula he was using for his studies. Rydberg rewrote Balmer's formula in terms of wavenumbers and developed a formula that provided the wavelengths of all of the hydrogen emission lines:

$$\frac{1}{\lambda} = R \left(\frac{1}{n_1^2} - \frac{1}{n_2^2} \right) \tag{1.7}$$

For integers n_1 and n_2 with $n_1 < n_2$, R (Rydberg constant) $= 1.097 \times 10^7$ m^{-1} Eq. (1.7) can be modified for all "hydrogen-like" nuclei (atoms with only one electron interacting with the nucleus such as He$^+$, Li^{2+}, Be^{3+}) by replacing the Rydberg constant R in Eq. (1.7) with $Z^2 R$, where Z is the atomic number of the element.

a) Show that the Rydberg formula reduces to Balmer's formula when $Z = 1$, $n_1 = 2$, and $n_2 = 3, 4, 5, \ldots$

b) How are h in Eq. (1.6) (not to be confused with Planck's constant) and R in Eq. (1.7) related when $Z = 1$?

c) If the indexes n_1 and n_2 are limited to $n_1 = 1$ and $n_2 = 2$ and the modified Rydberg formula is further modified by replacing Z with $Z - 1$, the formula can be used to estimate values for the K-alpha lines (analogous to the Lyman-alpha line transition for hydrogen). Use this formula to predict the frequency of the K-alpha line for Tungsten. Describe the type of radiation produced by this electron transition.

1.3 The Bohr Atomic Model

a) Derive an equation for the electron orbit radii using the Bohr model and Coulomb's law. Calculate the Bohr radius (orbit radius for the lowest electron energy state for hydrogen, $n = 1$).

b) Relativistic effects cause a splitting of some electron energy states. This leads to a splitting of lines in the spectrum for hydrogen. The fine structure constant determines the relativistic corrections to the Bohr energy levels. Calculate the fine structure constant for hydrogen

$$\alpha = \frac{v_1}{c} \tag{1.8}$$

$v_1 = $ electron velocity for $n = 1$ orbit, $c = $ speed of light

c) The electron volt (eV) is the preferred energy unit in particle physics. An electron volt is the energy acquired by and electron moving freely through a 1 V potential difference. Electrons moving in space have kinetic energy according to their velocity but no potential energy if they are not under the influence of an electric field. Electrons in orbit around a nucleus are said to be in a potential well and have a potential energy:

$$PE_n = -\frac{k_0^2 Z^2 e^4 m}{n^2 \hbar^2} \tag{1.9}$$

$k_0 = 8.98755 \times 10^9$ $N\,m^2/C^2$ ($= 1/4\pi\varepsilon_0$, $\varepsilon_0 = 8.85418782 \times 10^{-12}$ F/m).
The lowest electron energy for an atom occurs for the $n = 1$ orbit, or ground-state. Calculate this energy (answer in eV) for hydrogen and helium.

d) What is the significance of the negative sign in Eq. (1.9)? At what distance from a nucleus will and electron have a potential energy of 0?

References

It is not our objective to provide an exhaustive set of references, which in any case would take up too much space. We have provided a representative set of references that are of historical interest and will provide a good starting point for further study.

1 Maxwell, J.C. (1904). *A Treatise on Electricity and Magnetism*. Oxford: Clarendon Press.

2 Crookes, W. (1878). On the illumination of lines of molecular pressure, and the trajectory of molecules. *Philosophical Transactions of the Royal Society A* 170: 405–408.

3 Thomson, J.J. (1897). Cathode rays. *The Electrician* 39: 104.

4 Landwehr, G. and Hasse, A. (eds.) (1997). *Röntgen Centennial: X-Rays in Natural and Life Sciences*. World Scientific.

5 Rutherford, E. (1899). Uranium radiation and the electrical conduction produced by it. *Philosophical Magazine* 47 (284): 109–163.

6 Bohr, N. (1913). On the constitution of atoms and molecules. *Philosophical Magazine* 26, 1–25, 476–502, 857–875.

7 Chadwick, J. (1932). The existence of a neutron. *Proceedings of the Royal Society A: Mathematical, Physical and Engineering Science* 136 (830): 692.

8 Einstein, A. (1905). On the electrodynamics of moving bodies. *Annalen der Physik* 17 (10): 891–921.

9 Einstein, A. (1905). Does the inertia of a body depend upon its energy content? *Annalen der Physik* 18 (13): 639–641.

10 Hansen, C. (1995). *US Nuclear Weapons Histories, Swords of Armageddon: US Nuclear Weapons Development Since 1945*. Chukelea Publications.

11 Groves, L. (1962). *Now It Can Be Told: The Story of the Manhattan Project*. Harper & Row.

12 Fermi, E. (1946). The development of the first chain reaction pile. *Proceedings of the American Philosophical Society* 90: 20–24.

13 Foerstner, A. (2007). *James Van Allen: The First Eight Billion Miles*. University of Iowa Press.

14 Baker, D.N., Kanekal, S.G., Hoxie, V.C. et al. (2013). A long-lived relativistic electron storage ring embedded in Earth's outer Van Allen belt. *Science* 340 (6129): 186–190.

15 NASA Goddard Space Flight Center (1976) AP-8 Trapped Proton Environment for Solar Maximum and Solar Minimum. *Tech. Rep. NSSDC/WDC-A-R/S 76–06*.

16 USDOD/Los Alamos Scientific Lab (1962) A "Quick Look" at the Technical Results of Starfish Prime. Sanitized Version. *Tech. Rep. ADA955411*.

17 Clarke, A.C. Extra-terrestrial relays – can rocket stations give worldwide radio coverage? *Wireless World* 1945, 51 (10): 305–308.

18 Shockley, W. (1949). The theory of PN junctions in semiconductors and PN junction transistors. *Bell System Technical Journal* 28 (3): 435–489.

19 Bardeen, J. and Brattain, W.H. (1948). The transistor: a semiconductor triode. *Physics Review* 74: 230.

20 Hall, E.H. (1879). On a new action of the magnet on electric currents. *American Journal of Mathematics* 2: 287–292.

21 Lilienfeld, E.J. (1930). U.S. Patent 1745175, Method and Apparatus for Controlling Electric Currents.

22 Kilby, J. S. (1964) Miniaturized Electronic Circuits. U.S. Patent 3,138,743, filed Feb. 6, 1959 and issued June 23, 1964.

23 Moore, G.E. (1965). Cramming more components onto integrated circuits. *Electronics* 38 (8): 114–117.

24 Wanlass, F. (1967). Low Stand-By Power Complementary Field Effect Circuitry. U.S. Patent 3,356,858, filed June 18, 1963 and issued on Dec. 5, 1967.

25 Oldham, T.R. and McLean, F.B. (2003). Total ionizing dose effects in MOS oxides and devices. *IEEE Transactions on Nuclear Science* 50 (3): 483–499.

26 Benedetto, J.M., Boesch, H.E. Jr., McLean, F.B., and Mize, J.P. (1985). Hole removal in thin-gate MOSFETs by tunneling. *IEEE Transactions on Nuclear Science* 32 (6): 3916–3920.

27 Wallmark, J.T. and Marcus, S.M. (1962). Minimum size and maximum packing density of nonredundant semiconductor devices. *Proceedings of the IRE* 50: 286–298.

28 Binder, D., Smith, E.C., and Holman, A.B. (1975). Satellite anomalies from galactic cosmic rays. *IEEE Transactions on Nuclear Science* 22 (6): 2675–2680.

29 Pickel, J.C. and Blandford, J.T. (1978). Cosmic ray induced errors in MOS memory cells. *IEEE Transactions on Nuclear Science* 25 (6): 1166–1171.

30 Wyatt, R.C., McNulty, P.J., Toumbas, P. et al. (1979). Soft errors induced by energetic protons. *IEEE Transactions on Nuclear Science* 26 (6): 4905–4910.

31 Guenzer, C.S., Wolicki, E.A., and Allas, R.G. (1979). Single event upset of dynamic RAMs by neutrons and protons. *IEEE Transactions on Nuclear Science* 26 (6): 5048–5052.

32 May, T.C. and Woods, M.H. (1979). Alpha-particle-induced soft errors in dynamic memories. *IEEE Transactions on Electron Devices* 26 (1): 2–9.

2

Radiation Environments

2.1   Introduction

This chapter provides a brief introduction to the various radiation environments that may be encountered by electronic systems and components. In the first part of the chapter, some basic sources of radiation are reviewed, including atomic and nuclear processes such as radioactive decay. Sections 2.2 and 2.3 provide a good physics background for the deeper dive into fundamental radiation effects discussed in Chapters 3 and 4. We then discuss the three predominant natural sources of radiation in the space environment and the man-made environments found in commercial nuclear reactors, nuclear weapons scenarios, and high-energy physics facilities.

2.2   X-Rays, Gamma Rays, and the Atom

As a first step in understanding actual radiation environments, we examine some of the physical processes that give rise to radiation. These include atomic as well as nuclear scale events like radioactive decay. An obvious prerequisite for the following discussion is a good working knowledge of the atomic shell structure and the quantum mechanics behind this structure. The introduction provided in Chapter 1 as well as the problems at the end of Chapter 1 are good starting points for this knowledge.

2.2.1   X-Rays

The discovery of X-rays, attributed to Wilhelm Röntgen in 1895, came after numerous researchers noted odd effects while experimenting with cathode ray tubes. Crookes and Hitton noted fogging of photographic plates near their cathode ray apparatus in the 1880s, but many experimenters from about 1877 up through 1895 studied X-rays and noted such effects. The name "X-ray" comes from Röntgen's naming of the radiation, denoting its unknown nature.

   The most familiar sources of X-rays are modern versions of early tubes, such as in Figure 1.2. An example output from an X-ray tube is shown in Figure 2.1. The X-rays produced consist of a continuous spectrum of wavelengths with the shortest wavelength (cutoff frequency) corresponding to the applied anode voltage. The broad range of X-ray wavelengths from an X-ray tube is the result of electron bremsstrahlung. Bremsstrahlung, or braking radiation, results when an electron comes close enough

Integrated Circuit Design for Radiation Environments, First Edition.
Stephen J. Gaul, Nicolaas van Vonno, Steven H. Voldman and Wesley H. Morris.
© 2020 John Wiley & Sons Ltd. Published 2020 by John Wiley & Sons Ltd.

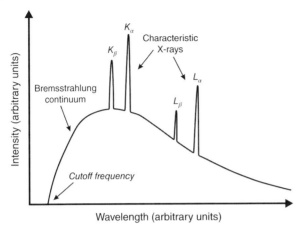

Figure 2.1 Output from an X-ray tube showing characteristic and bremsstrahlung components.

to a nucleus of an atom to cause a change in the electron velocity. As explained more thoroughly in Chapter 3, an accelerating electron will radiate. In an X-ray tube, the applied anode voltage establishes the maximum electron energy so that the shortest wavelength X-ray (highest energy) is produced when the electron loses all of its kinetic energy to bremsstrahlung.

The spectrum in Figure 2.1 also includes several characteristic X-ray spectral lines that depend on the anode material of the X-ray tube. The characteristic spectral lines are from X-ray fluorescence. In X-ray fluorescence, electrons with enough energy can displace an inner-shell electron from an anode metal atom and the resulting vacancy is filled by a higher-energy-shell electron. The energy difference between the shells is released as an X-ray photon with a characteristic frequency (Eq. (1.2)). The K_α and K_β lines are from $L \to K$ and $M \to K$ transitions, respectively, and are the most prominent lines. There may also be L_α and L_β lines in the X-ray tube output and these are X-rays from $M \to L$ and $N \to L$ shell transitions (either cascade fill-in from prior K-shell vacancies or L-shell electrons removed by an incident electron).

The displaced shell electron is called a secondary electron. Incident and secondary electrons will undergo elastic and inelastic interactions in the anode, losing their energy through fluorescence and bremsstrahlung radiation until they thermalize. Figure 2.2 provides a diagram for the X-ray fluorescence and bremsstrahlung radiation production. The removal of the k-shell electron e_k by an incident electron e_1 shown in Figure 2.2a leads to the emission of X-ray photon γ_1 in Figure 2.2b. The ejected electron is called a secondary electron. Also shown in Figure 2.2b is an incident electron e_2 that emits a photon γ_2 due to bremsstrahlung. Depending on the anode material, transitions involving the filling of empty L-shell as well as empty K-shell electron states may also produce characteristic X-rays. Elements with higher atomic number have more tightly bound inner shell electrons as shown in Figure 2.3 where the electron binding energy versus atomic weight is plotted with electron shell as a parameter. The 1s line corresponds to the K-shell electron. The characteristic X-ray photon energy produced by various anode materials is shown in Table 2.1.

In addition to the familiar man-made variety, there are also many sources of X-rays in space, including stars, galaxy clusters, quasars, black holes, and supernova remnants.

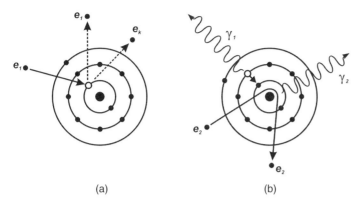

Figure 2.2 X-rays are produced by X-ray florescence and bremsstrahlung. (a) An incident electron e_1 removes a K-shell electron e_k. (b) A higher shell electron fills the K-shell vacancy, emitting a characteristic X-ray photon γ_1. Alternatively, an incident electron e_2 comes close enough to the nucleus to change direction (accelerate) and releases a bremsstrahlung (braking radiation) photon γ_2.

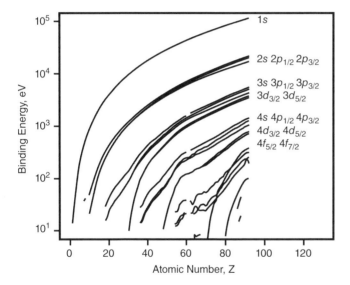

Figure 2.3 Binding energies for shell electrons. Source: Adapted from Thompson 2009 [1], Table 1-1, pp. 1-2 to 1-7.

Some of these are covered in more detail in "The Space Environment" section later in this chapter.

X-rays can be classified by their energy. Soft X-rays range from about 100 eV to 1000 eV whereas hard X-rays have energies from 1000 eV to 200 keV. X-rays with the highest energies (hex-ray) range in energy from 100 keV to 1 Mev. The dividing line between soft and hard X-rays is around 5 keV – 10 keV. While there is some overlap in energy when distinguishing X-rays from gamma rays, as seen in Figure 2.4, there are other factors to consider. For example, X-rays can be considered as originating from interactions at the atomic level and gamma rays from nuclear interaction but this classification ignores stellar bremsstrahlung events that produce gamma rays. There

Table 2.1 K-shell emission line energy for various X-ray tube anode materials.

Anode material	Atomic number	K-shell photon energy (keV)	
		$K_{\alpha 1}$	$K_{\beta 1}$
Cu	29	8.05	8.91
Ga	31	9.25	10.26
Mo	42	17.5	19.6
Ag	47	22.2	24.9
In	49	24.2	27.3
W	74	59.3	67.2

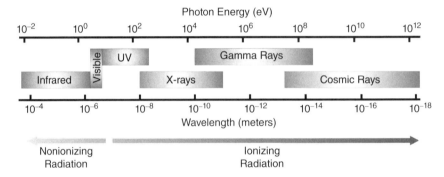

Figure 2.4 Portion of the electromagnetic spectrum showing the energy ranges of ionizing radiation.

may also be an historical aspect for the classification of sources as either X-ray or gamma ray, for example, when a detector did not register wavelength information. In general, the division can be in terms of energy, but one term or another can be used to preserve the historical context.

2.2.2   X-Ray Absorption

It is well known that X-rays have an ability to penetrate materials that are opaque to lower energy radiation. This ability is due to the small wavelength of X-ray radiation. The photon energy of less energetic radiation such as visible, ultraviolet, and infrared is absorbed through an energy transfer to electronic, vibrational, and rotational energy states in the material, but X-ray photons have energy levels that are several orders of magnitude larger than visible radiation. They are very unlikely to interact through low energy transfer processes. If they interact electronically with the material, it is with the tightly bound K- or L-shell electrons in the atoms of the heavier elements, so clearly X-rays will readily penetrate most materials.

The absorption of X-ray photons in a material can be derived for the simple case of a beam of monotonic X-rays that are incident on an absorber, as shown in Figure 2.5. A reduction in intensity occurs as the X-ray beam traverses the material that is proportional to the thickness of the material as well as the absorption property of the material.

Figure 2.5 X-ray absorption diagram.

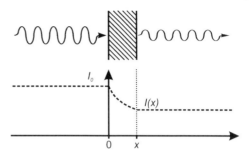

Figure 2.6 Comparison of the mass attenuation coefficients for lead and silicon. Source: Adapted from Hubbell and Seltzer 2004 [2], http://physics.nist.gov/xaamdi.

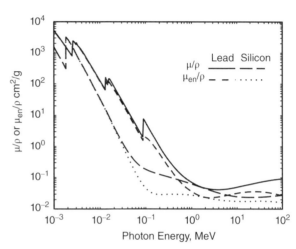

Over a small increment dx the intensity will be reduced by

$$dI = -\mu I dx. \tag{2.1}$$

Solutions to Eq. (2.1) have the form of an exponential decay

$$I = I_0 e^{-ux} \tag{2.2}$$

where I_0 is the initial intensity and μ is usually called the linear absorption coefficient (but macroscopic absorption coefficient or linear attenuation coefficient are also used). It is common to normalize the linear absorption coefficient to the density of the material, μ/ρ, and this quantity is called the mass attenuation coefficient.

The quantities μ and μ/ρ are dependent on the frequency of the X-rays because the absorption in a material will depend on the energy states of the bound electrons. Light elements may have only K-shell electrons that can absorb an X-ray photon's energy, with an uptick in the absorption at a relatively low photon energy corresponding to the K-shell to L-shell electron transition. Heavier elements will have several electron shell transitions that can contribute to the absorption of X-ray photon energy. Figure 2.6 provides a comparison of the mass attenuation coefficients for lead and silicon as a function of the X-ray photon energy. The mass energy absorption coefficient μ_{en}/ρ is also shown in Figure 2.6. This quantity removes the fraction of the energy of secondary charged particles that is lost to bremsstrahlung in the material from the calculation for the absorption coefficient.

The sharp transition in absorption occurs for photon energies that are equal to the binding energy of an electron shell in the absorber material. These transitions are called absorption edges or critical excitation energies. The primary absorption mechanism for lower-energy X-rays is the photoelectric effect, but other absorption processes also occur at low photon energy, including Compton scattering (or incoherent scattering) and Thomson scattering. At higher photon energies, pair production and photodisintegration contribute to photon energy loss. These effects are explored more thoroughly in Chapter 3.

2.2.3 Auger Electrons

A variant of the K-shell ejection process shown in Figure 2.2 produces an electron instead of a photon. The electron is called an Auger electron and is named after Pierre Auger, a French physicist who shares his discovery of the effect in 1923 with Austrian-Swedish physicist Lise Meitner, who independently discovered the effect in 1922. As shown in Figure 2.7, an electron dislodges an inner shell electron of an atom, but the energy released when the shell vacancy is filled is transferred to another electron in the same atom. This electron is ejected from the atom with energy that is equal to the orbital binding energy of the electron and the difference in the inner shell electron energies:

$$E = E_2 - E_1 + E_{core}. \tag{2.3}$$

The resultant atom has two electron vacancies in its electron shell.

The Auger effect has been used since the mid-1950s to analyze the chemical composition of surfaces and has wide application in the microelectronics industry. Sometimes the energy release from an inner electron shell vacancy filling is shared between the emission of an X-ray photon and the simultaneous promotion of an electron into either a bound or a continuum state. This process is called the radiative Auger effect and was discovered by F. Bloch and P. A. Ross in 1935.

2.2.4 Nuclear Structure and Binding Energy

We now turn to radiation that has its origins in the nucleus of atoms. Scattering experiments suggest that nuclei are roughly spherical and appear to have the same density

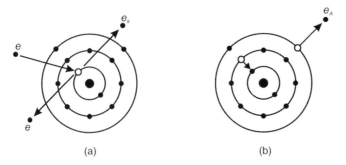

(a)   (b)

Figure 2.7 Diagram of the Auger effect. (a) An incident electron e scatters a K-shell electron e_k. (b) An L-shell electron fills the K-shell vacancy and an outer electron e_A is ejected.

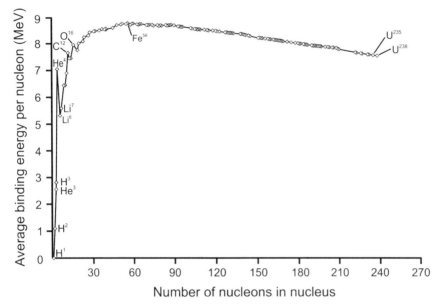

Figure 2.8 Nuclear binding energy versus nucleon.

regardless of atomic mass number A. The nuclear radius is given by the Fermi model

$$R = r_0 A^{1/3}, \tag{2.4}$$

where $r_0 = 1.25$–1.5 fm.

The energy required to remove one nucleon from a nucleus is called the average binding energy per nucleon. Measurements of this energy per nucleon indicate a range between 7 MeV to about 9 MeV for most elements, as shown in Figure 2.8. This is a million times higher than the ionization energy, the energy required to move a bound electron to infinity, or 25 eV or less for most elements [3]. Nuclear binding energies are highest near iron, making the nuclei of ^{56}Fe, ^{58}Fe, and ^{62}Ni the most stable of all elements. In fact, if one assumes the reaction must be exoergic, fission and fusion events are sharply divided around iron with fusion energetically favored for elements lighter than iron and fission favored for elements heavier than iron.

When nucleons combine into a nucleus, they lose energy so that the resultant nucleus has a smaller mass than its constituent nucleons. The missing mass, known as the mass defect, represents the energy released when the nucleus is formed and is a measure of the binding energy of the nucleus. For this reason, the plot in Figure 2.8 is sometimes called the mass defect curve. The mass defect was first noted by British chemist and physicist Francis William Aston in 1920. If $M(N, Z)$ is the nuclear mass of an element and N and Z are the neutron number and atomic number, respectfully,

$$M(N,Z) = ZM_H + NM_n - E_B \tag{2.5}$$

where M_H is the mass of the hydrogen nucleus, M_n is the mass of the neutron, and E_B is the binding energy of the nucleus. Note that E_B in Eq. (2.5) is a measure of the total binding energy and not the binding energy per nucleon as in Figure 2.8.

2.2.4.1 Models of the Nucleus

Over the years, various models have been proposed to describe nuclear structure, but there is no one model that fits all of the observed nuclear properties. While it is understood that there is strong binding between nucleons related to the strong force, this interaction is provided in some models via a central nuclear field or potential so that there is weak or no nucleon–nucleon interaction as such. These models can be referred to as central field or microscopic models where the nucleons rarely interact with each other like the molecules in a gas. Other models provide for strong nucleon–nucleon interaction so that the nucleus behaves like a liquid or a solid. These models are called collective or strong nuclear force models.

2.2.4.1.1 Liquid Drop Model The liquid drop model was first proposed by George Gamow in 1928 [4], and Gamow's analysis inspired Heisenberg (1933) and Weizsacker (1935) to study the mass defect curve (Figure 2.8). Gamow's nuclear model grouped the nucleons into N alpha particles, but in a more modern version of the liquid drop model, the semiempirical mass formula, protons and neutrons interact with each other at a short range through the strong force. This is similar to the way that water molecules are attracted to each other under the short-range van der Waals force. The liquid drop model uses this similarity to describe the nucleus as a classical fluid made up of protons and neutrons with an internal repulsive electrostatic force proportional to the number of protons [5]. In Bohr's concept of the compound nucleus in nuclear reactions, interactions between the particles in the nucleus are strong so that particles do not behave independently of each other. The energy of an incident particle that is captured by the nucleus is quickly shared by the other nucleons [6]. Furthermore, captured particles have a mean free path that is much smaller than the diameter of the nucleus.

Semiempirical mass formulas describe the binding energy E_B in Eq. (2.5) as the sum of a series of terms, each of which represent some aspect of the nucleus:

$$E_B = E_V + E_S + E_C + E_A + E_P + \ldots \tag{2.6}$$

While some mass formulas may have hundreds of terms, the first several terms are most important and provide an accurate model for the binding energy. The first term in Eq. (2.6), E_V, represents a volume energy or exchange energy. It is the energy lowering that occurs due to the potential well created by the attractive strong force and is expected to be proportional to the number of nucleons:

$$E_V = a_0 A. \tag{2.7}$$

The constant a_0 is evaluated empirically. For very large A, such as inside a neutron star where $A \to \infty$ and there are few protons, E_V would be the only major term in Eq. (2.6). Nuclei are small, however, and bounded by a surface where nucleons see lower exchange energy with their neighbors. We expect a deficit of binding energy for these surface nucleons and deduct a correction term E_S for nucleons at the nuclear surface

$$E_S = -a_1 A^{2/3}. \tag{2.8}$$

As with the volume term, the constant a_1 in the surface term is evaluated empirically. Coulomb repulsion between protons is the only long-range force in the nucleus. We can regard the total nuclear charge of Ze as spread uniformly throughout the nuclear volume. This leads to a loss in binding energy E_C due to the Coulomb repulsion:

$$E_C = -a_2 \frac{Z(Z-1)}{A^{1/3}} \tag{2.9}$$

with the constant a_2 evaluated from empirical data and noting that the Coulomb interaction is between a single proton in the nucleus and the $Z-1$ remaining protons.

According to Raoult's law, in any two-component liquid with nonpolar attractive forces, the minimum energy occurs when the two components are in equal concentrations. This effect is modeled in nuclei by an asymmetry term E_A given by

$$E_A = -a_3 \frac{(A - 2Z)^2}{A} \tag{2.10}$$

where the constant a_3 is evaluated from empirical data. The asymmetry term takes into account a composition effect in the nucleus. The maximum stability is achieved when there are an equal number of protons and neutrons. When there are excess neutrons, they occupy higher energy states than if paired with a proton, so excess neutrons are not as tightly bound in the nucleus. The E_A term subtracts from binding energy in Eq. (2.6) when there is an imbalance between proton and neutron number.

A final term in Eq. (2.6), E_P, represents an energy correction due to pairing of protons and pairing of neutrons. Pairing of proton with proton lowers energy through the Pauli exclusion principle when the number of spin-up protons is equal to the number of spin-down protons. The same is true for neutrons in the nucleus. The pairing term has an effect like the asymmetry term but has three possible settings. When there are an even number of protons and an even number of neutrons (e−e), the pairing term is additive to Eq. (2.6). If there is an even number of protons but the neutron number is odd, the term is zero. The term is also zero if neutron number is even and proton number is odd (e−o and o−e). When both proton and neutron number are odd, the term subtracts from Eq. (2.6) (lowers binding energy). The pairing term EP in Eq. (2.6) is given by

$$E_P = -a_4 \frac{\delta}{A^{1/2}}, \quad \text{where } \delta = \begin{bmatrix} +1 \text{ for e} - \text{e} \\ 0 \text{ for e} - \text{o and o} - \text{e} \\ -1 \text{ for o} - \text{o} \end{bmatrix} \tag{2.11}$$

As with the other terms, there is a fitting parameter, a_4, which is determined empirically. The final form of the semiempirical mass formula is

$$E_B = a_0 A - a_1 A^{2/3} - a_2 \frac{Z(Z-1)}{A^{1/3}} - a_3 \frac{(A - 2Z)^2}{A} - a_4 \frac{\delta}{A^{1/2}}. \tag{2.12}$$

Equation (2.12) provides a prediction of binding energy for heavier nuclei and can be accurate to about 1% for nuclei with $A > 40$. Some published values for the fitting parameters $a_0 - a_4$ in the semiempirical mass formula are provided in Table 2.2 and

Table 2.2 Comparison of fitting parameters for the semiempirical mass formula.

Reference	Volume	Surface	Coulomb	Asymmetry	Pairing
1936 Bethe and Bacher [7]	13.86	13.2	0.58	19.5	
1945 Fermi [8, 9]	14.0	13.0	0.583	19.3	$33.5/A^{3/4}$
1955 [5]	14.1	13	0.595	19.0	(Graph)
2004 [10]	15.19	16.51	0.675	20.72	12.53

Source: Adapted from Evans 1955 [5], table 3.3, p. 383.

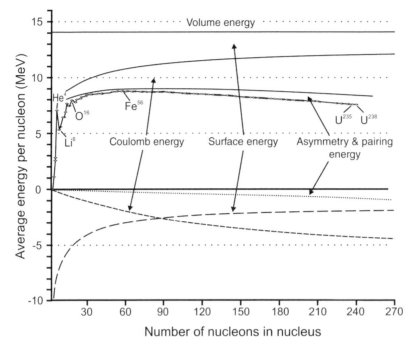

Figure 2.9 Nuclear binding energy showing the various energy components in the semiempirical mass formula. The net binding energy from the formula (not shown for clarity) aligns approximately with the actual binding energies for the heavier nuclei. The surface, Coulomb, and asymmetry energies subtract from the volume energy and are plotted separately in the lower part of the graph.

can also be found in [11–14]. Figure 2.9 shows the relative contributions of the various energy terms (with $a_0 = 14.1$ MeV) to the binding energy. Note that the influence of the surface and Coulomb terms are primarily responsible for the peak in binding energy around iron.

In spite of the different approaches taken by the various contributors, Eq. (2.12) gives an accurate match for the behavior of many elements and isotopes. For lighter nuclei, Eq. (2.12) is not a good predictor of binding energy, especially ^{4}He, mainly because the model doesn't consider internal structure of the nucleus and fails to predict the observed nuclear shell behavior. Modeled as an incompressible fluid, the ground state energies for the nucleus are accurately predicted, but excited states must be visualized as vibrational modes or nuclear surface vibrations corresponding to periodic deformations of the nuclear "droplet." The lowest permissible mode of surface vibration corresponds to excitation energy many times greater than the lowest observed excitation energy of the nucleus [15]. Still, the liquid drop model and nuclear models based on the concepts of the liquid drop model provide valuable insights into the possible energy states that could lead to nuclear fission and other nuclear events [16].

2.2.4.1.2 Fermi Gas Model The Fermi gas model was an early attempt to add quantum mechanics into the nuclear model, mainly through a statistical treatment. Protons and neutrons are fermions, so they will obey Fermi-Dirac statistics in the nucleus. Fermions obey the Pauli exclusion principle, so energy states in the nucleus are filled from the lowest energy state upward with each energy state accommodating a spin-up and spin-down particle. Nucleons move freely in a nuclear potential well, which is due to the influence of all nucleons and has as a first estimate a rectangular shape. Neutrons and protons experience different potential wells in the nucleus, however.

In addition to the potential created by the strong force, protons are affected by the Coulomb force. This causes the neutron potential well to be deeper than the proton potential well so that protons are less tightly bound in the nucleus than neutrons.

Since the nucleons are noninteracting and have only kinetic energy, the model is that of two Fermi gases, one proton and one neutron, occupying the same volume. A diagram of the Fermi gas model potential well is shown in Figure 2.10. The energy of the highest occupied state is the Fermi energy and protons and neutrons can be viewed as having separate Fermi energies. Note the Fermi energies are aligned for neutrons and protons in Figure 2.10 and this is the situation for all stable nuclei. If the proton and neutron Fermi energies were not aligned, the nucleus would undergo beta decay (Section 2.2.5.2) in order to establish a more stable condition.

We can estimate the Fermi energy by first calculating the number of nucleon states up to the Fermi energy E_F. Looking at a single dimension we have from the Heisenberg uncertainty principle, the one-dimensional volume of a particle in phase space:

$$\Delta x \Delta p \geq \frac{1}{2}\hbar \Rightarrow 2\pi\hbar. \tag{2.13}$$

The number of nucleon states in a volume V is then just the total number of states in phase space divided by the volume in phase space for one particle:

$$n = \frac{\iint d^3r d^3p}{(2\pi\hbar)^3} = \frac{V \cdot 4\pi \int_0^{P_F} p^2 dp}{(2\pi\hbar)^3} = \frac{V \cdot p_F^3}{6\pi^2\hbar^3}. \tag{2.14}$$

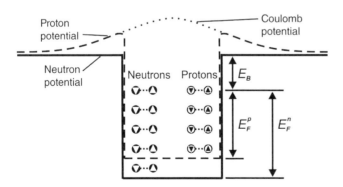

Figure 2.10 Fermi gas model diagram, note that E_B is the binding energy per nucleon.

Note that in Eq. (2.14) the integration over real space yields the volume V whereas the integration over momentum has been converted into spherical coordinates and represents a differential volume of a sphere of radius p from radius p to $p + dp$. Such a region represents an equal energy surface regardless of the direction of p. Given that each state of Eq. (2.14) can hold two fermions of the same species, we have

$$N = \frac{V \cdot (p_F^n)^3}{3\pi^2 \hbar^3} \qquad \text{for neutrons and} \tag{2.15}$$

$$Z = \frac{V \cdot (p_F^p)^3}{3\pi^2 \hbar^3} \qquad \text{for protons.} \tag{2.16}$$

In Eqs. (2.15) and (2.16) the Fermi momentum for protons and neutrons is designated by the addition of a p or n superscript, respectively. For a nucleus with $N = Z = A/2$ the proton and neutron Fermi energies will be equal. Using Eq. (2.4) we have

$$n = \frac{A}{2} = 2 \cdot \frac{V \cdot p_F^3}{6\pi^2 \hbar^3} = 2 \cdot \frac{4\pi}{3} r_0^3 A \cdot \frac{p_F^3}{6\pi^2 \hbar^3} = \frac{4Ar_0^3 p_F^3}{9\pi \hbar^3} \tag{2.17}$$

and

$$p_F = \left(\frac{9\pi}{8}\right)^{1/3} \cdot \frac{\hbar}{r_0}. \tag{2.18}$$

Using the values of r_0 from Eqs. (2.4) and (2.18) provides a momentum of 240–250 MeV/c for nucleons at the Fermi level, indicating that nucleons move in the nucleus with a high momentum. The corresponding Fermi energy is

$$E_F = \frac{p_F^2}{2M} \approx 33 \text{ MeV} \quad M = 938 \text{ MeV}. \tag{2.19}$$

Given the average binding energy of 7–9 MeV for most elements, Eq. (2.19) predicts a potential well depth of about 41 MeV so that nucleons are, in fact, lightly bound in the nucleus.

Estimating the average nucleon energy can provide some additional insight into elements with unequal proton and neutron numbers, but the Fermi gas model provides no information on the individual nucleon energy levels other than the pairing of fermions in each energy state.

2.2.4.1.3 Shell Model There is a lot of evidence that nuclei have a shell configuration and that nuclei have greater stability when shells are filled or closed. In particular, the stability of atomic nuclei that have either neutron number N or atomic number Z with the magic numbers 2, 8, 20, 28, 50, 82, and 126 has been noted. The influence of these magic numbers is seen in the relative abundance of the Earth's elements as in Figure 2.11 and also can be seen when the lifetime of the various elements and isotopes are plotted versus neutron and atomic number as in Figure 2.12. This plot also shows a deviation away from $Z = N$ for stable elements indicating a need for more neutrons to dilute the effect of the Coulomb repulsion between protons as the nucleus becomes larger. When the neutron cross sections for the various nuclei are examined, those with magic neutron numbers have much lower cross sections by about two orders of magnitude, indicating these nuclei are much less likely to absorb an additional neutron. The electric quadrupole moments for nuclei with magic numbers of neutrons or protons are found to be nearly zero, indicating they are spherically symmetric. It is also worth noting that of the four decay series (Section 2.3.2), three terminate in elements that are singly magic and the fourth terminates in a doubly magic isotope.

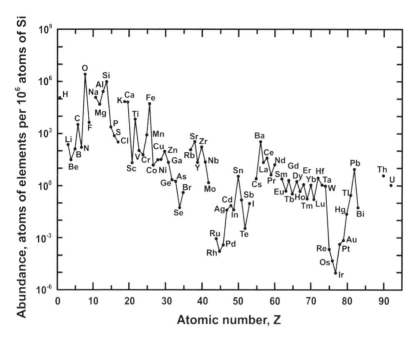

Figure 2.11 Relative abundance of the chemical elements in Earth's upper continental crust (atoms per 10^6 atoms of Si). Source: Adapted from Haxel et al. 2002, U.S. Geological Survey Fact Sheet 087-02, https://pubs.usgs.gov/fs/2002/fs087-02, Figure 4 [17].

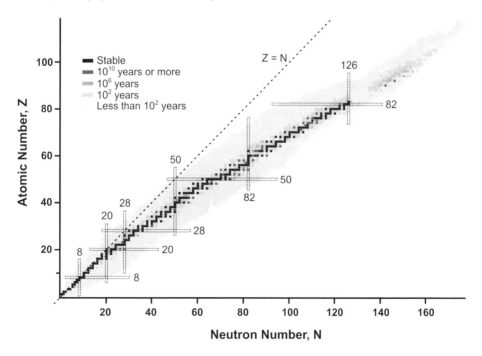

Figure 2.12 Half-life of elements and their isotopes. Source: Adapted from Chart of Nuclides, National Nuclear Data Center, http://www.nndc.bnl.gov/chart.

The observation of magic numbers for protons and neutrons and the success of quantum mechanical models for electron shells in atoms provided clues for a similar quantum mechanical treatment of the nucleus. The shell model of the nucleus, developed in 1949, was the result of independent development of earlier shell models (Dmitry Ivanenko and E. Gapon in 1932) by several physicists, including Eugene Paul Wigner, Maria Geoppert-Mayer, and J. Hans D. Jensen. In fact, these three physicists shared the Nobel Prize in 1963 for their contributions to the nuclear shell model.

A shell configuration in the nucleus implies that nucleons are moving in a potential well created by the forces of all of the other nucleons, but what is this potential? We know it isn't the strong force, as this force acts between the quarks in each nucleon. The energy required to separate two quarks far exceeds the creation energy for a quark/anti-quark pair so attempts to isolate individual quarks end up creating other hadrons (either two-quark particles called mesons or three-quark particles called baryons). Quarks carry a color charge, so unlike a Coulomb field that decreases in strength when two electrically charged particles are separated, the force-carrying gluons create a color field that stretches into a narrow tube or string when quarks are separated so that the strength of the force or the color field remains constant. The action of the color field is called color confinement as it keeps the color field from being observed.

What we observe as the binding force between nucleons is a residual strong force that can be thought of as a kind of London dispersion force. While the binding force is much weaker than the strong force binding quarks, it is still much stronger than the Coulomb repulsion and it also retains some features of the strong force. It is highly attractive in the range of 1 fm but becomes repulsive at about 0.7 fm in keeping with the Pauli exclusion principle. The potential drops off rapidly at a distance of about 2.5 fm. In the shell model, one seeks an average or mean potential that models the collective influence of the nucleus on a single nucleon. Simple potentials such as a finite spherical well or a harmonic oscillator well can be used, but the shape of the potential has a large effect on the prediction of the energy states. A typical choice for the potential is the Woods-Saxon potential:

$$V_{WS}(r) = -\frac{V_0}{1 + e^{(r-R)/a}}, \tag{2.20}$$

where the nuclear radius R is provided by Eq. (2.4), the potential well depth $V_0 \approx 60$ Mev and the surface thickness $a \approx 0.65$ fm. This potential has a nearly constant potential near the center when the atomic mass number A is large. For nucleons on the surface of the nucleus (i.e. $r \approx R$) there is an attractive force toward the center of the nucleus and the potential drops to zero as r goes to infinity ($r - R \gg a$). As with the Fermi model, neutrons and protons have independent states that are the result of the different potentials wells they occupy. So for protons, a Coulomb potential is added to Eq. (2.20), which extends beyond the nucleus and provides a net repulsive force for protons outside the nucleus.

The solution to Schrodinger's equation using potential fields such as Eq. (2.20) follows the usual route using separation of variables in the proposed wave function:

$$\Psi(r, \theta, \phi) = R(r)\Theta(\theta)\Phi(\phi). \tag{2.21}$$

The spherical symmetry of the potential field makes separation of variables possible, dividing the wave function into radial and harmonic parts. The wave function Ψ represents the physical probability for the location of the particle. As such, it must be single valued and continuous and finite everywhere. This restricts the harmonic functions $\Theta(\theta)$ and $\Phi(\phi)$ to the form

$$\Theta(\theta) = P_l^{(m)}(\cos\theta) \tag{2.22}$$

$$\Phi(\phi) = e^{im\phi} \tag{2.23}$$

where $P_l^{(m)}$ are the associated Legendre polynomials of order l in $\cos\theta$ and l, m are integers with $l \geq 0$ and $|m| \leq l$. It can be shown (see, e.g. any text on spherical harmonics and [18, 19]) that l and m are related to the orbital angular momentum L of the particle (about the center of the nucleus) with the magnitude of L given by

$$L = [l(l+1)]^{\frac{1}{2}}\hbar \tag{2.24}$$

and its z-axis component

$$L_z = m_l\hbar. \tag{2.25}$$

The result is a shell arrangement that is similar to the atomic shell numbering, so that a similar shell labeling can be used. The energy levels are determined by the orbital angular momentum quantum number l using the s, p d, f,... letters to denote the values of $l = 0, 1, 2, 3,...$ respectively. The primary quantum number n does not have a physical meaning in the nuclear shell model nor does the value of n limit the allowed values of l as it does in the atomic shell model. This leads to the appearance of some shells such as 1f, 1g, etc. that are not seen in the atomic shell labeling.

The energy states of the bound nucleons can be calculated by solving the radial wave equation for a given orbital angular momentum l:

$$-\frac{\hbar^2}{2m}\frac{d^2u}{dr^2} + \left[\frac{l(l+1)\hbar^2}{2mr^2} + V(r)\right]u = Eu. \tag{2.26}$$

Note that Eq. (2.26) uses the substitution $u(r) = rR(r)$ to simplify the form of the equation. Comparing Eq. (2.26) to the classical conservation of energy law with momentum p and potential energy V

$$\frac{p^2}{2m} + V = E \tag{2.27}$$

reveals that the bracketed portion of Eq. (2.26) represents the total potential energy of the nucleon. The interpretation is that the orbital angular momentum acts like an additional potential that adds to the central potential $V(r)$. Note that the angular momentum term, like the original potential field, is spherically symmetric, so solutions of Schrodinger's equation are still of the form of Eq. (2.21). The effect of the orbital angular momentum term in Eq. (2.26) is to move the wave function further away from the origin for increasing values of l. This influence can be seen by examining the bracketed expression in Eq. (2.26) for a finite potential well $V(r)$ function as shown in Figure 2.13. The orbital angular momentum represents an increase in potential energy over the negative potential energy of the finite spherical well, creating a potential barrier that is strongest near the center of the well ($r = 0$). Instead of a spherical well, the wave function is confined to a spherical shell when $l > 0$.

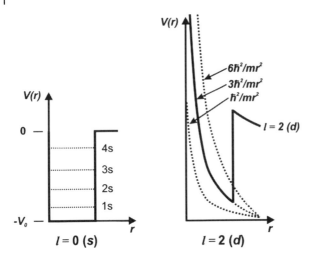

Figure 2.13 Superposition of the nuclear orbital angular energy in a finite spherical potential well. The $l = 0$ state is shown at the left, with the energy levels indicated (not to scale). At the right is the $l = 2$ state, but $l = 1$ and $l = 3$ orbital momentum components are also shown for reference.

Solutions to Schrodinger's equation are not easy to accomplish for most potential fields, but if the potential field is the harmonic oscillator potential:

$$V(r) = \frac{m\omega^2 r^2}{2} \tag{2.28}$$

closed forms for the energy states can be derived. No matter what potential $V(r)$ is used, the energy levels predicted by solutions of the radial function do not predict the shell closures for magic numbers above 20.

Around 1949, Maria Geoppert-Mayer, J. Hans D. Jensen, D. Haxel, and H. Suess independently realized that the nuclear potential is altered by the spin-orbit coupling of the nucleons. The nucleon orbital l and intrinsic spin $s = \pm\frac{1}{2}$ angular momenta interact strongly so that the total angular momentum j is given by

$$j = l + \frac{1}{2} \text{ or } j = l - \frac{1}{2}. \tag{2.29}$$

This interaction can be modeled by adding a coupling term to Eq. (2.20):

$$V(r) = V_{WS}(r) + V_{SO}(r)(L \cdot S) \tag{2.30}$$

that consists of a central potential $V_{SO}(r)$ multiplied by the dot product of the orbital and spin angular momentum operators L and S, respectively. The new momentum operator J has the same correspondence to j as was used in Eq. (2.24) for the orbital momentum operator L and the quantum number l. Likewise, the same can be stated for the nucleon spin operator S and the quantum number s. The dot product in the coupling term further complicates the problem and destroys the spherical symmetry of the potential field. By noting that $J = S + L$ and also that

$$L \cdot S = (J^2 - S^2 - L^2)/2, \tag{2.31}$$

solutions to the radial equation can be expressed in terms J (j) and J_z (m_j). The J_z can be visualized as the addition of either the parallel or anti-parallel combinations of L_z

and S_z leading to Eq. (2.29). The spin-orbit interaction causes a spitting of energy states and becomes quite pronounced as the orbital angular momentum l increases. This effect causes a reordering of the shell energies so that the magic numbers above 20 are correctly predicted.

The shell model is useful for predicting nuclear spin and spin states, parity, and magnetic moments, but is better when applied to closed-shell or nearly closed-shell nuclei. The shell model, like the Fermi gas model, provides for a filling of states that is consistent with fermions. Momentum states are singly occupied and fill from lowest energy to highest energy. The total spin of the nucleus, denoted by I, can be determined because as nucleons are added to a nucleus, opposite spin states pair with each other to reduce energy. If it is assumed that the nucleons in closed shells do not contribute to the nuclear spin, the spin will be given by the nucleons that are outside closed shells. For this reason, even A nuclei will have an integer spin, whereas odd A nuclei will have a half-integer spin. For odd A nuclei, the spin is determined by the last nucleon that is odd. Nuclei with even Z and even N have a nuclear spin of zero. For odd Z and odd N nuclei, the last neutron couples to the last proton with their spins in parallel orientation.

Each of the nuclear energy states is labeled in a manner that is somewhat parallel to the labeling of the atomic energy states. The label indicates the orbital momentum state as well as the spin state, $l \pm \frac{1}{2}$. The spin state for the nucleus can be predicted. For example ^{16}O has magic numbers for both protons and neutrons and hence has a nuclear spin of zero; however, if a neutron is added, it will fill the next energy level, leaving ^{17}O with a spin of 5/2 and a spin state of $(1d_{5/2})^1$. The parity of the spin state is found by examining the momentum state. The s, d, g, etc. momentum states are even parity, whereas the p, f, h, etc. momentum states are odd parity. For ^{17}O the momentum state is d or $L = 2$ and therefore is even parity. Taking a neutron away from ^{16}O to get ^{15}O results in odd parity, as the unpaired neutron is in a p state ($L = 1$).

While the shell model is predictive of observed nuclear behavior like shell closures, nuclear spin, and magnetic moments, the accuracy of the model is poor when the nucleus does not have closed or nearly closed shells. There are some aspects that are poorly modeled or missing altogether. Collective movement of nucleons is not modeled at all due to the starting assumption in the model for the treatment of nucleons as noninteracting point objects. Rotational and vibrational states as well as fission, particle capture, and emission cannot therefore be predicted.

There are also some isotopes that simply do not conform to the predicted shell behavior. For every element there is a point where no more neutrons can be added to the nucleus. In terms of Figure 2.12, the collection of these points is to the left of the stable isotopes and is called the neutron drip line. Some isotopes near the neutron drip line have very lightly bound neutrons that are not in the nucleus at all but occupy a halo around the nucleus. Examples of such halo nuclei include ^{6}He, ^{8}He, ^{11}Li, ^{11}Be, ^{14}Be, ^{17}B, ^{19}B, ^{19}C, and ^{22}C. One of the more scrutinized halo nuclei is ^{11}Li. First discovered in 1966, ^{11}Li was found to have a halo in 1985 by Isao Tanihata and his collaborators at Lawrence Berkeley Laboratory. It consists of a ^{9}Li core with a halo comprised of two neutrons. The seventh neutron in beryllium-11 surrounds the beryllium-10 core like a halo at a considerable distance from the core. Experiments on beryllium-12 indicate that the number of eight neutrons in beryllium isotopes is not a magic number [20]. In a like manner, there are isotones beyond which no more protons can be added to the nucleus. These can be represented by a proton drip line that is to the right of the stable isotopes in

Figure 2.12. There are some isotopes along the proton drip line that have lightly bound outer protons that form a halo around the nucleus. The known examples are ^{8}B, ^{17}Ne, ^{26}P, and ^{27}S. These exotic nuclei are the result of high-energy physics experiments and have very short lifetimes. They are detected in part because their cross sections indicate a nuclear size that is larger than predicted by formulas such as Eq. (2.4). While it might be tempting to think of halo nuclides as orbiting a core nucleus it is more realistic to consider them as low binding energy states near the top of the nuclear potential well with a halo that is caused by tunneling.

2.2.4.1.4   Other Nuclear Models   The preceding discussion of the liquid drop, Fermi, and shell models for the nucleus provide the background for understanding various approaches to modeling the nucleus. As stated at the beginning of this section, nuclear models can be classified as central field or microscopic if they have weakly interacting nucleons. The shell model is an example of a central field model. On the other hand, the individual nucleons in the liquid drop model interact strongly with each other so this is an example of a collective or strong nuclear force model.

Independent Particle Model   Ideally, the wave function for each nucleon in a nucleus could be calculated by taking into account all of the nucleon–nucleon interactions that are possible. Unfortunately, such an approach for n nucleons becomes increasingly complex, using either classical or quantum mechanics even for small values of n. Independent particle models assume the nucleons in the nucleus produce a stable field with a potential energy function. The shell model is an example or an independent particle model where the potential energy function is a relatively simple function. More advanced approaches approximate the n-body interaction of the nucleons with n single-body interactions. An initial trial potential energy function is used to solve for the nucleon wave functions and an iterative approach is used to refine the potential function to achieve a result closer to experimental observations. Like the shell model, this approach is not generally consistent with scattering experiments, which indicate a shorter nucleon mean free path than predicted.

Collective Model   When a nuclear transition involves many or all of the nucleons acting together, the transition is called *collective*. Examples of collective transitions or movements in a nucleus include vibrations and rotations. The collective model was developed in 1953 by Aage Bohr and Benjamin Mottelson using prior work by James Rainwater. The model treats the surface of the nucleus like the liquid drop model with internal motions of the nucleons influencing the shape of the nucleus. The shape of the nucleus is not in all cases spherical, there are cases where the shape is an ellipsoid. Bohr and Mottelson found that the shape of nonspherical nuclei can be explained in terms of nuclear rotation. The collective model can be considered as an extension of the liquid drop model.

Cluster Model   This model is also called the substructure model. It is based on the idea that certain combinations of nucleons produce stability. The pairing and asymmetry terms in the liquid drop model are examples of corrections due to nucleon clustering. In the former, like nucleons are energetically favored to form pairs, while in the latter, pairing between neutrons and protons is considered. The alpha (α) particle model is probably the best-known example of a cluster model appearing first as a type of liquid drop model

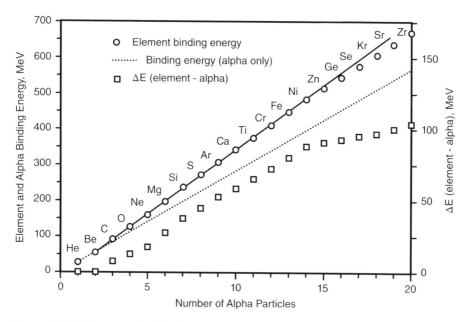

Figure 2.14 Binding energies of nuclei comprised of an integer number of alpha particles.

proposed by George Gamow in 1928 [4]. In this model, protons and neutrons when added to the nucleus will, whenever possible, form into alpha particle subunits. The compelling evidence for this model is alpha particle emission (Section 2.2.5.1) but a plot of the binding energies for nuclei comprised of an integer number of alpha particles, as shown in Figure 2.14, indicates a consistent binding energy for each additional alpha particle. The alpha particle model starts to fall apart when neutrons start to outnumber protons, so the model is best applied to low A nuclei.

Interacting Boson Model The interacting boson model (IBM) was created by A. Arima and F. Iachello [21] in 1974. In this model the protons and neutrons in the nucleus pair up and each pair acts like a single particle with the properties of a boson and angular momentum of 0, 2, or 4. While this model is based on the shell model, it reduces the complexity of the shell model by the combination of the nucleons in pairs, making the model attractive for describing nuclei with $A > 50$. The complexity of the shell model for nuclei in this atomic weight range limits its usefulness for predicting various nuclear states.

Lattice Model Lattice models try to reconcile the nuclear density of the liquid drop model with the shell structure of the shell model. They do this by exploring the properties of specific arrangements of nucleons or their constituent quarks in a given space or volume. This wasn't a popular direction for nuclear models until the 1960s following the discovery of quarks and neutron stars. Prior to this time, the lack of nuclear diffraction and the success of quantum mechanics and the uncertainty principle kept researchers away from investigating solid-phase nuclear models. The nucleon arrangements in lattice models resemble crystal lattices and are named accordingly. For example, there is a face-centered-cubic (FCC) arrangement proposed by Cook and Hayashi [22] and a

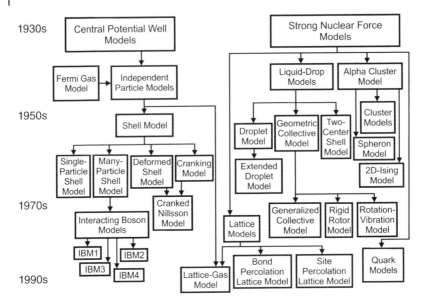

Figure 2.15 Evolution of nuclear models in the years since the 1930s. Source: Cook et al. 1999, Figure 1, p. 55. © 1999 IEEE. Reprinted, with permission, from [24].

body-centered-cubic (BCC) arrangement proposed by Nasser [23]. One of the hallmarks of lattice models is the massive amount of computation used to verify the models. For each arrangement of nucleons, a Monte Carlo approach is used simulate the result of a neutron collision and the corresponding fragmentation of the nucleus. The results are compared to experiment to fine tune the nuclear structure.

Lattice models provide answers to many of the observed nuclear properties. They provide a constant nuclear density like the liquid drop model but also offer insight for asymmetric fission. Lattice models have a shell structure because there is a relationship between lattice positions of nucleons and the quantum number in the shell model. Lattice models can also explain alpha particle emission.

The evolution of nuclear models in the years since the 1930s (Figure 2.15) provides a perspective for the effort extended to reconcile the good points about each classification of model with the observed behavior of nuclei.

2.2.5 Alpha and Beta Decay

An atom is said to decay when it changes into another element by emitting a particle or particles. Alpha and beta decay are two types of atomic decay that result in a transmutation of the original atom into another element. Alpha decay is a type of radioactive decay in which an atomic nucleus emits an alpha particle. The nucleus reduces both its neutron and proton count by two in this process. Beta decay consists of two complementary processes that allow an atom to obtain the optimal ratio of protons and neutrons. In beta decay, a proton is transformed into a neutron or a neutron is converted into a proton. The two types of decay are known as beta minus (β−) and beta plus (β+). Figure 2.16 illustrates the effect each of these decay processes will have on the neutron and proton

Figure 2.16 Illustration showing the effect that alpha and beta decay have on neutron and proton count in the nucleus.

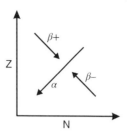

count. Figure 2.16 can be compared to Figure 2.12, noting that beta decay brings unstable nuclei closer to the stable isotopes.

2.2.5.1 Alpha Decay

Alpha decay typically occurs only in heavy elements such as uranium, thorium, and radium. These elements have neutron-rich nuclei, which makes the emission of an alpha particle more likely. The process starts when several protons at the surface of the nucleus become somewhat separated from the rest of the nucleus. This can happen because of an imbalance between the short-range strong force and the longer-range repulsive Coulomb force between protons. Once there is sufficient separation, a more stable clump of nucleons, an alpha particle, is formed with enough energy that it can tunnel out of the nucleus. This process is shown schematically in Figure 2.17. The energies of two possible alpha states are shown. The higher-energy alpha state has a smaller barrier to tunneling and so has a higher probability of emission from the nucleus. It is also emitted with the same energy as its state while confined in the nucleus. This leads to a dependency of the emission probability on the emitted energy of the alpha particle, so that higher-energy alphas are emitted by isotopes with small decay lifetimes.

The relationship between alpha particle energy and the isotope decay constant is called the Geiger-Nuttall law and was first formulated by Hans Geiger and John Mitchel Nuttall

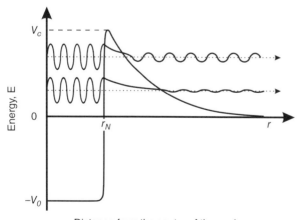

Figure 2.17 The emitted alpha particle energy is dependent on the emission probability (decay lifetime of the isotope).

[25]. Its modern form is

$$\ln \lambda = -a_1 \frac{Z}{\sqrt{E}} + a_2 \tag{2.32}$$

where λ is the decay constant, E is the total kinetic energy (alpha particle and recoil energy of the decaying isotope) and a_1 and a_2 are constants. The exponential relationship in Eq. (2.32) means a wide range in half-life results in a very narrow range in alpha particle kinetic energy. While this law was empirically derived in 1911 by Geiger and Nuttall, the theoretical cause due to tunneling was provided in 1928 by the Russian theoretical physicist George Gamov.

2.2.5.2 Beta Decay

Beta decay is a type of radioactive decay in which a beta particle (either an electron or a positron) is emitted from an atomic nucleus. This type of decay consists of two complementary processes, as indicated in Figure 2.16, that serve to move less-stable isotopes toward a more stable ratio of protons and neutrons. Both processes are mediated by the weak force, a force that seems to tear things apart rather than keep things together like the other forces. The weak interaction is responsible for the radioactive decay of subatomic particles and also plays an important role in nuclear fission. In β− decay a neutron decays into a proton emitting an electron and an electron anti-neutrino. Likewise, in β+ decay a proton decays into a neutron, emitting a positron and neutrino in the process. Both processes are best understood by an inspection of the Feynman diagrams for β decay in Figure 2.18. In β− decay, the weak interaction converts a negatively charged (−1/3e) down quark to a positively charged (+2/3e) up quark with the emission of a W− boson, which decays into an electron and an electron antineutrino.

It is worth noting that β− decay is energetically favored because the mass of a neutron is greater than the mass of a proton. This means that an isolated neutron will always decay and has a half-life of about 10.5 minutes. Isolated protons and hydrogen nuclei, however, do not appear to decay and have a predicted lifetime of 10^{32} years or higher. In the nucleus, it is the change in binding energy that results in either β− or β+ decay. The emitted particles from β− decay have a continuous energy spectrum, ranging from 0 to the maximal available energy. This spectrum is continuous because there is energy

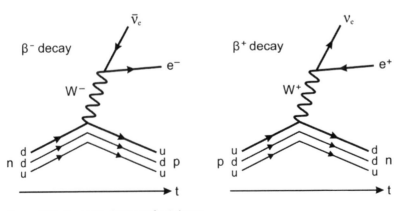

Figure 2.18 Feynman diagrams for β decay.

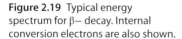

Figure 2.19 Typical energy spectrum for β− decay. Internal conversion electrons are also shown.

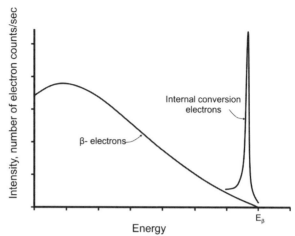

sharing between the beta particle, the neutrino, and the rebound of the nucleus. Typical energies are around 1 MeV, but the range in energies can be from a few keV to 10 MeV or more so that the most energetic beta particles are ultra-relativistic.

There is a variation to the typical β+ decay called electron capture. In a nucleus where β+ decay is energetically allowed, the nucleus can capture an atomic electron emitting only a neutrino. This process is sometimes called K-capture, as it refers to the capture on an inner K-shell electron. There is also a process called bound-state β− decay, which is only applicable for fully ionized atoms. In this process, the electron emitted during β− decay is captured into a low-lying atomic bound state.

The typical beta decay energy spectrum indicated by the solid line shown in Figure 2.19 provides insight into the decay process. The difference in mass energies of the original neutron and the resultant proton, electron, and the electron antineutrino is distributed to the three particles as kinetic energy so the electron can have any energy up to the maximum available energy E_β. Another related process, internal conversion, is also shown in Figure 2.19. The kinetic energy imparted to the proton causes the nucleus to enter an excited state. This energy is usually released as a gamma ray photon (see below), but it can in some cases be imparted to orbital electrons with wave functions that overlap the volume of the nucleus. In this case, the energy of the electron is less than the energy of a gamma ray because the binding energy of the electron is subtracted from the available energy.

2.2.6 Gamma-Ray Emission or Gamma Decay

When a nucleus emits an alpha or beta particle, the nucleus can be left in an excited state. The emission of a gamma ray photon allows the nucleus to drop to a lower energy state (this is similar to the emission of photons by atomic electrons when they change to a lower energy state/orbit). The speed at which the gamma ray photon is emitted is on the order of 10^{-12} seconds, or nearly instantaneous with the alpha/beta decay emission.

A good example of beta decay followed by gamma photon emission is provided by Cobalt-60. Cobalt-60 is a radioactive isotope of cobalt that is produced artificially in nuclear reactors. It is often used for experimental purposes, including radiation testing of electronic parts. It has a half-life of about five years and decays into nickel-60

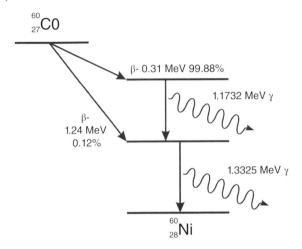

Figure 2.20 Energy diagram for Cobalt-60 β decay.

by β− decay. A decay diagram for cobalt-60 (Figure 2.20) shows that most of the time, beta decay is followed by the emission of two gamma photons with energies of 1.17 and 1.33 MeV.

2.2.7 Other Types of Nuclear Radiation

Spontaneous fission is a type of radioactive decay where an atom's nucleus splits into two smaller nuclei and one to several extra neutrons. It is usually seen in atoms with atomic numbers above 90 and is a very slow process when compared to other types of decay. For example, uranium-238 decays by alpha decay with a half-life of about 10^9 years but decays by spontaneous fission with a half-life of 10^{16} years.

Proton emission and neutron emission can occur in a nucleus whenever there is an excess of protons or neutrons. In terms of nuclear stability, there are maximal extents to the proton/neutron ratio called drip lines. When a nuclide has too many protons or neutrons the excess particles can leak or drip from the nucleus until at least a somewhat stable nucleus can be established. Figure 2.21 illustrates a portion of a nuclide chart showing the decay process. Proton emission and neutron emission are the prominent processes at edges of the plot. The emission of a proton moves the nuclide down vertically in proton count toward a more stable nuclide. Likewise, the emission of a neutron moves the nuclide toward the left and to a more stable nuclide. Proton or neutron emission can be viewed more clearly in terms of the shell or the Fermi gas models for the nucleus. There simply are no more states available in the nucleus for a nucleon to occupy and no binding energy available. Protons that have a negative binding energy can tunnel through the Coulomb barrier and out of the nucleus similar to alpha decay.

Radioisotopes that decay predominantly by alpha emission can sometimes emit a fragment larger than an alpha particle in a process called *cluster decay*. As with alpha particle emission, the emitted particle or cluster tunnels through the nuclear Coulomb barrier. The emitted cluster is not quite large enough to be considered the result of fission.

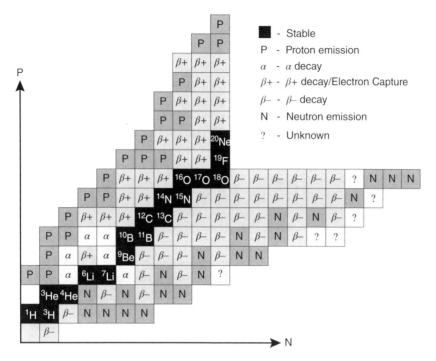

Figure 2.21 Portion of the nuclide chart showing decay mode. Source: Adapted from Chart of Nuclides, National Nuclear Data Center, http://www.nndc.bnl.gov/chart.

2.3 Natural Radioactivity

The discovery of natural radioactivity by Henri Becquerel a year after the discovery of X-rays led to a broader understanding of radioactivity and radioactive substances. It was Becquerel's experiments with radioactive elements and magnetic fields that helped to classify alpha, beta, and gamma radiation. The naming convention, however, can be attributed to Ernest Rutherford, who also discovered the differences in penetrating power between the three radiation types. Natural radioactivity has many uses today, including providing some of the best simulation sources for space and other total dose environments (Chapter 5). At the heart of natural radiation is nuclear decay, as discussed in the previous sections. The next few sections discuss how decay is measured and present the four decay series for the heavy radioactive elements.

2.3.1 Exponential Decay

Radioactive decay can be characterized in terms of exponential decay because the decay rate is proportional to the amount of material that has yet to decay. The relationship can be described mathematically by the first-order differential equation:

$$\frac{dN}{dt} = -\lambda N \tag{2.33}$$

with the solution

$$N(t) = N_0 e^{-\lambda t}. \tag{2.34}$$

The quantity N_0 is the original amount of material and λ is called the decay constant. The mean lifetime $\tau = 1/\lambda$ can be substituted in Eq. (2.34) and represents the average lifetime of an isotope before it decays. It is a simple exercise to derive the more familiar *half-life*, which represents the time it takes for a quantity of radioactive material to decay to half of its initial quantity. Often, there are several decay processes going on in the same sample of material. In this case, the decay constant can be treated as the summation of the various constituent decay constants. A more complicated situation arises when an isotope decays to another by one process and then into a third isotope by yet another process.

2.3.2 Decay Series

Elements with $Z > 83$ found in nature are radioactive (except for Bi 209, which is stable) and decay by alpha or beta emission until a stable isotope is achieved. The decay toward nuclear stability is usually the result of many intermediate beta or alpha decay steps. Noting that beta decay does not change the nucleon count in the nucleus, but that alpha decay decreases nucleon count by four leads to the possibility of four separate decay paths, which are illustrated in Figures 2.22–2.25. Three of these paths or decay series have long-lived isotopes at their starting points, so that these radioactive elements can still be found in nature. For example, the thorium series starts in nature with thorium 232 with a half-life of about 14 billion years. Likewise, the uranium and actinium series also begin in nature with the long-lived starting points of uranium 238 (half-life 4.4 billion years) and uranium 235 (704 million year half-life), respectively. The fourth decay

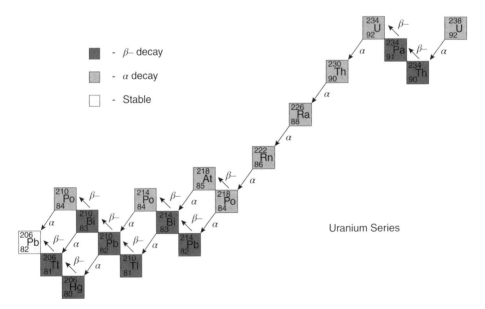

Figure 2.22 The uranium series.

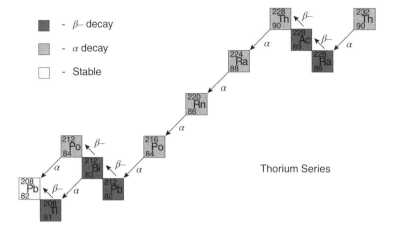

Figure 2.23 The thorium series.

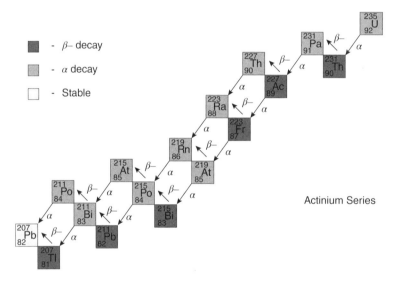

Figure 2.24 The actinium series.

series, the neptunium series, is comprised of mostly short half-life elements. Since the formation of the Earth products from this decay series have long since decayed away, leaving only the two last elements in the series, bismuth 209 and thallium 205, found in nature today.

Although Figures 2.22–2.25 are based on the natural starting points for each decay series, the decay chain for each series can be extended backward from the natural starting point to include elements not normally or easily found in nature. For example, americium 241 is used in smoke detectors and is normally produced by neutron irradiation of uranium or plutonium in nuclear reactors. It decays by alpha emission with a half-life of about 430 years to neptunium 237. The ending points for each decay series are also noteworthy. Stable elements at the end of the naturally occurring radioactive series all

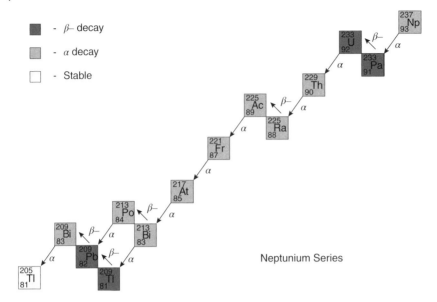

Figure 2.25 The neptunium series.

have magic number of neutrons or protons; the uranium series ends with stable 206/82 Pb, the thorium series ends with 208/82 Pb, and the actinium series ends with 207/82 Pb.

A list of selected radionuclides and their properties is provided in Table 2.3. The list includes both naturally occurring and man-made isotopes and provides the decay energies from the most likely decay path. Medicinally important isotopes like ^{68}Ga, ^{99m}Tc, and ^{103}Pd are made from precursor isotopes using specially made generators or cyclotrons while ^{60}Co, ^{238}Pu, ^{239}Pu, and ^{241}Am are the result of neutron irradiation. Naturally occurring ^{3}H and ^{14}C are the result of atmospheric neutrons (Section 2.4.4) interacting with nitrogen in the atmosphere

$$^{3}\text{H} : {}_{7}^{14}N + n \rightarrow {}_{6}^{12}C + {}_{1}^{3}H, \tag{2.35}$$

$$^{14}\text{C} : {}_{7}^{14}N + n \rightarrow {}_{6}^{14}C + p. \tag{2.36}$$

Nuclear reactors produce ^{239}Pu from neutron irradiation of ^{238}U and also ^{90}Sr and ^{137}Cs as fission products. Water-moderated reactors will also slowly build up ^{3}H, although ^{2}H has a very low cross section for neutrons.

2.4 The Space Environment

The radiation environment in space is highly complex and dynamic but can be roughly classified into solar radiation, trapped radiation, and cosmic rays. Solar radiation includes radiation and particle flux from the Sun, while trapped radiation includes radiation in the well-known van Allen radiation belts circling the Earth. Cosmic rays are actually high-energy particles that come from the Sun or from outside the solar system.

Table 2.3 Properties of naturally occurring and man-made radionuclides.

Nuclide	$T_{1/2}$	Decay process	Photon energy	Source	Usage
^{3}H	12.32 yr	β− (18.6 keV)	—	NO, NR	Tracer for sewage and liquid wastes, nuclear weapons
^{14}C	5730 yr	β− (16 keV)	—	NO	Dating of carbon-containing items
^{36}Cl	301 000 yr	β− (16 keV)	—	NO	Used to measure sources of chloride and the age of water
^{60}Co	5.3 yr	β− (0.31 MeV) (1.48 MeV)	1.173 MeV 1.333 MeV	^{59}Co NI	Irradiation, sterilization
^{68}Ga	67.6 mo	β+ (1899 keV)	—	^{68}Ge	PET and PET-CT imaging
^{90}Sr	28.6 yr	β− (546 keV)	—	NR, NW	RTGs, industrial gauging
^{99m}Tc	6 h	γ emission	140.5 keV	^{99}Mo	Bone, tissue imaging
^{103}Pd	17 d	e- capture	21 keV	^{102}Pd, ^{103}Rn	Brachytherapy
^{133}Xe	5.3 d	β− (346 keV)	81 keV	NR	Pulmonary studies
^{137}Cs	30.17 yr	β− (512 keV)	661.7 keV	NR, NW	Irradiation, sterilization
^{210}Pb	22.2 yr	β− (16.96 keV)	46.5 keV	NO	Sand, soil dating up to 80 yr
^{213}Bi	46 mo	α emission	440 keV	NR	Targeted alpha therapy (TAT)
^{235}U	703.8×10^6 yr	α emission	4.679 MeV	NO	Nuclear power and weapons
^{238}Pu	87.8 yr	α emission	5.593 MeV	^{237}Np NI	Pacemakers, RTGs
^{239}Pu	24 110 yr	α emission	5.245 MeV	^{238}U NI	Nuclear power and weapons
^{241}Am	232.2 yr	α emission	5.485 MeV	U, Pu NI	Smoke detectors

Key: NO, naturally occurring; NR, nuclear reactor; NI, neutron irradiation; NW, nuclear weapons.

2.4.1 Solar Radiation

The Sun is the dominant influence on the space environment within the solar system, with even energetic cosmic rays of extragalactic origin subject to the influence of the particles and magnetic fields of the solar wind. The Sun is an extremely active gaseous body at the center of the solar system, and both the particle and magnetic fluxes it emits are subject to large periodic and aperiodic variations. The 11-year solar activity cycle, for example, has been well known since first being identified in 1843 by the German

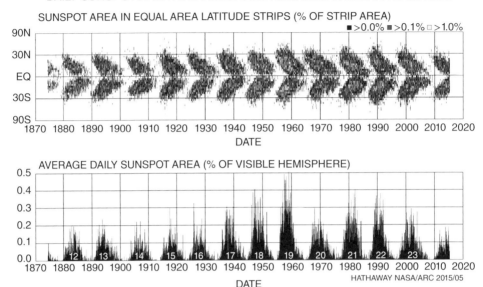

Figure 2.26 Daily sunspot area averaged over individual solar rotations. Source: http://nasa.gov.

astronomer Heinrich Schwabe. A plot of this cycle shown in Figure 2.26 covers sunspot observations over many years and provides sunspot size and latitudinal location (top plot) in what is sometimes called a *butterfly plot*.

The structure of the Sun includes several shells as indicated in Figure 2.27. The innermost core where nuclear fusion takes place is surrounded by a radiative shell, which is itself surrounded by a convection zone. The core is estimated have a radius of a quarter of the solar radius and a temperature higher than 15 million degrees Kelvin. The radiative shell is estimated to extend to about 70% of the solar radius and is about 5 million degrees Kelvin. Energy produced in the core is transferred primarily by radiation rather than convection in the radiative zone. Photons produced in the core can take over 150 000 years to reach the convection zone. The temperature in the convection zone is about 2 million degrees Kelvin. At this temperature the most efficient means of energy transfer is convection. The photosphere at the outer surface of the convection zone completes the visible areas of the Sun. It has a temperature between 4500 and 6000 K.

Two additional layers, the chromosphere and corona, are visible only during a total eclipse. The chromosphere is a low-density region about 2000 km deep above the photosphere but with a density of about 1/1000th that of the photosphere. The outermost layer of the Sun, the corona, extends out to several solar diameters and can be considered the rough equivalent of a planetary atmosphere. The corona is the origin of most of the solar particle and photon emission due to its very high temperature between 1–3 million degrees Kelvin and its extreme magnetic field intensities. In the corona, low mass particles such as electrons are accelerated to very high velocities sufficient to escape the Sun's gravitational field. The electron deficiency caused by the massive emission of negative charge results in intense localized charge imbalance conditions, creating high electric

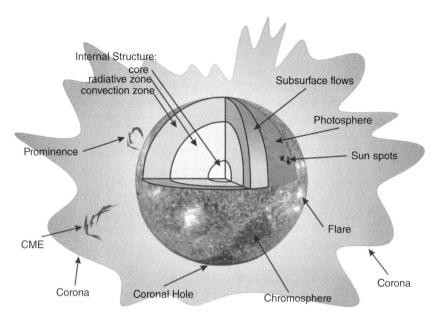

Figure 2.27 Diagram of the Sun. Source: http://nasa.gov.

fields, which, in turn, accelerate and eject other charged particles such as protons, alpha particles, and heavier ionized species.

The flow of electrons, protons, and other particles from the Sun is called the solar wind and extends into space well past the limits of the solar system, as shown in Figure 2.28. The influence of the Sun forms the heliosphere, an immense magnetic bubble, consisting of electrically neutral plasma moving away from the Sun at velocities ranging from 300 to 900 km/s and a temperature ranging from 1×10^4 to 1×10^6 K. The large volume of the heliosphere translates to a very low effective plasma density of 30 particles/cm^3 within this volume, which is very nearly a vacuum. These conditions effectively define one of

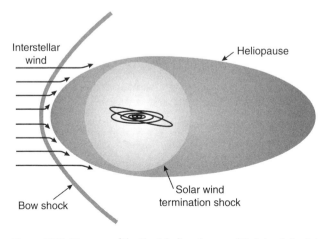

Figure 2.28 Diagram of the Sun's heliosphere and its interstellar interaction. Source: http://nasa.gov.

the basic properties of plasma. The solar wind consists largely of protons (95%), with positively charged He nuclei (alpha particles) and positively charged heavy ions making up the remaining 5%, and with enough electrons to provide charge neutrality within the overall system.

The boundary between the heliosphere and the interstellar medium is believed to be a structure somewhat similar to but much larger than the Earth's magnetosphere. This boundary is located some 10 billion kilometers from Earth, as estimated from *Voyager* data. It is dominated by the interaction of the interstellar medium with the solar wind, which creates a bow shock in the direction of the solar system's motion with respect to the rest of the local region of the Milky Way galaxy and an elongated tail in the opposite direction. An alternative to this classical interpretation is provided by recent *Cassini* mission data, which has suggested that the interaction between the solar wind and the interstellar medium is controlled by particle pressure and magnetic field energy density rather than by velocity with respect to the galactic environment, resulting in a more spherical region rather than the highly asymmetrical region predicted by classical theory.

The Sun displays other aperiodic phenomena, including the well-known solar flares as first described in 1859 [26], which are caused by the sudden release of magnetic energy stored in the solar corona. Flares appear as bright spots adjacent to sunspots, which are actually cooler regions, and affect all layers of the solar atmosphere. Solar flares release tremendous amounts of energy (to the order of 10^{25} J) over a broad range of the electromagnetic spectrum, ranging from gamma and X-ray radiation through visible light and radio frequency (RF) energy.

A solar flare will generally (but not always) trigger a coronal mass ejection (CME), which takes place in the chromosphere. CMEs consist of large volumes of plasma and entrapped magnetic fields, with a typical event ejecting some 10^{17} g (about 100 billion tons) of plasma into space. The plasma consists mostly of electrons and protons, but will also contain other species such as helium, oxygen, and some heavier elements. The correlation between solar flares and CMEs and the mechanisms behind CMEs are not yet well understood. It is believed that these phenomena originate as a result of disturbances in the coronal magnetic field, resulting in magnetic reconnection of opposing magnetic fields. This rearrangement results in a sudden release of energy stored in the magnetic fields, accelerating charged particles to relativistic velocities and causing the solar flare or CME. These solar phenomena occur more frequently in magnetically active regions of the Sun. The shock wave associated with CME events is responsible for large aperiodic increases in the solar wind velocity causing disturbances to the Earth's magnetosphere, including magnetic storms and radio interference. The analog of solar flares occurs on other stars, and these phenomena are known as stellar flares.

2.4.2   Trapped Radiation

As discussed briefly in Chapter 1, radiation belts surrounding the Earth were discovered in 1958 by James van Allen. The van Allen belts are comprised of electrons and protons that are trapped by Earth's magnetic field and are continuously fed and shaped by the solar wind. The cross section through the van Allen radiation belts was shown in Figure 1.5; Figure 2.29 more clearly indicates the relationship between the radiation belts and Earth's magnetic field. This magnetic field is a basic bipolar field ascribed to convection

Inner Proton Belt
1,000 to 6000 km
(610 to 3,700 mi)

Outer Electron Belt
13,000 to 60,000 km
(8,100 to 37,300 mi)

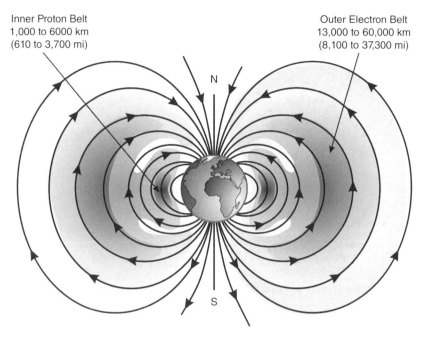

Figure 2.29 Cross-sectional view of the van Allen radiation belts showing their relationship to the Earth's magnetic field.

effects in the planet's metallic core, which is believed to be composed of liquid Ni-Fe. The magnitude and distribution of Earth's magnetic field are time variant, which results in periodic changes in the geographic location of the magnetic poles with respect to the rotational poles. The dipole component of Earth's magnetic field axis is not coincident with its rotational axis, but has an 11° offset [27, 28]. The magnetic North Pole is currently located at 76° North and 100° West and the magnetic South Pole is at 66° South and 139° East. The absolute value of the field in any fixed location also varies over a considerable range, from a few nanotesla (nT) at low latitude and high altitude to tens of thousands of nT at low altitude in the high-latitude polar regions. The field offset from the rotational axis also leads to variations in the altitude of the trapped radiation belts, mostly the inner (proton) belt, with the closest approach to the surface occurring in the South Atlantic Ocean off the South American coast. This is the well-known [29] South Atlantic Anomaly (SAA), which causes serious increases in proton and electron flux and which has historically resulted in persistent on-orbit system difficulties over this region. The SAA produces the majority of trapped radiation encountered in low Earth orbits (LEOs). Since these magnetic structures are at least somewhat symmetrical with respect to their axis (but not with respect to the Earth's axis), the corresponding Southeast Asia Anomaly on the opposite side of the Earth has the proton belt at a higher altitude, resulting in reduced trapped particle flux.

The majority of particles encountered in space and particularly in the near-Earth solar wind are electrons and protons. As these are both charged particles, they interact readily with electric and magnetic fields. The interaction between the solar wind plasma and the Earth's magnetic field distorts the field into a highly asymmetrical structure except

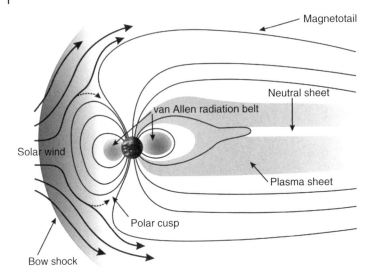

Figure 2.30 Structure of the Earth's magnetosphere under the influence of the solar wind. Source: http://nasa.gov.

in the near-Earth region of 4–5 Earth radii (R_e). Although not to scale, Figure 2.30 shows the distortion in the magnetosphere caused by the solar wind. The magnetosphere protects the Earth from the majority of solar wind particles and galactic cosmic rays; this geomagnetic shielding varies with latitude and is described quantitatively by the geomagnetic rigidity, which is defined as the resistance to charged particle deflection by the magnetic field. The magnetic rigidity of a charged particle is defined as its momentum per unit of particle charge, with the units working out to gigavolts (GV). The Earth's magnetic field offers the least protection at the poles, and the cosmic ray flux is the highest there. The cosmic ray flux ionizes air molecules in the upper atmosphere causing the Northern Lights or Aurora Borealis at the North Pole and the Southern Lights or Aurora Australis at the South Pole.

The solar wind plasma (mostly protons, with some heavy ions as well and electrons to achieve overall charge neutrality) and the magnetic fields embedded in the solar wind plasma interact with the magnetosphere and form it into a complex shape with several clearly defined regions. These regions include a hemispherical bow shock region facing the Sun with a radius of some 10–12 R_e (Earth radii) and a highly extended magnetotail facing away from the Sun; this region is roughly cylindrical with a diameter of approximately 40 R_e and can be as much as 300 R_e long. At the bow shock region, the solar wind is prevented from penetrating into Earth's atmosphere by the geomagnetic rigidity. This deflection mechanism results in the great majority of solar wind particles bypassing Earth entirely; estimates of the fraction of bypassed particles are as high as 99%. The bypassing mechanism also results in the stretching of the geomagnetic field into the magnetotail on the side of the Earth turned away from the solar wind. The ionosphere is the effective lower boundary of the magnetosphere, while the upper boundary is the interface between the magnetosphere and the solar wind plasma.

The magnetotail forms a first magnetic confinement system, in which solar wind particles are trapped; this region is called the neutral plasma sheet. A second trapping mechanism is responsible for most near-Earth trapped radiation. These trapped particles consist mostly of electrons and protons with energies up to a few MeV, as higher energy particles such as cosmic rays will be deflected around the magnetosphere but not trapped. The trapping mechanism is interesting and involves the constraint of the charged particles by Earth's magnetic field lines through the Lorentz force, given by

$$d\mathbf{p}/dt = q\mathbf{v} \times \mathbf{B}. \tag{2.37}$$

in which $\mathbf{p}$ is the particle momentum, q is the particle charge, $\mathbf{v}$ is the particle velocity, and $\mathbf{B}$ is the magnetic field intensity. The velocity is a vector quantity consisting of normal and parallel components to the magnetic field. The Lorentz force results in the charged particles oscillating (Figure 1.6) back and forth between the North and South Poles while performing a helical motion around the magnetic field lines and with a mirror point at each end of the trajectory. The period of this helicoid motion from pole to pole is to the order of a second for protons and perhaps 100 ms for electrons. Additionally, the magnetic field nonuniformity causes a slow precession of the particles, with the electrons drifting East and the protons drifting West. A complete azimuthal drift around Earth takes to the order of an hour, with a high degree of variability due to magnetic field disturbances. These multiple trapping and precessional effects result in clearly defined drift shells, or belts, of trapped (but not static; paradoxically, the particle needs to remain moving in order to remain trapped) particles. These charged particle belts were first discovered by James van Allen during the 1958 International Geophysical Year (IGY), and are appropriately known as the van Allen belts. The belts derive their long-term stability from four balancing phenomena: particle injection into the belt from the solar wind, particle acceleration due to the Lorentz force, particle diffusion, and particle loss. Short-term stability is driven by the solar wind's cyclic behavior superimposed on magnetic disturbances and solar rotation variations.

Not all planets have radiation belts. The minimum requirement for these belts to exist is driven by the minimum requirements for trapping charged particles; the planet's magnetic dipole moment must be sufficiently high in magnitude to stop the solar wind before it reaches the upper layers of the atmosphere, where deceleration will take place through collision (which is not a trapping mechanism) rather than through magnetic deflection. Within the solar system, Venus and Mars do not have magnetospheres and hence will not have trapped radiation belts, while Saturn, Uranus, and Neptune have magnetospheres but have weaker radiation belts. In contrast, the Jovian magnetic field is well over an order of magnitude more intense than that of Earth, which results in particularly intense Jovian belts that present particular challenges [30] for missions venturing into this environment.

Under normal conditions there will be two van Allen belts, the inner and outer. The inner belt consists of electrons and protons in the several hundred MeV energy range and extends from near the surface to approximately $2.4\,R_e$. The outer belt extends from $2.8\,R_e$ to $12\,R_e$ and is composed mostly of energetic electrons up to 7 MeV energy, with some protons. The region between the belts is known as the slot. The belts vary in composition and altitude, and atmospheric disturbances such as magnetic storms or man-made phenomena such as high-altitude thermonuclear detonations have been known to greatly enhance the trapped particle flux in the belts and to even create a temporary third belt. A nuclear weapon detonation injects [31] energetic electrons from the beta decay of fission fragments into the near-Earth environment, and these particles are trapped through the same mechanisms as solar electrons. This injected electron component is most stable and causes most damage at LEO altitudes.

The van Allen belts have been extensively mapped by satellites such as the Combined Release and Radiation Effects Satellite (CRRES), which gained insight into not only the belt environments but their effects on spacecraft electronics [32]. A mathematical description is a key requirement for modeling these regions for radiation environment prediction for different satellite orbits. Analytical solutions have not been found, as you would expect for such a large, varying and chaotic structure, and numerical solutions are the only practical approach. Modeling codes include AP8 [33] for protons and AE8 [34] for electrons, in which the "A" recognizes the leadership role of the Aerospace Corporation in first developing these models. AP8 was released in 1976 and was followed by AE8 in 1983. These codes are periodically updated, including the recent development [35] of the Monte Carlo Radiative Energy Deposition (MRED) code by researchers at Vanderbilt University. MRED introduced Monte Carlo simulation technology into this key set of tools. The codes develop predictions of trapped particle fluxes for any set of orbital parameters and mission duration. The solar cycle is modeled as a simple set of solar maximum and solar minimum values for the predicted fluxes. Figures 1.7 and 1.8 provided examples of AP8 proton flux maps. It becomes clear from this contour plot that the radiation profile will vary greatly depending on the orbit, with low Earth and geosynchronous orbits relatively benign and the highly eccentric Molniya orbits a much more difficult environment.

2.4.3 Cosmic Rays

The term *cosmic rays* has historical origins but is somewhat misleading, as these are not rays at all, but the term persists in the popular literature as well as in science. Cosmic "rays" are energetic charged particles originating in the Sun as well as outside the solar system; these are termed solar and galactic cosmic rays, respectively, and differ from other charged particles in that they are not trapped in the magnetosphere, usually due to their much higher kinetic energy. This high energy arises from the higher mass of the particles and from their sometimes extreme velocity, which can approach relativistic values. Cosmic rays are composed of 90% protons, 9% alpha particles, and 1% electrons, with some other particles such as heavy ions and neutrons as well. Solar flares emit large quantities of energetic charged particles, and this solar cosmic ray fluence is composed almost entirely of protons, with some alpha particles, heavy ions, and electrons. The magnitudes of these events vary, with the most active ones weakly correlated with the solar maximum part of the solar cycle.

Galactic cosmic rays originate outside the solar system, as the name implies. A background ionizing radiation component at the Earth's surface was first described in the nineteenth century and was verified as originating beyond the Earth's atmosphere through manned high altitude balloon experiments conducted by the Austrian scientist Victor Hess in 1912, using ionization chambers to detect the unknown radiation. Cosmic ray particles vary widely in energy, ranging from a few KeV for low-energy protons to relativistic heavy ions in the GeV range. A good reference plot of the cosmic ray flux as a function of energy is provided by Swordy [36]. In Figure 2.31 we show a log-log plot of galactic cosmic ray flux as a function of energy from the Particle Data Group (figure created by P. Boyle and D. Muller, source of data can be found in refs. 2

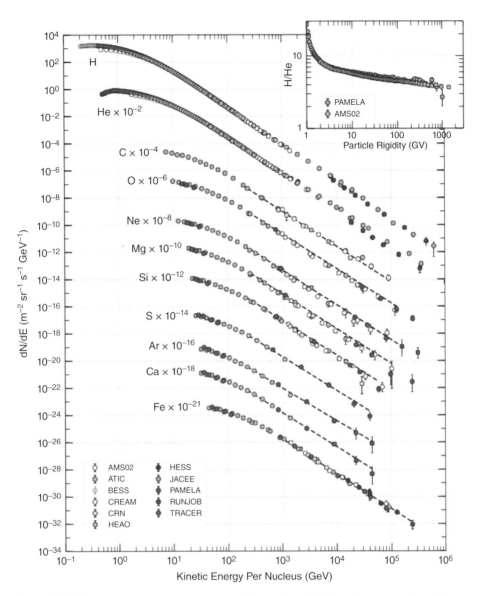

Figure 2.31 Galactic cosmic ray energy spectrum. Fluxes of nuclei of the primary cosmic radiation in particles per energy-per-nucleus are plotted vs energy-per-nucleus. Plot was created by P. Boyle and D. Muller, references for data can be found in [37]. Source: Tanabashi et al. (Particle Data Group) [37] 2018, figure 29.1.

through 13 in [37]). It is important to note that the tail of this distribution extends out to nearly arbitrarily high energies; the flux here is very low, but these few ions still cause fundamental difficulties in the cosmic ray hardening of satellite systems because of shielding ineffectiveness. Reasonable Al shielding thicknesses in the 100 mil (2.5 mm) range as commonly used in satellite airframes are effective at removing the low energy proton component, but have no effect at all on the high-energy end of the

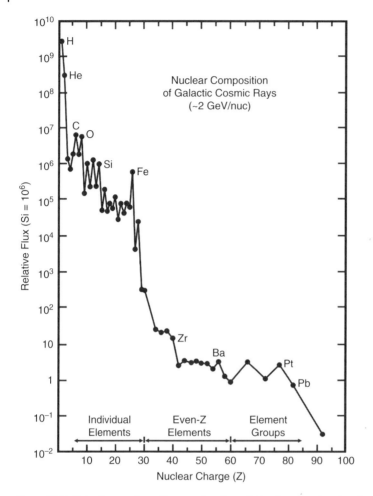

Figure 2.32 Galactic cosmic ray flux as a function of atomic mass, showing the dominant species below the Ni/Fe cutoff "knee." Source: Mewaldt 1988 [38], Figure 1, p. 122.

distribution. Cosmic ray hardening must thus be carried out at the component level, through redundancy and error correction, as shielding is ineffective. The abundance distribution shown in Figure 2.32 plots the cosmic ray flux as a function of particle species [38] and has a similarly long tail, this time in the high atomic number species, which runs all the way to the transuranic elements and isotopes.

Characterizing cosmic rays requires very large detectors and long integration times to collect enough particles for even reasonable statistical significance, and the low flux at the higher-energy end of the spectrum makes this an inefficient procedure. As an example, the Heavy Nuclei Experiment (HEAO-3) was launched in 1979 to detect heavy and ultraheavy nuclei in cosmic radiation and consisted of two large argon-methane ionization chambers and Cerenkov counters read out by photomultipliers. HEAO-3 observed [39] over 100 high atomic number cosmic ray species between element 75 and element 87 in almost a year and a half of flight, and was much larger than most scientific instruments flown by NASA today. Much cosmic ray research today is performed on

Earth's surface using large arrays [40] of detectors, and a huge number of experiments are in progress. These can detect neutrons or charged particles through the light given off by the interaction between cosmic rays and gas molecules or liquid-based Cerenkov detectors. None of these arrays image incoming cosmic rays directly, but sense the secondary particle showers generated as the high energy particle interacts with the upper atmosphere. Most of the data agree with the theoretical upper limit to cosmic ray energy, the so-called Greisen-Zatsepin-Kuzmin (GZK) limit, caused by interaction of cosmic rays with the microwave background radiation. The highest energy cosmic rays appear to be of extragalactic origin but physicists puzzle over what process could produce such high energy particles.

In general, the origin of galactic cosmic rays is not well understood. The particle flux is isotropic, and as the charged particles are influenced by magnetic fields no clear point source can be determined. Low- to medium-energy cosmic rays with energies up to 10^{18} eV are believed to originate in the local Milky Way galaxy, with higher-energy particles of likely extragalactic origin. A number of theories have been advanced [41, 42], including an interpretation that galactic cosmic rays are accelerated in intense blast waves occurring as a side effect of supernova explosions and the interaction of supernova remnants with the interstellar medium. The remnants are expanding clouds of gas and magnetic field, and the violent interaction with their environment accelerates particles created in the supernova explosion. Particles are believed to oscillate within the magnetic fields, resulting in gradual acceleration up to the escape velocity, at which they escape the supernova remnant; this escape velocity depends on the volume of the acceleration region and the magnetic field strength. However, cosmic rays have been observed at much higher energies than can be theoretically predicted for supernova remnants to generate, and where these ultra-high energies come from is an active research topic. Leading theories include massive black holes at galactic centers, quasar activity, or the motion of galaxies through the intergalactic medium.

2.4.4 Atmospheric Neutrons

Atmospheric neutrons are a secondary effect of solar and galactic cosmic rays and arise from the interaction of galactic cosmic rays with the gaseous upper layers of the Earth's atmosphere, as discussed previously. The energetic cosmic ray particles collide with atomic nuclei and cause nuclear reactions, which result in a cascade of secondary particles including neutrons, protons, and pions. Recall that there is a small but nonzero flux of extremely high energy ions in the cosmic ray spectrum, and the interaction of these particles can lead to large cascades ("showers" of billions of secondary particles) of multiple energetic nuclear reactions. A diagram showing a portion of a cosmic ray shower or cascade is shown in Figure 2.33.

The incident particle in a cosmic ray shower or cascade shatters a nucleus in an air molecule, releasing numerous fragments, which go on to interact with other air molecules or decay in some way. The incident particle can also go on to interact with many more air molecule nuclei. The secondary particles from such interactions create various components of the cosmic ray cascade including electromagnetic, nucleonic/hadronic, and mesonic components. For example, π^0 mesons decay into two photons, which if energetic enough can produce one or more electron/positron pairs though pair production. Electrons and positrons can release photons through a

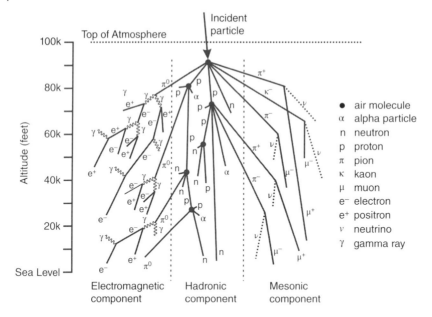

Figure 2.33 Diagram of a cosmic ray cascade.

process called bremsstrahlung, and bremsstrahlung photons can, if energetic enough, release energy to pair production. Such interactions comprise the electromagnetic component. The π^+ and π^- mesons contribute to the mesonic component and decay into muons and neutrinos. Muons can interact with nuclei or release bremsstrahlung photons if they have enough energy. The neutrinos are mostly noninteracting. Proton and neutron fragments, including alpha particles, can continue to interact with the atmosphere causing a continuance of the cascade until the energy imparted from the incident particle is dissipated.

Neutrons from cosmic ray cascades, called atmospheric neutrons, can have energies [42] ranging from 1 to 100 MeV, with the 1 MeV flux some three orders of magnitude higher than the 100 MeV flux. A plot of the atmospheric neutron flux as a function of altitude is shown in Figure 2.34. The maximum neutron flux of 1.2 n/cm^2 s is at 60 000 ft (20 km) and drops off by 60% at 30 000 ft and to 0.25% at sea level, at which there is still quite significant neutron flux. At commercial aircraft cruising altitude (10 km) the radiation is still several hundred times the ground-level intensity. Practical cosmic ray detection is carried out by detecting these nuclear reaction byproducts at the Earth's surface using detector arrays, as discussed previously.

Atmospheric neutrons are a relatively recent research topic but have been shown to cause a number of important effects, including primary radiation exposure to aircrew in high-altitude commercial aircraft. Neutrons represent about half the incurred dose in this environment, and significant research has been done in characterizing it; much early neutron mapping data was developed using the CREAM detector on transatlantic Concorde flights at up to 60 000 ft. These neutrons also cause single-event effects (SEEs) in electronics; since these are not charged particles, the mechanism is again one of nuclear reactions resulting in high energy ions and progressing to the ionizing effects of these ions, a process known as indirect ionization. A great deal of experimental work has

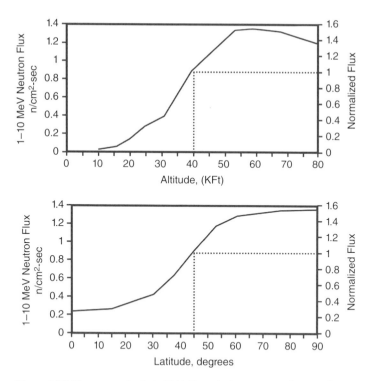

Figure 2.34 The atmospheric 1–10 MeV equivalent neutron flux as a function of altitude (top plot) and latitude (bottom plot). Source: Adapted from Taber and Normand 1995 [43], Figures 7-2 and 7-3, pp. 7-3 and Figure 7-4.

gone into characterizing high-altitude avionics systems for neutron SEEs [44], which is routinely addressed by hardware redundancy and error-correcting software. Upset rates of about 1 per 200 hours have been reported for the Boeing 777 autopilot system, which is orders of magnitude higher than the tolerances required by the manufacturers' requirements of 10^{-6} upsets per hour. Neutron SEE is also a threat in low-altitude atmospheric applications and is an important emerging threat in sea-level applications, such as server farms, and in high-density components such a field-programmable gate arrays. In particular, for the higher-density processes such as deep submicron CMOS, the critical charge required for an upset is greatly reduced, and this makes the problem worse as the flux at lower neutron energies increases rapidly. Server farms can have 50 000 individual servers, and the statistics become very difficult as downtime in these installations is not well tolerated. Formally radiation-hardened components are not of much use here due to cost constraints, which dictate the use of commercial off-the-shelf (COTS) parts.

2.5 The Nuclear Reactor Environment

Nuclear reactors for commercial power generation present a much different but equally challenging and specialized environment. The reactor core itself presents too intense

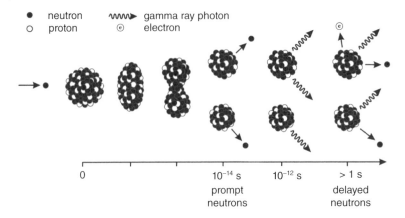

Figure 2.35 Fission reaction showing timescale and decay of the fission fragments.

an environment for the operation of nearly all practical electronic systems, but reactor safety, monitoring, and control systems are used outside the primary reactor shield. These environments are still orders of magnitude more intense than those encountered in space, with gamma radiation levels as high as 10 Mrad(Si) and 1 MeV equivalent neutron levels as high as 10^{15} n/cm^2. Much of the electronics for these applications use discrete semiconductors, with high-temperature wide bandgap materials, such as SiC and GaN provide an attractive solution.

Commercial nuclear power installations are built around a reactor core using, for example, ^{235}U or ^{239}Pu fission material to sustain a continuous but controlled nuclear reaction [45]. This process is based on neutron-induced fission of the fuel material as indicated in Figure 2.35. This figure provides a time-based diagram of the fission event as well as decay that happens in the fission fragments. A neutron hitting the nucleus causes it to enter a vibrational mode that is reminiscent of those from the liquid drop model. After several femto-seconds, the nucleus splits approximately in two and releases a few free neutrons as well. The free neutrons, coming from the initial fission event, are called prompt neutrons. As the fission fragments settle, they emit gamma ray photons. Many fragments are short-lived and undergo beta decay and/or emit neutrons to achieve stability. Neutrons emitted by fragments are called delayed neutrons and occur about a second or longer after the original fission event.

In a nuclear chain reaction, prompt and delayed neutrons contribute to further fission reactions, which lead to more neutrons feeding the reaction. A chain reaction is said to be critical if every fission event produces a neutron that also causes fission. If a reaction is critical due to the contribution of prompt neutrons alone any additional neutron production would quickly take the reaction out of control. This is due to the prompt nature of these neutrons – no control loop can react quickly enough to their production. In practice, nuclear reactions in power stations are kept subcritical with respect to prompt neutrons with the control loop geared toward the criticality of the delayed neutrons. Neutron absorption in control rods and reactor coolant are used to control the degree of criticality of the reaction, avoiding the uncontrolled chain reaction on which nuclear weapons are based. The control rods are composed of materials of a high neutron capture cross-section that efficiently absorb neutrons without themselves

Table 2.4 Comparison of the fission critical energy to the neutron binding energy for various isotopes used in nuclear reactors.

Nucleus	Critical energy (Mev)	Binding energy of last neutron (Mev)
^{239}Pu	5.0	6.6
^{238}U	7.0	5.5
^{235}U	6.5	6.8
^{233}U	6.0	7.0
^{232}Th	7.5	5.4

becoming unstable; typical materials include AgInCd alloys and B-based alloys. The physical position of the control rods within the core determines the amount of neutrons absorbed and hence the power output of the reactor. In case of emergency, the rods can be rapidly dropped into the core to immediately reduce criticality and stop the chain reaction. The fission of the ^{235}U or ^{238}Pu atoms releases neutrons, heat and gamma rays, and the absorption of neutrons reduces their energy in a process known as thermalization. The intense neutron environment activates other materials rendering them radioactive as well. These activated materials also produce further energy by decay of unstable isotopes.

Very few radioactive isotopes can be used as fuel in a nuclear reactor. The viable candidates are listed in Table 2.4 along with critical energies needed for fission and binding energies for the last neutron in each nucleus. For ^{233}U, ^{235}U, and ^{239}Pu, the critical energies are lower than the binding energy for another neutron. This means that these nuclei are fissile or easily fissionable, even with very slow neutrons. On the other hand, ^{238}U and ^{232}Th will likely absorb a slow neutron, with fission occurring only in ^{238}U if the neutron can provide the energy difference between the critical energy and the neutron binding energy. While ^{238}U nuclei are fissionable with fast neutrons, a chain reaction cannot be sustained because of inelastic scattering of fast neutrons to below the critical energy. However, both ^{238}U and ^{232}Th are fertile – they can be converted into fissile nuclei by absorbing neutrons in a breeder reactor:

$$n + {}^{232}_{90}Th \rightarrow {}^{233}_{90}Th \xrightarrow{\beta-} {}^{233}_{91}Pa \xrightarrow{\beta-} {}^{233}_{92}U, \tag{2.38}$$

$$n + {}^{238}_{92}U \rightarrow {}^{239}_{92}U \xrightarrow{\beta-} {}^{239}_{93}Np \xrightarrow{\beta-} {}^{239}_{94}Pu. \tag{2.39}$$

The fission of ^{235}U releases a large amount of energy – around 200 MeV. This energy is shared by the fission products and fission neutrons. Fission neutrons are thus very energetic and are called fast neutrons because most of them have energy greater than 1 MeV. An energy spectrum for prompt neutrons from ^{235}U fission is shown in Figure 2.36. The neutron energy in this spectrum, sometimes called a Watt distribution, is a calculated spectrum by David Madland [46] under the assumption of first-chance fission only that shows the dependence of the spectrum on the kinetic energy of the incident neutron.

The Watt distributions as in Figure 2.36 can be described by the empirical probability function:

$$P(E) = 0.4865 \sinh(\sqrt{2E})e^{-E} \text{ MeV}^{-1}. \tag{2.40}$$

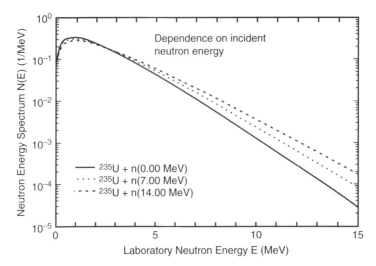

Figure 2.36 Dependence of the prompt fission neutron spectrum on the kinetic energy of the incident neutron, for the fission of ^{235}U. Source: Madland 1982 [46], Figure 2, p. 5.

While it is not obvious due to the linear energy scale of Figure 2.36, Eq. (2.40) indicates that very few prompt neutrons have energy in the thermal range (a few eV) that is favorable for fission of ^{235}U. For this reason, nuclear reactors utilize materials that moderate the energy of neutrons to increase the number of thermal neutrons. Moderating materials include water, heavy water, and carbon as these materials slow down neutrons through inelastic collisions providing more neutrons for fission.

The prediction from the liquid drop model for equally sized fission fragments is not born out in nature. Fission of ^{235}U by thermal neutrons usually results in two fragments of unequal size centered at $A = 95$ and $A = 137$ along with several fission neutrons. These fragments, in turn, decay, usually through beta decay, to more stable elements.

The reactor core thus releases large amounts of energy through various nuclear reactions; as mass and energy are related by the well-known Einstein equation, nuclear reactions are a great deal more energy-intensive than chemical reactions and nuclear materials have the highest energy density of any fuel. One kilogram of ^{235}U processed through a nuclear reactor will yield 7.2×10^{13} J, while burning the same quantity of coal releases 2.4×10^7 J, which is a six order of magnitude difference. These large amounts of thermal energy are removed from the core by a primary cooling loop, using either water or, alternatively, molten metals or salts, transferring the energy as well as preventing the core from melting down. This entire loop is highly radioactive and it, the core, and peripheral equipment are enclosed in a concrete and steel containment structure. The heated coolant is circulated through another heat exchanger, transferring the energy to a secondary water loop that is not radioactive and that is then used to drive turbines with superheated steam. The radiation environment within the containment facility consists of neutrons, gamma radiation released by nuclear reactions between neutrons and the surrounding materials, and (under accident conditions) protons and electrons. Table 2.5 summarizes the representative radiation environments found in a nuclear power plant. Unlike the weapons environment, there are no transient effects to be considered in reactor applications, as the dose rates are moderate. Compared to

Table 2.5 Representative radiation environments in a nuclear power plant. Both the normal and loss of coolant accident (LOCA) levels are shown.

Environment	40-year normal operation level	Peak level, accident	Units
Neutrons, 100 KeV, dose	1×10^9 to 1×10^{14}	—	n/cm^2
Neutrons, 100 KeV, dose rate	1×10^1 to 1×10^5	—	n/cm^2 s
Gamma, dose	1×10^3 to 1×10^8	1×10^7	rad(Si)
Gamma, dose rate	1×10^{-3} to 1×10^2	1×10^6	rad(Si)/h
Electrons and protons	—	1×10^8	rad(Si)

Source: Adapted from Johnson et al. 1983 [47], table 2, p. 4359.

space applications, the levels encountered are very high, although the dose rates and operational times are somewhat comparable.

Hardening of nuclear power plant electronics is accomplished by using critical control functions designed as simply as possible and by using a combination of shielding and radiation-hardened parts. Hardened parts for this environment present a serious procurement and obsolescence problem, as sources and parts selection are very limited. Shielding is an attractive method and widely used, as the stringent weight considerations found in space and mobile applications are of no immediate concern in reactor applications. Hardness requirements are also heavily driven by accessibility, as the electronic systems in the containment structure cannot be directly serviced on a routine basis. Much of the work within these structures involves the handling of highly radioactive materials, which necessarily needs to be carried out by robots, and mobility requirements for these machines are stringent and use multiplexed control signals rather than direct heavy cable bundles.

2.6 The Weapons Environment

Weapons spectra were part of the very earliest studies of radiation effects in electronics. These environments are driven by the detonation of nuclear weapons of both the fission and fusion type, which we euphemistically called devices in the Cold War years. In Chapter 1, we discussed the very limited experience with which the nuclear powers undertook their testing programs, driven by the pressures of war, and the long learning curve that eventually enabled engineers to understand the effects of the various types of radiation on electronics. Weapons spectra are in nearly all cases much different from the space environments we're now familiar with and are largely classified, and will be discussed only in general terms.

A fission weapon uses a nuclear reaction in which heavy atoms such as ^{235}U or ^{239}Pu are split into lighter species, releasing the mass differential as energy as discussed above. In these weapons, two subcritical masses of fissionable material are driven together by external means or a single subcritical mass is compressed into a single supercritical mass, which then initiates an uncontrolled chain reaction and explodes (Figure 2.37). Early fission weapons used a simple gun-type arrangement to drive the two masses together, while later devices used an implosion driven by precisely machined shaped

Figure 2.37 The Crossroads Baker nuclear event at Bikini Atoll. This was a 23 kt device exploded as part of Operation Crossroads, which investigated the effects of fission weapons on naval assets; note the silhouettes of ships near the event site. Source: US DoD.

charges. In contrast, a fusion or thermonuclear device uses a similarly uncontrolled reaction between light isotopes of hydrogen, which combine into helium accompanied by the release of large amounts of energy. A practical fusion weapon uses a fission device as a primary source of energy to drive the fusion reaction of hydrogen isotopes, a nuclear reaction that is entirely similar to those driving the processes in the Sun. Both classes of devices result in large amounts of released energy, which consists of blast (shock waves), thermal radiation, ionizing radiation, and neutrons. Our interest is in the ionizing and neutron components of the released energy.

Figure 2.38 provides the time dependence of gamma ray output from a hypothetical nuclear explosion [48]. Gamma radiation is produced directly by the nuclear reactions that constitute the explosion, and indirectly by interaction of prompt neutrons with the atmosphere and the decay of residual nuclear reaction products, which include fission debris and radioactivity in materials as induced by neutron activation. The explosion emits products in a definite sequence, starting with a short (about 20–100 ns) burst of gamma rays and neutrons occurring before the device has fully disintegrated. The intensity of these emissions depends very heavily on the type of weapon and the specific design and burn rate. This process is followed by a burst of fast neutrons, which are very penetrating and are not absorbed.

A fusion weapon will produce more neutrons per kiloton of yield, and these neutrons are generally more energetic than fission neutrons; a fission weapon can thus be optimized for maximum neutron yield and minimum blast, minimizing physical damage and maximizing lethality. The fast neutron burst is also a source of secondary gamma radiation through energy loss and activation, as the neutrons are absorbed by the atmosphere. The time constant for this process is to the order of microseconds. A thermonuclear detonation neutron energy spectrum [49] has many of the same properties as the Watt distribution for fission of ^{235}U. There is a peak flux around

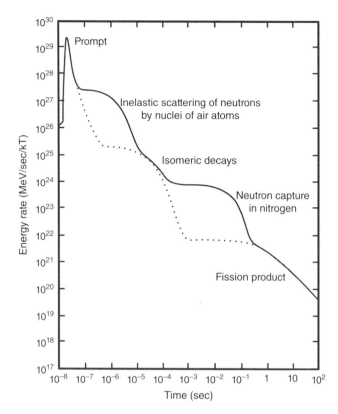

Figure 2.38 Calculated time dependence of the gamma ray energy output per kiloton energy yield from a hypothetical nuclear explosion. The dashed line refers to an explosion at very high altitude. Source: Glasstone and Dolan 1977 [48], Figure 8.14, p. 328.

0.8 MeV for fission produced neutrons with an average energy of about 2.0 MeV. There is also a peak at 14 MeV that is associated with fusion-produced neutrons.

The neutron fluence from a nuclear explosion can be estimated from the empirical equation [50]

$$\Phi_n = \frac{2 \cdot 10^{25} Y}{r^2} e^{-(\rho r/2.38 \cdot 10^4)} \, \text{n/cm}^2 \tag{2.41}$$

where Y is the yield of the detonation in megatons, r is the distance from the detonation in cm, and ρ is the density of the air in grams per liter ($\rho = 1.1$ for sea-level air). Equation (2.41) includes and attenuation by $1/r^2$ due to the geometry of the explosion and an exponential term to take into account the atmospheric attenuation.

The explosion also generates substantial amounts of fission products with very short half-lives (milliseconds to minutes), including the remnants of the bomb's structure. The rapid decay of these isotopes generates debris gamma radiation but is essentially complete within 10 seconds. The relative proportion of these gamma ray and neutron sources depends on the device yield. Tactical devices with yields to the order of a few kilotons can generate up to 25% of the gamma dose from direct gamma and neutron reactions, while this contribution is essentially zero for larger thermonuclear explosions. References on these weapons environments are expectedly uncommon in the open literature, but health physics journals are a resource. For a discussion of the physics of weapons effects on electronics the very early Nuclear and Space Radiation Effects Conference (NSREC) Short Course notes (1980, 1982, and 1984) provide [51–53] a reasonable background.

Survivable electronic systems in the nuclear weapon environment present an entirely different set of challenges than space or reactor applications. The key difference is the dose rates involved. Gamma radiation dose rates in space are generally low, as we've seen, to the order of a few millirad(Si)/s with important exceptions in such places as the Jovian radiation belts. Industrial acceptance testing of hardened parts is commonly performed within the 50–300 rad(Si)/s, a few orders of magnitude above that. A nuclear event however has dose rates in the 10^9–10^{11} rad(Si)/s range, with 20–100 ns pulse widths. The time integral of the pulse is the total dose, which can run in the hundreds of kilorads even for these short pulse widths. The high dose rate drives a correspondingly high rate of electron–hole pair generation, which reduces the semiconductor material resistivity to a very low value, and also generates high photocurrents (amperes) in p-n junctions. These effects usually latch the device or result in a low-resistance path between the supplies. In the absence of protective measures, the part will be destroyed through excessive current leading to damage to interconnect metallization and bond wires. Protection can be implemented at the system level through a nuclear event detector, which is a PIN diode-based transient gamma detector that generates a signal used to crowbar the power supplies of critical functions. Protection can also be done at the part level, using a combination of on-chip current limiting resistors, photocurrent compensation diodes, and nonlatching dielectrically isolated processes, and this has been the historical approach in US strategic systems.

It should also be noted that neutron levels in nuclear weapons environments are correspondingly high, with fluences as high as 1×10^{15} n/cm^2. This leads to displacement damage effects in which one or more silicon atoms are displaced from the lattice, creating a Frenkel pair. This model is simplistic, as the actual process involves a great many lattice sites and results in a defect cluster [54]. The lattice damage results in reduced minority carrier lifetime, with actual resistivity shift (in the direction of increased resistivity) occurring at very high levels. This radiation damage affects mostly bipolar junction transistors, which are a minority carrier device, with less of an effect on MOSFET devices. Hardening is accomplished through higher doping levels, narrower base width, and shallow junctions, and as a general rule higher F_T devices will be harder in the neutron environment. All parts hardened to weapons environments are highly application specific and closely export controlled, and the use of commercial components in strategically hardened systems is uncommon.

2.7 The Environment in High-Energy Physics Facilities

This is another interesting man-made environment that is specific to the large particle accelerator facilities used for the experimental aspects of high-energy particle physics. The leading accelerator facility in the current high-energy field and the one defining the state of the art in this field is the Large Hadron Collider (LHC) at the Conseil Européen pour la Recherche Nucléaire (CERN) facility, located on both sides of the French-Swiss border outside Geneva, Switzerland. The facility and experiments have a number of physics objectives including confirming the existence of the Higgs boson and alternative symmetry-breaking mechanisms and the exploration of alternatives to the standard model of particle physics. LHC is located in a 27 km circumference tunnel, which is reused from the Large Electron-Positron Collider (LEP) project and accelerates two counter-rotating proton beams to a peak energy of 7 TeV, resulting in highly energetic proton–proton collisions. In order to create a sufficient number of collisions, LHC is capable of reaching an unprecedented peak luminosity of $10^{34}/cm^2$ s at two high luminosity interaction points. At this peak luminosity, LHC will produce nearly 1×10^9 p–p collisions per second, creating an extremely hostile radiation environment near the beam line, the deflection magnets, and at the experiments. These are the most intense radiation fields found in any large-scale particle physics experiment and require specifically radiation-hardened design and hardware for the control electronics and the experiments.

An experiment in this context is a very large (perhaps 20 m in diameter) vertex detector consisting of a number of cylindrical detector arrays of several types, all coaxially surrounding the beam interaction region and operating in an externally applied magnetic field of to the order of 3T. The collision products are tracked by the detectors, and the resulting signals are routed into massive data acquisition systems of tens of millions of channels each. Since the particles are charged, the mass and momentum can be derived from track curvature and magnetic field strength. The resulting amount of data is massive, and a substantial amount of real-time data filtering is performed that allows only "interesting events" to be further processed and discards most others. Even with data filtering the amount of data generated by LHC is enormous, adding up to some 15 PB per year (15 million gigabytes or nearly 2 million dual-layer DVDs). The data is processed by a distributed computing network operating in 34 countries, not a surprising approach from the organization that pioneered the World Wide Web as a means of scientific communication. After initial processing, the data are distributed to 11 primary computer centers in Canada, France, Germany, Italy, the Netherlands, Scandinavia, Spain, Taiwan, the United Kingdom, and the United States. These primary centers then distribute the data to some 160 secondary centers for analysis. A backup of all data is retained on tape at CERN.

There are seven principal experiments at LHC, including the "Compact Muon Solenoid" (CMS), the "Toroidal LHC Apparatus" (ATLAS) and the "Large Ion Collider Experiment" (ALICE), all addressing different issues in particle physics. These vertex detectors surround an interaction region between the two beams and are subjected to the intense radiation environments generated by the collisions; since the interaction region is imaged directly there is no opportunity to shield the sensors and front-end processing electronics. In addition, the acceleration and deflection magnet environments have their own set of radiation environments, in which the beam interacts with gases and materials. Collisions between the proton beam and residual gas molecules in the vacuum beam line are responsible for a large percentage of the radiation environment in these locations. The radiation environment varies greatly with location, but maximum levels can range as high as 5×10^5 rad(Si) (5000 Gy) ionizing dose and 1×10^{14} n/cm^2 per year in some limited areas. Many areas have significant muon fluences as well; this negatively charged particle has approximately 200 times the mass of an electron, is only produced in high-energy nuclear interactions and has a relatively high lifetime of 2.2 µs, which renders it very penetrative. Muon effects are an issue in high-energy physics but also play an emerging role in atmospheric SEE in electronics. The diversity of radiation environments at LHC is huge, and electronic components used range from COTS parts to specifically designed radiation-hardened parts.

2.8 Summary and Closing Comments

This chapter examined the atomic sources of radiation through a review of some basic nuclear physics. This included the sources of X-rays as well as X-ray absorption. Also covered were basic and historical models of the nucleus that provided context for following sections about alpha and beta decay. Various radiation environments were also discussed, including natural radioactivity, space, nuclear reactors, nuclear weapons, and high-energy physics facilities. The discussion of natural radioactivity introduced the idea of exponential decay and the four decay series. The space environment includes discussion of solar radiation, cosmic rays, and the radiation belts, as well as atmospheric neutrons. Aspects of the nuclear reactor environment were discussed, including basic fission reaction concepts. Finally a brief discussion was provided about the nuclear weapon and high-energy physics facility environments.

Problems

2.1 Characteristic X-rays
The wavelength of the characteristic line for a given electron transition and element is given by Moseley's law

$$\sqrt{f} = k_1(Z - k_2) \tag{2.42}$$

where k_1 and k_2 are constants.
a) Derive Moseley's law from Bohr's formula (Eq. (1.2)). Assume that the brightest spectral line (K-alpha) is from an L- to K-shell transition and substitute $Z - 1$ for Z. Why is the substitution of $Z - 1$ for Z reasonable?
b) Moseley's law indicates that heavier elements will produce characteristic radiation with shorter wavelengths. Discuss the suitability of the various elements in Table 2.1 as well as other elements for use as X-ray anodes.

2.2 Absorption
a) An X-ray source produces a continuous spectrum of radiation (due to bremsstrahlung) up to about 75 MeV. There is thick lead shielding used as protection against stray radiation. What is the most likely frequency (highest relative intensity) of radiation to be detected when testing the shielding for stray radiation?
b) After an examination of Figure 2.6 one might be tempted to state that lead is about 100 times more effective than silicon at attenuating 100 keV X-rays. Is this correct?
c) According to Figure 2.6, as photon energy increases, the attenuation in lead and silicon decreases. Why is this?

2.3 Liquid Drop Model

For the following liquid drop model problems, use the nuclear radius given by the Fermi model (Eq. (2.4)).

a) Derive the surface energy term in the liquid drop model.

b) Derive the Coulomb energy term in the liquid drop model.

2.4 The Fermi Gas Model

a) Calculate the average nucleon energy in the Fermi gas model.

2.5 The Shell Model

a) Derive the spin-orbit interaction dependence on l.

2.6 Alpha Decay

a) Li^6 has an average binding energy that is less than He^4 (see Figure 2.8) so why doesn't Li^6 undergo spontaneous alpha decay?

2.7 Natural Radioactivity

a) Elements from three of the decay series are found in nature but only the last few elements from the Neptunian series can be found in nature. Why is this?

2.8 Radiation Environments

Discuss the electronic design challenges for the various radiation environments introduced in Chapter 2.

a) Are some environments more challenging to electronic design?

b) Total hardness will often be a combination of shielding, radiation hardening (IC process), and circuitry redundancy. Discuss the effectiveness of each for the various types of radiation as well as how the various radiation environments may influence tradeoffs in total hardness design.

c) Some radiation environments are not available to the designer (e.g. space and nuclear weapons environments). Discuss the various ways that radiation environments could be simulated using available resources such as radioactive elements, linear accelerators, etc.

References

There is a robust literature on radiation environments and radiation testing. It is not our objective to provide an exhaustive set of references, which in any case would take up entirely too much space. We have provided a representative set of references that are of historical interest or provide a good starting point for further study. The IEEE Transactions on Nuclear Science are the journal of record in America and include the conference proceedings for the two leading radiation effects conferences, the IEEE Nuclear and Space Radiation Effects Conference (NSREC) and its European counterpart, the [conference on] Radiations et ses Effets sur Composants et Systèmes (RADECS). The Short Courses presented at each of these conferences are an excellent resource as well.

1 Williams, G.P. (2009). Electron binding energies. In: *X-Ray Data Booklet*, 3e (ed. A.C. Thompson), 1-2–1-7. Berkeley, CA Table 1-1: Lawrence Berkeley National Laboratory, University of California.

2 Hubbell, J.H. and Seltzer, S.M. (2004). *Tables of X-Ray Mass Attenuation Coefficients and Mass Energy-Absorption Coefficients* (version 1.4), http://physics.nist.gov/xaamdi (accessed 18 Aug 2017). Gaithersburg, MD: National Institute of Standards and Technology.

3 Kramida, A., Ralchenko, Y., Reader, J., and NIST ASD Team (2018). *NIST Atomic Spectra Database* (ver. 5.5.6), http://physics.nist.gov/asd (accessed 19 April 2018). Gaithersburg, MD: National Institute of Standards and Technology.

4 Stuewer, R.H. (1997). Gamow, alpha decay, and the liquid-drop model of the nucleus. In: *George Gamov Symposium, ASP Conference Series*, vol. 129 (eds. E. Harper, W. Parke and D. Anderson), 29–43.

5 Evans, R.D. (1955). *The Atomic Nucleus*. New York: McGraw-Hill.

6 Bohr, N. (1937). Transmutations of atomic nuclei. *Science* 86 (2225): 161–165.

7 Bethe, H.A. and Bacher, R.F. (1936). Nuclear physics A. stationary states of nuclei. *Reviews of Modern Physics* 8: 82–229.

8 Fermi, E. (1946). A Course in Neutron Physics. Los Alamos Document *LADC-255*. Notes by I. Halpern on Los Alamos lecture series on neutron physics.

9 Fermi, E. (1950). *Nuclear Physics*. Notes compiled by J. Orear, A.H. Rosenfeld, and R.A. Schluter. Chicago: University of Chicago Press.

10 Daley, C. (2004). An improved mass formula. B.Sc. thesis. University of Surrey.

11 Feenberg, E. (1939). On the shape and stability of heavy nuclei. *Physics Review* 55 (5): 504.

12 Bohr, N. and Wheeler, J.A. (1939). The mechanism of nuclear fission. *Physics Review* 56 (5): 426.

13 Feenberg, E. (1947). Semi-empirical theory of the nuclear energy surface. *Reviews of Modern Physics* 19: 239.

14 Green, A.E.S. (1954). Coulomb radius constant from nuclear masses. *Physics Review* 95: 1006.

15 Blatt, J.M. and Weisskopf, V.F. (1952). *Theoretical Nuclear Physics*. New York: Wiley.

16 Royer, G. (2012). Generalized liquid drop model and fission, fusion, alpha and cluster radioactivity and superheavy nuclei. 4th International Conference on Current Problems in Nuclear Physics and Atomic Energy. Kiev, Ukraine.

17 Haxel, G. B., Hedrick, J. B., and Orris, G. J. (2002). Rare earth elements – critical resources for high technology. U.S. Geological Survey Fact Sheet 087-02. https://pubs.usgs.gov/fs/2002/fs087-02 (accessed 18 May 2017).

18 Meyerhof, W.E. (1967). *Elements of Nuclear Physics*. New York: McGraw-Hill.

19 Cohen, B.L. (1971). *Concepts of Nuclear Physics*. New York: McGraw-Hill.

20 Krieger, A., Blaum, K., Bissell, M.L. et al. (2012). Nuclear charge radius of ^{12}Be. *Physical Review Letters* 108: 142501.

21 Arima, A. and Iachello, F. (1975). Collective nuclear states as representations of a SU(6) group. *Physical Review Letters* 35: 1069–1072.

22 Cook, N.D. and Hayashi, T. (1997). Lattice models for quark, nuclear structure and nuclear reaction studies. *Journal of Physics G: Nuclear Physics* 23 (9): 1109.

23 Nasser, G.A. (2014). Body-centered-cubic (BCC) lattice model of nuclear structure. *Journal of Nuclear Physics*.

24 Cook, N.D., Hayashi, T., and Yoshida, N. (1999). Visualizing the atomic nucleus. *IEEE Computer Graphics and Applications* 19 (5): 54–60.

25 Geiger, H. and Nuttall, J.M. (1911). The ranges of the α particles from various radioactive substances and a relation between range and period of transformation. *Philosophical Magazine* 22 (130): 613–621.

26 Carrington, R.C. (1859). Description of a singular appearance seen in the Sun on September 1, 1859. *Monthly Notices of the Royal Astronomical Society* 20: 13–15.

27 Stassinopoulos, E.G. and Raymond, J.P. (1988). The space radiation environment for electronics. *Proceedings of the IEEE* 76 (11): 1423–1442.

28 Barraclough, D.R., Harwood, R.M., Leaton, B.R., and Malin, S.R.C. (1975). A model of the geomagnetic field at epoch 1975. *Geophysical Journal of the Royal Astronomical Society* 43: 645–659.

29 Stassinopoulos, E.G. and Staffer, C.A. (2007). *Forty-Year "Drift" and Change of the SAA*. Greenbelt, MD: NASA Goddard Space Flight Center.

30 Fillius, R.W. and McIlwain, C.E. (1974). Measurements of the Jovian radiation belts. *Journal of Geophysical Research* 79 (25): 3589–3599.

31 Teague, M. J. and Stassinopoulos, E. G., (1971). A model of the Starfish flux in the inner radiation zone. *NASA/GSFC report X-601-72-487*.

32 Gussenhoven, M.S., Mullen, E.G., and Brautigam, D.H. (1996). Improved understanding of the Earth's radiation belts from the CRRES satellite. *IEEE Transactions on Nuclear Science* 43 (2): 353–368.

33 Sawyer, D.M. and Vette, J.I. (1976). *AP-8 Trapped Proton Environment for Solar Maximum and Solar Minimum. NASA report NASA-TM-X-72605, NSSDC/WDC-A-R/S-76-06*, . Greenbelt, MD: NASA/Goddard Space Flight Center.

34 Vette, J.I. (1991). *The AE-8 Trapped Electron Model Environment. NASA report NSSDC/WDC-A-R/S-91-24*, . Greenbelt, MD: NASA/Goddard Space Flight Center.

35 Weller, R.A., Mendenhall, M.H., Reed, R.A. et al. (2010). Monte Carlo simulation of radiation effects in microelectronics. In: *IEEE Nuclear Science Symposium Conference Record, Nuclear Science Symposium*, 1262–1268. IEEE.

36 Swordy, S. (2001). The energy spectra and anisotropies of cosmic rays. *Space Science Reviews* 99: 85–94.

37 Tanabashi, M., Hagiwara, K., Hikasa, K. et al. (Particle Data Group) (2018). Review of particle physics. *Physical Review D* 98 (030001): 447.

38 Mewaldt, R.A. (1988). Elemental composition and energy spectra of galactic cosmic rays. In: *Proc. of Conf. on Interplanetary Particle Environment*, JPL Publication 88-28, 121–132.

39 Binns, W.R., Fickle, R., Waddington, C.J. et al. (1981). The heavy nuclei experiment on HEAO-3. *Proceedings of the International Astronomical Union* 94: 91–92.

40 Morrison, P., Olbert, S., and Rossi, B. (1954). The origin of cosmic rays. *Physics Review* 94: 440.

41 Ginsburg, V.L. and Pluskin, V.S. (1976). On the origin of cosmic rays: some problems in high-energy astrophysics. *Reviews of Modern Physics* 48: 161.

42 Gordon, M.S., Goldhagen, P., Rodbell, K.P. et al. (2004). Measurement of the flux and energy spectrum of cosmic-ray induced neutrons on the ground. *IEEE Transactions on Nuclear Science* 51 (6): 3427–3434.

43 Taber, A.H. and Normand, E. (1995). *Investigation and Characterization of SEU Effects and Hardening Strategies in Avionics*, Report No. DNA-TR-94–123. Alexandria, VA: Defense Nuclear Agency.

44 Johnson, C.W. and Holloway, C.M. (2007). The dangers of failure masking in fault-tolerant software: aspects of a recent in-flight upset event. In: *2007 2nd Institution of Engineering and Technology International Conference on System Safety*, 60–65. London.

45 Glasstone, S. (1955). *Principles of Nuclear Reactor Engineering*. New York: van Nostrand.

46 Madland, D.G. (1982). *New Fission Neutron Spectrum Representation for ENDF*, Technical Report LA—9285-MS. Los Alamos National Laboratory, DOE.

47 Johnson, R.T. Jr.,, Thome, F.V., and Craft, C.M. (1983). A survey of aging of electronics with application to nuclear power plant instrumentation. *IEEE Transactions on Nuclear Science* 30 (6): 4358–4362.

48 Glasstone, S. and Dolan, P.J. (eds.) (1977). *The Effects of Nuclear Weapons*. Washington, D.C.: United States Department of Defense and the Energy Research and Development Administration.

49 Messenger, G.C. and Ash, M.S. (1986). *The Effects of Radiation on Electronic Systems*. New York: Van Nostrand Reinhold Company.

50 Brode, H.L. (1969). Close-in weapon phenomena. *Annual Review of Nuclear Science* 8: 153–202.

51 Gover, J. E., Johnston, A. H., Halpin, J. and Rudie, N. J. (1980). Radiation effects and systems hardening. IEEE Nuclear and Space Radiation Effects Conference (NSREC) Short Course.

52 Srour, J. R., Longmire, C. L., Raymond, J. P. and Allen, D. J. (1982). Radiation effects and systems hardening. IEEE Nuclear and Space Radiation Effects Conference (NSREC) Short Course.

53 Gover, J. E., Rose, M. A., Tigner, J. E. et al. (1984). Radiation effects and systems hardening. IEEE Nuclear and Space Radiation Effects Conference (NSREC) Short Course.

54 van Lint, V.A.J. (1987). The physics of radiation damage in particle detectors. *Nuclear Instruments and Methods in Physics Research Section A: Accelerators, Spectrometers, Detectors and Associated Equipment* 253 (3): 453–459.

3

Radiation Effects in Semiconductor Materials

3.1 Introduction

Particles and photons interact with matter, and this chapter provides the introductory physics behind these interactions. An understanding of the basic effects in semiconductor materials caused by particles and photons (e.g. X-rays and gamma rays) will lead to a better understanding of more complex or higher-order effects on devices and circuits, for example the change in beta of a bipolar transistor or the dependence of operational amplifier open-loop gain on total dose of radiation accumulated.

An understanding of the interaction of radiation on semiconductor materials and structures also guides semiconductor process development, whether specifically aimed at designing radiation hardened processes or used during the development of conventional foundry processes. The ion implantation process used to form doped regions in semiconductor wafers is a good example of a conventional process step where the effects of a radiation source (ionized boron, phosphorous, arsenic, etc.) are finely calculated. The energy of the ionized specie and the dose of ions are chosen to achieve a desired distribution of dopant in the semiconductor wafer. The implant angle is chosen to avoid channeling effects along certain crystal lattice directions. Lattice damage caused by the ion implant step is repaired using a high-temperature anneal ($>700\,^\circ$C).

While analogous to an ion implantation processing step, the incident conditions of radiation particle sources are less controlled. Most obviously, there is no high temperature anneal to repair crystal damage in the radiated integrated circuit, only normal operating temperatures. Depending on the radiation source, the energy and angle of incidence with the integrated circuit can vary. The effect that the incident radiation particle has on devices in the integrated circuit can depend on voltages applied to gates, junctions, and interconnect layers. Operation of the integrated circuit during the radiation event can be affected both by transient effects and accumulated dose effects.

The discussion in this chapter is aimed at providing a working knowledge of the physics for some basic radiation effects in materials. Chapter 6 reviews some of the effects that are seen in semiconductor devices, along with some direction for developing simulation models.

Integrated Circuit Design for Radiation Environments, First Edition.
Stephen J. Gaul, Nicolaas van Vonno, Steven H. Voldman and Wesley H. Morris.
© 2020 John Wiley & Sons Ltd. Published 2020 by John Wiley & Sons Ltd.

3.2 Basic Effects

When photons or particles travel through space containing atoms or molecules, they lose energy through interactions with the surrounding matter. This interaction can take several forms, depending on the incident photon or particle energy. Some interactions, like the photoelectric and Compton effects, are specific to photons. For incident particles, the mass and charge are also of interest. Some of the more fundamental ideas about this interaction, like electronic stopping (Section 3.2.1.1) are specific to particles and have theoretical origins in the early twentieth century.

The discussion of basic effects starts with a focus on the interaction of heavy, charged particles with matter. These include protons and any ionized element or molecule. Electrons and neutrons are separated into their own sections as each has unique interactions with matter. Photons are discussed in Section 3.2.4 and include X-rays and gamma rays.

3.2.1 Heavy Charged Particles

This section discusses the interaction of heavy charged particles in matter. The section starts with a discussion of stopping power, a concept that is useful in Sections 3.2.2 through 3.2.4 as well.

3.2.1.1 Stopping Power

When a particle interacts with matter, it loses energy in a number of ways, depending on the energy, mass, and charge state of the particle. The energy loss can come from electronic stopping, the slowing that is caused by Coulomb interactions between the particle and electrons in the matter. The moving charged particle exerts an electromagnetic force on atomic electrons, resulting in energy transfer that can be sufficient to remove the electron from the atom or leave the atom in an excited, nonionized state. There is also the possibility of some type of nuclear stopping, an interaction with the nuclei of atoms that can be inelastic, elastic, or result in a fission event. The particle can also emit electromagnetic radiation.

A quantity can be defined for the total average linear rate of energy loss of a heavy charged particle in a medium

$$S(E) = -\frac{dE}{dx}. \tag{3.1}$$

In Eq. (3.1) E is the energy of the particle and x is the path length. This quantity is called the stopping power of the medium for the particle, and is of fundamental importance in radiation physics and dosimetry. Note that stopping power is a property of the medium and not the particle. The negative sign in Eq. (3.1) ensures stopping power will be a positive quantity. It is useful to consider the electronic, nuclear, and radiative energy components to the total stopping power separately:

$$-\frac{dE}{dx} = \left(-\frac{dE}{dx}\right)_e + \left(-\frac{dE}{dx}\right)_n \tag{3.2}$$

where $(-dE/dx)_e$ is the electronic energy loss due to Coulomb interactions and $(-dE/dx)_n$ is the nuclear energy loss. The nuclear energy loss component includes losses due to emission of bremsstrahlung or Cerenkov radiation, and nuclear interactions.

Figure 3.1 Bragg curve for 5.49 MeV alpha particles in air.

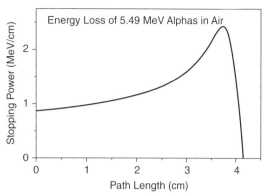

Figure 3.1 shows the stopping power or Bragg curve for 5.49 MeV alpha particles in air. These particles are typical for radiation from naturally occurring radon gas. As the particles lose energy, the likelihood of an interaction increases with a corresponding increase in stopping power up to a maximum called the Bragg peak, named for William Henry Bragg, who discovered it in 1903.

The $(-dE/dx)_e$ term in Eq. (3.2) is called the linear energy transfer (LET). LET focuses on the energy transfer to the media and is useful for modeling radiation effects in electronic circuits

$$L_\Delta = -\frac{dE_\Delta}{dx}. \tag{3.3}$$

Since the energy transfer of interest is usually closely aligned with the track of an incident particle, LET calculations will sometimes exclude secondary electrons with energies larger than Δ, as these electrons are likely to travel far from the primary particle track. As Δ approaches infinity, LET becomes identical to the electronic stopping power term in Eq. (3.2).

Sometimes LET is defined with respect to the material density ρ, changing the units from MeV/cm to MeV-cm^2/mg, for example. The mass stopping power S is defined as

$$S = \frac{-dE/dx}{\rho}. \tag{3.4}$$

Yet another way to describe the energy loss of a particle is specific ionization. Specific ionization (SI) is the number of ion pairs formed per unit distance traveled by the charged particle

$$SI = \frac{-dE/dx}{w} \tag{3.5}$$

where w is the average energy required to create an ion (or electron–hole pair in a semiconductor). Table 3.1 provides ion pair generation energy for several common media.

Although it is not possible to model exactly what will happen to a particular particle during its transit of matter, it is important to estimate the energy loss. We now examine each of the mechanisms that contribute to stopping power.

3.2.1.2 Electronic Stopping

For high-energy heavy charged particles, the energy loss is primarily due to electronic stopping. When such particles enter matter, they interact with the electrons in the outer

Table 3.1 Ion pair generation energy for various materials.

Material	Ion pair generation energy w (eV/ion pair)
Air	34 eV
Silicon (Si)	3.6 eV
Germanium (Ge)	2.8 eV
Gallium-arsenide (GaAs)	3.8 eV
Silicon dioxide (SiO$_2$)	17 eV

shells of nearby atoms. The energy loss at each encounter with an electron through this Coulomb interaction is small, however, due to the mass difference between the particle and electrons. Each encounter with an electron will affect the speed and direction of the particle only a little, so the path taken by the high-energy particle is fairly straight.

3.2.1.2.1 Q_{max} Calculation Consider a heavy charged particle of mass M and initial velocity V_i hitting an electron of mass m_e so that the maximum energy transfer takes place. For maximum energy transfer, assume the particle and electron are confined to motion in one dimension only. Also, the particle is moving rapidly compared to the electron so that the energy transfer to the electron is large enough to ignore the binding energy of the electron in the atom. For conservation of energy, we have

$$E_i = \frac{1}{2}MV_i^2 = \frac{1}{2}MV_f^2 + \frac{1}{2}m_e v_f^2 \tag{3.6}$$

And for conservation of momentum

$$MV_i = MV_f + m_e v_f. \tag{3.7}$$

In Eqs. (3.6) and (3.7), V_f and v_f are the post-collision velocities of the particle and electron, respectively. Solving Eqs. (3.6) and (3.7) for the post-collision particle velocity V_f yields

$$V_f = \frac{(M - m_e)V_i}{M + m_e}. \tag{3.8}$$

Now the maximum energy transfer to the electron for the particle due to this collision with an electron can be calculated. It is just the difference in the incident and final kinetic energies of the particle

$$Q_{max} = \frac{1}{2}MV_i^2 - \frac{1}{2}MV_f^2 = \frac{4m_e M E_i}{(M + m_e)^2}. \tag{3.9}$$

Note that the special case where $M = m$ results in $Q_{max} = E_i$, indicating that all of the energy of the particle is transferred to the electron. If the particle is a proton, then $M = 1000\,m$ and Eq. (3.9) can be simplified and rewritten to represent the ratio of energy transfer to incident energy

$$\frac{Q_{max}}{E_i} = \frac{4m(1000m)}{(1001m)^2} \cong \frac{4}{1001} \cong 0.004. \tag{3.10}$$

Clearly, many such collisions must occur between a heavy charged particle and surrounding electrons before the particle will slow appreciably. Once the particle slows down enough, it is more likely to collide with nuclei, eventually coming to a stop. In Eqs. (3.6) through (3.10) a nonrelativistic analysis is followed for the purpose of illustrating the energy transfer of such collisions. It is also possible to repeat this analysis, taking into account relativistic velocities.

3.2.1.2.2 Bohr Analysis of Electronic Stopping

A formula for stopping power was derived by Niels Bohr in 1913 [1]. The following derivation is along the same lines as Bohr's original formulation.

Consider the interaction of a heavy, charged particle with a single electron as diagramed in Figure 3.2. A particle with charge ze and velocity V travels near an electron and the resulting Coulomb interaction exerts a force F on the electron. For this simple analysis we ignore the binding energy of the electron and assume its velocity is zero. We also assume the velocity of the particle is unchanged by the interaction with the electron. A portion of the particle's momentum is thus imparted to the electron but note that the x-axis component of this force, when integrated over the path of the particle, sums to zero

$$|-F_x| = |+F_x|. \tag{3.11}$$

Only the F_y component of the Coulomb force contributes to the net momentum imparted to the electron. The imparted momentum can be calculated by integrating F_y over time

$$P = \int_{-\infty}^{\infty} F_y dt = \int_{-\infty}^{\infty} F \cos\theta dt = k_0 z e^2 \int_{-\infty}^{\infty} \frac{\cos\theta}{r^2} dt. \tag{3.12}$$

If the $t = 0$ point is made to coincide with $x = 0$, the integration in Eq. (3.12) is symmetric about $x = 0$:

$$P = k_0 z e^2 \int_{-\infty}^{\infty} \frac{\cos\theta}{r^2} dt = 2k_0 z e^2 \int_{0}^{\infty} \frac{b}{r^3} dt. \tag{3.13}$$

Noting that

$$r = \sqrt{V^2 t^2 + b^2} \tag{3.14}$$

Figure 3.2 Particle interaction with a single electron.

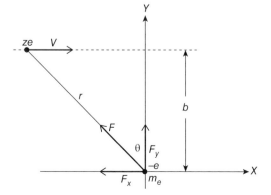

we have

$$P = 2k_0 z e^2 \int_0^\infty \frac{dt}{(V^2 t^2 + b^2)^{3/2}} = 2k_0 z e^2 b \left[\frac{t}{b^2 (b^2 + V^2 t^2)^{1/2}} \right]_0^\infty = \frac{2k_0 z e^2}{Vb}. \quad (3.15)$$

We can now calculate the energy transfer to the electron using the imparted momentum in Eq. (3.15)

$$Q = \frac{P^2}{2m_e} = \frac{2k_0^2 z^2 e^4}{m_e V^2 b^2}. \quad (3.16)$$

Equation (3.16) provides the energy transfer to a single electron, but we need to take into account all of the electrons in a material that are close enough to the particle track to calculate the electronic stopping power. The particle's closest approach to a given electron is called the impact parameter and corresponds to the distance b in Figures 3.2 and 3.4. A maximum value for this parameter, b_{max}, would correspond to the most distant electron that could still be affected by the particle's passage. Such an electron would require a minimum amount of energy to become ionized. We set this minimum energy to the mean excitation energy I for the material and calculate b_{max}:

$$b_{max} = \frac{k_0 z e^2}{V} \sqrt{\frac{2}{m_e I}}. \quad (3.17)$$

The mean excitation energy is specific to the material as seen in Figure 3.3, but for target materials heavier than sulfur, a good approximation is given by [2]

$$I = 10 eV \cdot Z. \quad (3.18)$$

The smallest impact parameter b_{min} can be calculated by assuming the maximum energy Q_{max} is transferred to the electron. Rewriting Eq. (3.9) assuming the mass of the particle is much larger than the electron mass, we have

$$E_{max} = 2V^2 m_e. \quad (3.19)$$

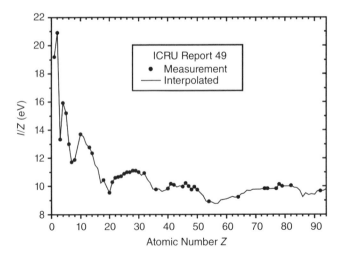

Figure 3.3 Mean excitation energy (relative) versus atomic number.

Figure 3.4 Stopping power calculation diagram.

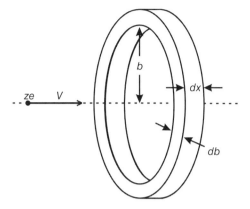

Solving for b using Eqs. (3.16) and (3.19) provides the estimate for b_{min}:

$$b_{min} = \frac{k_0 z e^2}{m_e V^2}.$$ (3.20)

Now the calculation of the electronic stopping power can proceed. Figure 3.4 provides a diagram showing an incremental volume of electrons that will interact with the passing particle. For matter of density ρ, with atomic weight A and Z electrons per atom, the electron density is $N_e = N_A Z \rho / A$ where N_A is Avogadro's number. So according to Figure 3.4, the passing particle encounters $2\pi N_e b db$ electrons for each unit distance dx traveled. Using Eq. (3.16) the energy loss per unit distance traveled is $2\pi N_e Q b db$. Integration over all possible impact parameters will give the total linear rate of energy loss:

$$-\frac{dE}{dx} = 2\pi N_e \int_{b_{min}}^{b_{max}} Q b db = \frac{4\pi k_0^2 z^2 e^4 N_e}{m_e V^2} \int_{b_{min}}^{b_{max}} \frac{db}{b} = \frac{4\pi k_0^2 z^2 e^4 Z N_A}{m_e A V^2} \rho \ln \frac{b_{max}}{b_{min}}.$$ (3.21)

Using Eqs. (3.17) and (3.20) we have the classical formula for energy loss:

$$-\frac{dE}{dx} = \frac{4\pi k_0^2 z^2 e^4 Z N_A}{m_e V^2 A} \rho \frac{1}{2} \ln \frac{2m_e V^2}{I}.$$ (3.22)

We see in Eq. (3.22) that only the particle's charge ze and velocity V affect the stopping power. For the material, the electron density $Z N_A \rho / A$ is an obvious factor, but only as a multiplicative influence.

3.2.1.2.3 *Bethe Formula* Hans Albrecht Bethe (1906–2005) was a German American nuclear physicist and Nobel laureate in physics, who in 1930 devised a nonrelativistic formula for the energy loss of heavy, high-energy charged particles using a quantum mechanical derivation. This nonrelativistic version of the formula is sometimes called the Bethe-Bloch formula, as it utilized an approximation for the mean ionization energy of the atom (Eq. (3.18)) attributed to Bloch. Bethe derived a relativistic version of the formula in 1932 [3]:

$$-\frac{dE}{dx} = \frac{4\pi k_0^2 z^2 e^4 Z N_A}{m_e \beta^2 c^2 A} \left[\ln \frac{2m_e c^2 \beta^2 \gamma^2 T_{max}}{I^2} - \beta^2 - \frac{\delta}{2} - \frac{C}{Z} \right]$$ (3.23)

where $\beta = V/c$ and $\gamma = \sqrt{1/(1-\beta^2)}$. Note that Eq. (3.23) corresponds to the mass stopping power as in Eq. (3.4), so the units will be MeVV-cm^2/g for example. When comparing the first term in square brackets in Eq. (3.23) to the classical value of the same term in Eq. (3.22), we see that the quantum mechanical treatment has resulted in a doubling of the strength of this term. One also notes that the mass of the particle is Lorenz contracted at relativistic speeds. The value used for the maximum kinetic energy transferred T_{max} depends on the speed of the particle (Problem 3.1). A good summary of formulas for calculating corrections to the stopping power has been provided by Fano [4]. The Bethe-Bloch formulation in Eq. (3.23) also includes several correction factors provided by Fano:

$\delta(\beta)$ – A polarization (density effect) correction that accounts for reduced participation by distant atoms due to the Coulomb field reduction caused by the dipole of nearby atoms. This is an important correction for heavy charged particles at relativistic energies, but is also important for light charged particles at all energies.

C/Z – A shell correction accounting for nonparticipation of K-shell electrons at low particle energies.

The Bethe-Bloch formula becomes invalid at low energies, for example when the velocity of the charged particle is comparable with K-shell electrons (10 MeV/A). The charged particle can also pick up electrons from the medium, reducing its charge. This charge exchange is significant below particles energies of about 0.2 MeV/A. At very low particle energies (approx. 0.01 MeV), nuclear interactions become significant.

3.2.1.3 Nuclear Stopping

As heavy, charged particles slow down in a material, particle energy loss due to interaction with the material atomic nuclei becomes more important. This dependence on particle speed/energy is due to the fact that faster particles have less interaction time with the atomic nuclei.

Nuclear stopping, like the electronic stopping, can be modeled using the Coulomb potential, except the screening effect of the surrounding electrons must also be modeled. The repulsive Coulomb field is combined with a screening function that is dependent on the impact parameter. The screening function can provide a unity or multiplicative value for the smallest impact parameters and a value approaching zero for complete shielding at impact parameters that are larger than the outer electron shell radius of the target atoms, for example. An example potential function $V(r)$ for nuclear stopping has the form

$$V(r) = \frac{k_0 zZe^4}{r}\Psi(r) = \frac{k_0 zZe^2}{r}e^{\left(-\frac{r}{a}\right)} \tag{3.24}$$

where $\psi(r)$ is the screening function and a is some screening distance from the nucleus.

Figure 3.5 shows a graph of nuclear and electronic stopping power for protons in silicon. The general characteristics of nuclear and electronic stopping power are shown. Nuclear stopping is greater at low energies with increasing atomic number in the target material, leading to greater nuclear stopping. The electronic stopping shows the Bragg peak, then a gradual rise with increasing energy after the Bragg peak due to relativistic effects and at the highest energies, the energy loss due to bremsstrahlung. Losses due to bremsstrahlung are important only for very light particles such as electrons, positrons, and muons, however.

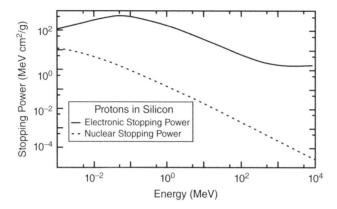

Figure 3.5 Nuclear and electronic stopping power for protons in silicon. Source: Adapted from Berger et al. 2005 [5], http://physics.nist.gov/Star.

3.2.2 Electrons

3.2.2.1 Electromagnetic Radiation

All charged particles can lose energy through "braking radiation" or bremsstrahlung, although the lightest charged particles like electrons can lose substantially more energy to bremsstrahlung due to their low rest mass. Bremsstrahlung is electromagnetic radiation produced when a charged particle is accelerated in an electric field. This is the case, for example, where an electron is deflected toward a nucleus in the target material or is decelerated when approaching the electrons around a nucleus. Figure 3.6 shows an example where an electron approaches a nucleus. As the electron velocity changes, radiation is produced. Clearly there are colinear and perpendicular components to acceleration, each of which can produce radiation.

The general expression for the total radiated power for a light particle of charge q is [6]

$$P = \frac{q^2 \gamma^4}{6\pi\varepsilon_0 c}\left(\dot{\vec{\beta}}^2 + \frac{(\vec{\beta}\cdot\dot{\vec{\beta}})^2}{1-\beta^2}\right) \tag{3.25}$$

Figure 3.6 Bremsstrahlung (braking radiation) for an electron.

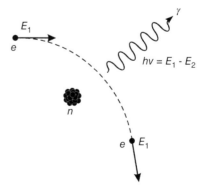

where $\overrightarrow{\dot{\beta}}$ represents the time derivative of β. When the particle acceleration is colinear with its velocity, the radiated power is proportional to γ^6 whereas for acceleration perpendicular to velocity radiated power is proportional to γ^3. This leads to a dependency on mass of the particle of m^{-6} and m^{-4}, respectively, for the two cases, so that bremsstrahlung is a significant energy loss mechanism for only the lightest particles including electrons, positrons, and muons.

It is very likely that you have encountered bremsstrahlung – it is the mechanism behind the production of X-rays used for commercial purposes and was introduced in Chapter 1. In an X-ray tube, electrons are accelerated in a vacuum by an applied electric field (typically >10 kV) toward a target made of some high-Z metal, for example tungsten. As the electrons slow in the target, a continuous spectrum of X-rays is produced as shown in Figure 2.1. The spectrum has a minimum wavelength cutoff that is set by the voltage V applied to the X-ray tube because there is a small probability that an electron could lose all of its incident energy in one collision (corresponding to $E_2 = 0$ in Figure 3.6):

$$\lambda_{\min} = \frac{hc}{eV}. \tag{3.26}$$

It is more likely that the electron will emit photons to lose energy in several collisions as it slows in the material.

Physicist and mathematician Sir Joseph Larmor (1857–1942) first calculated the rate of energy radiation from an accelerating electron, but the following derivation is along the lines of one used by the discoverer of the electron, J. J. Thomson in 1906 and 1907. A rigorous derivation based on Maxwell's equations and retarded potentials can also be done, but it doesn't illustrate the physics as clearly as J. J. Thomson's simpler derivation.

Figure 3.7 helps to illustrate the steps taken in Thomson's approach. Start with a charge q at the origin O at time $t = 0$. The charge is accelerated for a short time interval Δt, causing a change Δv in its velocity. After a time, t we can find a transition region in the shape of a hollow sphere of radius $r = ct$ and thickness $c\Delta t$ centered on the origin where the field lines from the perturbed charge meet the original $t = 0$ field lines outside

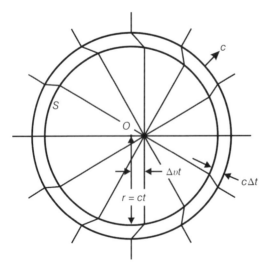

Figure 3.7 Diagram for Larmor formula derivation.

Figure 3.8 Diagram for Larmor formula derivation (detail).

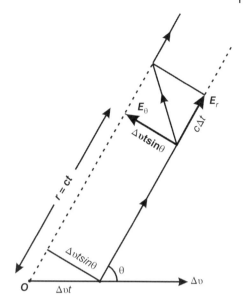

reference frame S as illustrated in Figure 3.7. We assume that $\Delta v \ll c$ so that the electric field lines are radial at $t = 0$ and also at time t in the frame of reference S. Inside the transition region, the electric field has both E_r and E_θ components, representing a pulse of electromagnetic energy that is traveling outward from O at the speed of light.

A portion of Figure 3.7 is replicated in Figure 3.8 to better illustrate the transition region and the relationship between E_r and E_θ. The electric field line is traced through the transition region where it is seen that the ratio of E_θ to E_r is found by an inspection of the geometry in Figure 3.8:

$$\frac{E_\theta}{E_r} = \frac{\Delta v t \sin \theta}{c \Delta t}. \tag{3.27}$$

E_r is given by Coulomb's law:

$$E_r = \frac{q}{4 \pi \varepsilon_0 r^2} \text{ where } r = ct, \tag{3.28}$$

so that

$$E_\theta = \frac{q(\Delta v / \Delta t) \sin \theta}{4 \pi \varepsilon_0 c^2 r} \approx \frac{k_0 q (dv/dt) \sin \theta}{c^2 r} = \frac{k_0 q \dot{v} \sin \theta}{c^2 r}. \tag{3.29}$$

We see the expected r^{-2} dependence in the radial component of the electric field but note from Eq. (3.29) that the pulse of electromagnetic energy represented by E_θ decreases only as r^{-1}. The radial component is called the near field and E_θ is the radiation field. Unless the charge is accelerated, the radiation field falls off rapidly at large distances. But when the radiation field is present, it dominates over the near field far from the charge. Figure 3.9 shows the cross section of the dipole radiation field that results when a charged particle undergoes a small change in velocity Δv. It takes the shape of a donut with no radiation along the particle path.

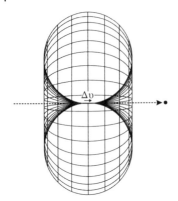

Figure 3.9 Representation of the dipole radiation field from a particle undergoing acceleration Δv.

We now calculate the energy flow per unit area per second at distance r using the Poynting vector $E \times H = \varepsilon_0 c E^2$ to quantify the energy lost due to radiation by the accelerated charge. Using the solid angle $d\Omega$ as the unit area and Eq. (3.29) we have

$$-\left(\frac{dE}{dt}\right)_{rad} d\Omega = \frac{\varepsilon_0 c k_0^2 q^2 \dot{v}^2 \sin^2\theta}{c^4 r^2} r^2 d\Omega = \frac{q^2 \dot{v}^2 \sin^2\theta}{16\pi^2 \varepsilon_0 c^3} d\Omega \tag{3.30}$$

The total radiated energy is found by integrating over the solid angle $d\Omega = 2\pi \sin\theta d\theta$

$$-\left(\frac{dE}{dt}\right)_{rad} = \int_0^\pi \frac{q^2 \dot{v}^2 \sin^2\theta}{16\pi^2 \varepsilon_0 c^3} 2\pi \sin\theta d\theta = \frac{q^2 \dot{v}^2}{6\pi \varepsilon_0 c^3}. \tag{3.31}$$

Equation (3.31) is known as the Larmor formula for nonrelativistic accelerated charge, and is the same result that is obtained from a more rigorous derivation based on Lienard-Weichert potentials (see, for example, chapter 14 of [7]). An easy relativistic extension to the Larmor formula leads to the Lienard result:

$$P = \frac{q^2}{6\pi \varepsilon_0 c} \gamma^6 [(\dot{\beta})^2 - (\beta \times \dot{\beta})^2]. \tag{3.32}$$

3.2.2.2 Stopping Power

Electrons behave differently when stopping in a target material because they can lose all of their kinetic energy in one collision to an orbital electron, although this is unlikely. This is clearly seen in the Q_{max} calculation of Eq. (3.9) by setting $m_e = M$. If we take into account relativistic effects for the energetic particle, Eq. (3.9) becomes

$$Q_{max} = \frac{2m_e M E_i (E_i + 2Mc^2)}{(M + m_e)^2 c^2 + 2m_e E_i}. \tag{3.33}$$

More typically electrons will take several hundred collisions to slow down completely, but they can scatter in any direction and lose large fractions of energy with each collision. Their path through matter is very erratic as a result.

When there is inelastic scattering between electrons (Moller scattering), the particle and target electron are indistinguishable if binding effects are ignored. The scattered electron could be the original incident electron or it could be the orbital electron. The convention is to identify the highest energy post-collision electron as the scattered electron. For this reason, the maximum energy lost in such collisions is $E_i/2$. The Feynman

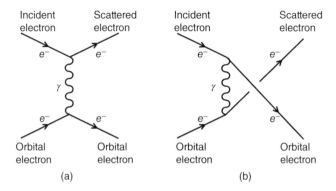

Figure 3.10 Two tree-level Feynman diagrams for the scattering between two electrons; (a) t-channel diagram, (b) u-channel diagram.

diagrams in Figure 3.10 illustrate the two possible outcomes to an electron/electron interaction.

Equation (3.33) is correct for all energies E_i and reduces to Eq. (3.9) when the kinetic energy E_i of the particle is smaller than its mass Mc^2, providing a good criterion for the energy regimes where Eq. (3.9) can be used instead of Eq. (3.33). Electrons have a low rest mass ($m_e c^2 = 0.511$ keV) and are often relativistic, however.

Electrons also radiate energy by bremsstrahlung, as discussed in Section 3.2.1.3 and Section 3.2.2.2.2, so the stopping power can therefore be expressed as the sum of ionization and radiative components:

$$-\frac{dE}{dx} = \left(-\frac{dE}{dx}\right)_{ion} + \left(-\frac{dE}{dx}\right)_{rad}. \tag{3.34}$$

3.2.2.2.1 Collisional Losses The Bethe-Bloch formula, Eq. (3.23), can be used to model electron and positron energy losses due to Coulomb interactions with orbital electrons, but the value of the maximum kinetic energy of the electron/positron T_e will depend on the type of collision. The maximum kinetic energy T_e will be provided by Eq. (3.33) for positron/electron interactions, but electron/electron interactions will use $T_e = Q_{max}/2$ because the two interacting particles are indistinguishable from each other. A modified version of the Bethe-Bloch formula describing the collisional losses is given by [8]

$$\left(-\frac{dE}{dx}\right)_{ion} = \frac{2\pi k_0^2 z^2 e^4 Z N_A}{m_e \beta^2 c^2 A}\left[\ln\frac{\tau^2(\tau+2)}{2(I/m_e c^2)^2} + F(\tau) - \delta(\beta\gamma) - \frac{2C}{Z}\right] \tag{3.35}$$

where $\tau = T_e/m_e c^2$ (T_e is the electron/positron kinetic energy) and

$$F(\tau) = 1 - \beta^2 + \frac{\frac{\tau^2}{8} - (2\tau+1)\ln 2}{(\tau+1)^2} \quad \text{for electrons} \tag{3.36a}$$

$$F(\tau) = 2\ln 2 - \frac{\beta^2}{12}\left[23 + \frac{14}{\tau+2} + \frac{10}{(\tau+2)^2} + \frac{4}{(\tau+2)^3}\right] \quad \text{for positrons.} \tag{3.36b}$$

Note that the logarithmic term in Eq. (3.35) resembles the Bethe-Bloch term from Eq. (3.23) and F(τ) in Eqs. (3.36a) and (3.36b) is proportional to $-\beta^2$. The remaining two terms in the square brackets of Eq. (3.35) provide the same corrections as for the Bethe-Bloch formula for heavy particles.

3.2.2.2.2 Bremsstrahlung Electrons are one of the few particles where radiation losses can contribute substantially to energy loss. The radiated power in Eq. (3.31) is proportional to the square of the acceleration, so for a given force F we have $\dot{v} = F/m$. The radiated power is thus proportional to $1/m^2$ making electrons much better at radiating than other particles due to their small mass. For example, compared to protons, electrons are better at radiating by a factor of 4×10^6! Even when compared to muons ($\mu \pm$ the next lightest particles) electrons are still better at radiating by a factor of 4×10^4.

Bremsstrahlung emission depends on the acceleration of the electron when passing near atomic nuclei in the target material, which leads to a dependency on the electric field felt by the electron. As with nuclear stopping power (Section 3.2.1.3) the electric field felt by the electron will depend on the screening from atomic electrons surrounding the nucleus. We can define a bremsstrahlung cross section $d\sigma/dE$ that depends on the energy E of the emitted photon and the atomic number Z of the target atom [9]:

$$\frac{d\sigma}{dE} = \frac{d\sigma_n}{dE} + Z\frac{d\sigma_e}{dE}.$$

(3.37)

Equation (3.37) separates the contributions to bremsstrahlung produced in the field of the screened atomic nucleus $d\sigma_n/dE$ and the bremsstrahlung produced in the field of the Z atomic electrons $Z(d\sigma_e/dE)$. Equation (3.37) can be rewritten to parameterize the influence of the atomic electrons:

$$\frac{d\sigma}{dE} = \left(1 + \frac{\eta}{Z}\right)\frac{d\sigma_n}{dE}$$

(3.38)

where η is the cross-section ratio:

$$\eta = \frac{d\sigma_e}{dE} \bigg/ \left(\frac{1}{Z^2}\frac{d\sigma_n}{dE}\right).$$

(3.39)

At very high energies, the electron–electron and electron–nucleus cross sections are nearly identical for an unscreened target of unit charge [10]. This provides a basis for an approximation that $\eta = 1$ that is often used in Eq. (3.38). However, there are a number of differences between electron–electron and electron–nucleus cross sections that counter this approximation, including differences in the screening of the fields due to atomic electrons and the nucleus, a low-energy bremsstrahlung produced by recoil electrons in electron–electron interactions, and a vanishing electron–electron cross section at low incident-electron energy caused by a lack of a dipole moment. Generally η is greater than unity at high energies because atomic electrons are less effectively screened than the nucleus. At lower energies, η is less than unity.

The radiative stopping power is usually written as [9]

$$-\frac{1}{\rho}\left(\frac{dE}{dx}\right)_{rad} = \frac{N_A}{A}\alpha r_e^2 Z^2(T_1 + m_e c^2)\Phi_{rad}$$

(3.40)

where ρ is the mass density, α is the fine structure constant, $r_e = k_0 e^2/m_e c^2$ is the classical electron radius, T_1 is the initial electron energy, $m_e c^2$ is the electron rest energy and

$$\Phi_{rad} = \int_0^{T_1} E\frac{d\sigma}{dE}dE/[\alpha r_e^2 Z^2(T_1 + m_e c^2)]$$

(3.41)

is a dimensionless, scaled, integrated bremsstrahlung energy-loss cross section. Like Eqs. (3.37) through (3.39), Φ_{rad} can be written as the sum of the electron–nucleus and electron–electron components

$$\Phi_{rad} = \Phi_{rad}^{(n)} + Z\Phi_{rad}^{(e)}. \tag{3.42}$$

Equation (3.42) is rewritten to parameterize the electron–electron component of the scaled energy-loss cross section

$$\Phi_{rad} = \left(1 + \frac{\bar{\eta}}{Z}\right)\Phi_{rad}^{(n)}, \tag{3.43}$$

where $\bar{\eta}$ is the scaled energy-loss cross-section ratio:

$$\bar{\eta} = \Phi_{rad}^{(e)} / \left(\frac{1}{Z^2}\Phi_{rad}^{(n)}\right). \tag{3.44}$$

There is a substantial body of theory on the bremsstrahlung cross section covering various approximations, limitations, and regions of applicability. A review of the analytical theories is given by Koch and Motz [11] and a more recent view by Pratt [12] provides discussion of low-energy theories.

Calculation of losses for electrons and other light particles can become complicated because of the production of photons as well as ionization losses. High-energy photons can lead to pair production (electron–positron), leading to further radiation losses by the product particles. The product particles create a cone of influence from the original electron, as interactions have a radial and angular distribution, resulting in an electron–photon shower or electromagnetic cascade. Calculating the energy loss for such a cascade involves many, many particles, photons, and interactions. Early work using computers and Monte Carlo methods to help deal with the complexity can be found in [13]. Figure 3.11 is based on these early calculations and shows the

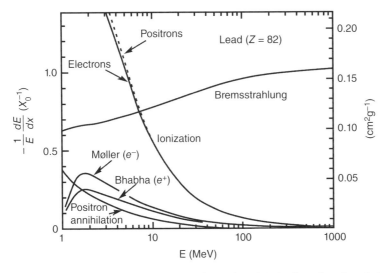

Figure 3.11 Fractional energy loss per radiation length in lead as a function of electron or positron energy Source: From Figure 3.2 Messel and Crawford, *Electron-Photon Shower Distribution Function Tables for Lead, Copper, and Air Absorbers*, Pergamon Press, 1970). Messel and Crawford use $X_0(\text{Pb}) = 5.82 \text{ g/cm}^2$ but the figure has been modified to reflect the value $X_0(\text{Pb}) = 6.37 \text{ g/cm}^2$. Source: Tanabashi et al. 2018 [14], Figure 33.11.

contributions from collisional and radiative losses for electrons and positrons stopping in lead.

3.2.2.2.3 Radiation Length

For high-energy electrons, the predominant energy loss mechanism is bremsstrahlung so we can define a radiation length X_0 as the mean distance over which the electron will lose all but $1/e$ of its initial energy:

$$E = E_0 e^{-x/X_0}. \tag{3.45}$$

The energy loss per unit length due to bremsstrahlung is then

$$-\frac{dE}{dx} = \frac{1}{X_0} E_0 e^{-x/X_0} = \frac{E}{X_0}. \tag{3.46}$$

The radiation length for a high-energy electron is usually measured in g/cm^2. Tsai [15] has calculated and tabulated radiation lengths, finding that

$$\frac{1}{X_0} = 4\alpha r_e^2 \frac{N_A}{A} [Z^2 (L_{rad} - f(Z)) + Z L'_{rad}]. \tag{3.47}$$

Values for Z, L_{rad} and L'_{rad} determined by Tsai for use in Eq. (4.47) are given in Table 3.2. The radiation length can also be applied to pair production from a high-energy photon and high-energy electromagnetic cascades (Section 3.2.4).

3.2.2.2.4 Critical Energy

When comparing the electron energy loss in a material due to collisions (ionization) with the energy loss due to bremsstrahlung, we see from Eqs. (3.35) and (3.40) that ionization losses are proportional to Z and $ln(E)$, whereas radiative losses are proportional to Z^2 and E. So as the electron energy increases, losses due to bremsstrahlung grow to a point where they are equal to or greater than the ionization loss. The energy where the energy losses are equal is called the critical energy, E_c, and is material dependent. The critical energy can be defined in two ways. The first is based on computer simulations using Monte Carlo methods (electron–gamma showers or EGS4 computer program, for example) to calculate energy loses and is defined simply as the energy where ionization losses equal radiative loses:

$$\left(\frac{dE}{dx}\right)_{rad} = \left(\frac{dE}{dx}\right)_{ion}. \tag{3.48}$$

A second definition from Rossi [16] defines critical energy as the energy at which the ionization loss per radiation length equals the electron's own energy. Rossi's definition

Table 3.2 Values of Z, L_{rad}, and L'_{rad} for calculating radiation length using Eq. (3.47).

Element	Z	L_{rad}	L'_{rad}
H	1	5.31	6.144
He	2	4.79	5.621
Li	3	4.74	5.805
Be	4	4.71	5.924
Others	>4	$ln(184.15Z^{-1/3})$	$ln(1194Z^{-2/3})$

Source: Tanabashi et al. 2018 [14], Table 33.2, p. 19.

Table 3.3 Radiation length and critical energy for integrated circuit and packaging materials.

Material	Radiation length $(g\text{-}cm^{-2}, cm)$	Critical energy Electrons (MeV)	Positrons (MeV)	Muons (GeV)
B	52.69, 22.23	93.95	91.41	1170
C	42.70, 12.13	80.17	77.94	1044
Al	24.01, 8.897	42.70	41.48	612
Si	21.82, 9.370	40.19	39.05	582
P	21.21, 9.639	37.92	36.84	552
Ti	16.16, 3.560	26.01	25.23	402
Co	13.62, 1.530	20.82	20.16	336
Cu	12.86, 1.436	19.42	18.79	317
Ga	12.47, 2.113	18.57	17.98	304
Ge	12.25, 2.302	18.16	17.58	297
As	11.94, 2.084	17.65	17.09	290
Ag	8.97, 0.8543	12.36	11.94	215
Sn	8.82, 1.206	11.86	11.46	207
Sb	8.73, 1.304	11.70	11.31	204
W	6.76, 0.3504	7.97	7.68	150
Pt	6.54, 0.3051	7.59	7.31	144
Au	6.46, 0.3344	7.53	7.26	143
Pb	6.37, 0.5612	7.43	7.16	141
SiO_2	27.05, 12.29	50.58	49.17	708
Al_2O_3	27.94, 7.038	50.18	48.74	706
GaAs	12.19, 2.296	18.11	17.53	297
CdTe	8.90, 1.436	11.84	11.44	208

Values are calculated using Eq. (3.47) and the parameters in Table 3.2.

can be reduced to Eq. (3.48). Table 3.3 provides values of radiation length and critical energy for some common semiconductor materials using Eq. (3.47) and the parameters in Table 3.2.

3.2.3 Neutrons

Neutrons were discovered in 1932 by English Nobel Laureate James Chadwick [17, 18] but were anticipated by Rutherford in 1920 [19], who proposed a neutral nuclear particle in order to explain the difference found between the atomic number of an atom and its atomic mass. In 1931, there was some study of a new type of radiation that occurred when alpha particles fell on light elements like beryllium, boron, and lithium [20, 21]. These rays were found to be more penetrating than gamma rays. When they were aimed at paraffin or other hydrogen-containing compounds in 1932, high-energy protons were ejected [22]. The mysterious rays were neutrons, as determined by Chadwick, and the way neutrons interact with matter as noted by early experimenters has since been used to advantage for probing and studying cellular and molecular structure.

When they traverse matter, neutrons behave quite differently from charged particles. They have zero electrical charge so they do not directly cause ionization and so penetrate further into materials than charged particles. Neutrons do have a small magnetic moment so they can interact with the magnetic moment of electrons ($\mu_B = e\hbar/2m_e = 9.274 \times 10^{-21}$ Erg-G^{-1}), but the neutron magnetic moment is about 1000 times smaller than μ_B. Most interactions are between the neutron and the nucleus. Ionization from neutrons is caused solely through secondary effects.

Neutrons are classified according to their energy, as shown in Table 3.4, although the dividing lines between the various classifications are sometimes fuzzy. For example, thermal neutrons at 20 °C with a Maxwellian distribution will have a most probable energy of 0.025 eV but the distribution would extend to about 0.2 eV. It is also common to speak of "slow" neutrons. These are neutrons that span in energy from just above thermal to 10 eV and sometimes higher.

3.2.3.1 Neutron Cross Section

It is very useful to speak in terms of cross section when discussing neutron interactions. Cross section simply measures the probability for the occurrence of a particular event. For example, the probability that a neutron will be absorbed by a thin layer of material is the ratio of the number of neutrons that do not emerge from the back of the layer to the number that are incident on the layer. When this probability is divided by the areal atom density (number of atoms per unit area of the layer) the result is the cross section σ for neutron absorption. This type of cross section is called the microscopic cross section because it describes the interaction with a single nucleus. One could look at the probability for a neutron interaction with a nucleus as the ratio of the target nuclei areas in a sample to the sample area, but this approach doesn't work for neutron cross sections because some interactions at the atomic level are not described by simple geometry. Still, most cross sections are on the order of the physical cross-sectional area of a large nucleus or about 1×10^{-24} cm^2, so the cross section σ can be viewed as an effective

Table 3.4 Neutron energy classification.

Classification	Energy	Comment
Cold	0–5 meV	
Thermal	5–100 meV	Neutrons in thermal equilibrium with their surroundings. E = 0.025 eV @ 20 °C.
Epithermal	100 meV–1 eV	
Resonant	1 eV–100 eV	Neutrons that are strongly captured in the resonance of U-238
Intermediate	100 eV–100 keV	
Fast	100 keV–10 MeV	
Ultra-fast	10 MeV–10GeV	
Relativistic	>10GeV	

cross-sectional area for the reaction. Working with such small numbers is inconvenient, so a new unit, the barn, is used. One barn (b) is equal to 10^{-24} cm^2. Neutron interaction cross sections can range between 0.001 b and 1000 b.

A macroscopic cross section can also be defined for a material by considering the transmission of a parallel beam of neutrons through a thick sample. By treating the sample as a series of thin layers, the microscopic total cross section σ_t can be used to develop a relationship between the intensity of the beam $I(x)$ at a depth x in the sample and the incident intensity I_0

$$I(x) = I_0 e^{-N\sigma_t x}. \tag{3.49}$$

In Eq. (3.49), N is the atom density in the sample so that the multiplier in the exponent represents the total macroscopic cross-section $\Sigma_t = N\sigma_t$ with dimensions of cm^{-1}. Note that the energy dependence for $I(x)$ is in the value for the microscopic cross section σ_t. Eq. (3.49) gives the intensity of neutrons that have had no reaction at depth x, so the actual number of neutrons present may be larger due to multiple scattering or multiplicative reactions. When considering compound materials the macroscopic cross sections of each of the constituent materials comprise the total macroscopic cross section of the material

$$\Sigma = \Sigma_1 + \Sigma_2 + \Sigma_3 + \dots \tag{3.50}$$

The atom density N_i for the each constituent material given by

$$N_i = \frac{\rho N_A n_i}{M} \tag{3.51}$$

where ρ and M are the density and molecular weight of the compound and n_i is the number of atoms of material i in the compound.

Equation (3.49) resembles Eq. (3.45) and in fact the reciprocal of the total macroscopic cross-section $\lambda = 1/\Sigma_t$ is the mean free path for neutrons in the material. The mean free path is the average distance that a neutron will travel between interactions in a material.

3.2.3.2 Interactions with Matter

Neutrons are slowed in matter through three types of nuclear interaction; elastic scattering, inelastic scattering, and absorption. Each of these interactions can be written in shorthand, as indicated in Table 3.5. In elastic scattering, the neutron bounces off of the

Table 3.5 Expressions for neutron interactions with nuclei.

Interaction	Expression	Comment
Elastic scattering	X(n,n)X	X is the target nucleus.
Inelastic scattering	X(n,n')X*	X* target nucleus in excited state.
Absorption	X(n,γ)Y	New nucleus Y, gamma ray emitted.
	X(n,α)Z	New nucleus Z, alpha particle emitted.

nucleus, losing some kinetic energy to the nucleus but without leaving the nucleus in an excited state. Inelastic scattering is the same as elastic scattering, except the nucleus is left in an excited state. The nucleus exits the state by emitting one or more gamma rays. In absorption, the neutron is absorbed by the nucleus causing either a gamma or alpha emission or causing the nucleus to undergo fission. Absorption causes the original nucleus, represented by X in the Table 3.5 expression, to change into a new element or isotope represented by Y or Z in the expression.

The dependence of total neutron cross section on neutron energy in a material is complicated by the interaction of the neutron with the quantum states of the nucleus. This is seen in the neutron cross section for silicon, as shown in Figure 3.12. The total neutron cross section includes contributions from all possible neutron reactions with the nucleus. In general, for neutron energies less than about 1 eV the cross section is inversely proportional to the neutron velocity. This is followed by a resonance region where neutrons of certain energies are more likely to interact with the nucleus.

3.2.3.2.1 Elastic Scattering It is useful to return to the elastic scattering Q_{max} calculation from Section 3.2.1.2 and apply it to elastic scattering of a neutron and nucleus. We use a more general approach where the paths of the neutron and nucleus are not confined to one dimension. We further introduce a common practice in particle physics and transform the frame of reference from the laboratory system to the center of mass system. In the laboratory frame of reference, the neutron is moving while the nucleus is assumed to be stationary. In the center of mass frame of reference, both the neutron and nucleus are moving so that the center of mass is stationary. This means that the speed of the elastically scattered neutron and nucleus are constant before and after the collision with only their direction of motion changing. These two reference frames are illustrated in Figure 3.13.

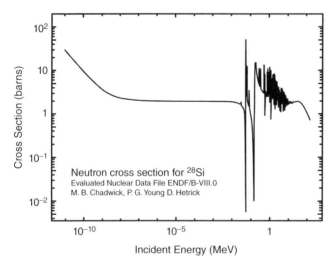

Figure 3.12 Total neutron cross section versus neutron energy for ^{28}Si. Source: Chadwick et al. 2018 [23], ENDF/B-VIII.0 evaluation, http://www-nds.iaea.org.

Figure 3.13 Laboratory (L) and Center of Mass (C) reference frames.

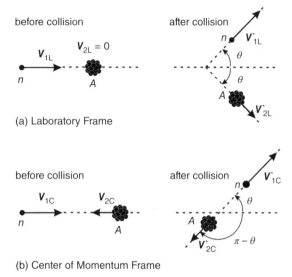

(a) Laboratory Frame

(b) Center of Momentum Frame

Starting with the assumption of a nucleus with mass number A at rest and a neutron velocity of V_1 in the laboratory frame, the speed of the center of mass V_m will be

$$V_m = \frac{V_1}{A+1}. \tag{3.52}$$

The maximum energy is transferred to the nucleus when the ratio of the final neutron energy to the initial neutron energy, E_2/E_1, is a minimum. It can be shown that this occurs when

$$\frac{E_2}{E_1} = \left(\frac{A-1}{A+1}\right)^2. \tag{3.53}$$

We see that neutrons lose more energy to lower A nuclei with the possibility of losing all of their kinetic energy to hydrogen nuclei ($A = 1$), assuming the mass of a neutron is approximately equal to that of a proton. This means that lower Z materials are more efficient at stopping neutrons, unlike the stopping of charged particles (and photons, Section 3.2.4) where higher Z materials are more effective. A good example of the effectiveness of low Z materials for stopping neutrons can be found in their use as moderators in nuclear reactors. Some commonly used materials include (light) water, carbon (graphite), and heavy water.

Another aspect of elastic neutron scattering is similar to electron–electron scattering. There is indeterminacy for exactly what happens during the collision between the neutron and nucleus because there are two possible types of interaction. In one scenario, called resonance scattering, the neutron is captured by the nucleus, causing it to gain enough energy to be at or near one of its excited states. The nucleus returns to its prior state by emitting a neutron and recoiling in a manner that conserves energy and momentum. The second scenario, called potential scattering, occurs when the neutron interacts with the nuclear potential well through the strong force.

3.2.3.2.2 Inelastic Scattering Inelastic scattering of neutrons occurs in a material when some of the neutron kinetic energy raises the energy state of the nucleus. During the collision, the neutron is absorbed by the nucleus but then escapes. The neutron loses some of its energy to the nucleus during the encounter, causing the nucleus to change its quantum state. The neutron is then released by the nucleus and the nucleus eventually returns to its prior state by releasing one or more photons (gamma rays). This interaction only occurs if the neutron has enough energy to alter the quantum state of the nucleus. Light nuclei have energy levels that are separated by several MeV, so only fast or higher-speed neutrons will interact in this way. Heavier nuclei like uranium have energy levels that are closer together with a separation around 0.1 MeV. These nuclei can be excited by slow neutrons, but the gamma rays emitted by the heavier nuclei after inelastic scattering will be less energetic compared to lighter nuclei.

3.2.3.2.3 Absorption There is the possibility that the neutron could be absorbed by the nucleus during a collision. When this happens, the nucleus enters an excited state and returns to its ground state by releasing a photon or a particle. When a photon is released, the process is called radiative capture. If the nucleus emits a particle, the process is called a charged particle reaction or particle ejection. Sometimes an absorbed neutron will cause the nucleus to break apart in a process called fission (see Chapter 2, Section 2.5). Like inelastic collisions, absorption occurs when the neutron has enough energy to raise the energy level of the nucleus. The newly formed compound nucleus can be stable after absorbing a neutron, but often the nucleus is radioactive and decays to a stable state by further photon/particle emissions. Each of these absorption processes has a cross section that is complicated by the quantum states and energies in the nucleus.

Generally, radiative capture is less important for lighter nuclei and thermal to intermediate energy neutrons but is the predominant reaction for heavy nuclei for these neutron energies. Overall, radiative capture cross section decreases for faster neutrons. Charged particle reactions are more likely for lighter nuclei, with the cross section increasing for faster neutrons. The emitted charged particle must overcome the Coulomb barrier of the orbital electrons, which is much smaller for light nuclei.

3.2.4 Photons (X-Rays, Gamma Rays)

Like neutrons, photons are electrically neutral and do not steadily lose energy via coulombic interaction with atomic electrons like charged particles. Photons have zero mass and always travel the speed of light. When they interact with matter, they may transfer all of their energy and be absorbed or scatter elastically or inelastically. The depth to which a photon will penetrate a material depends on the material and the photon energy, but photons are more penetrating than charged particles of similar energy.

Photons interact with material in several ways, including photoelectric absorption, scattering, pair production and photonuclear reactions. Each of these processes will be treated in some detail in the following sections. For the scattering process, there are several possibilities, including Compton, Thomson, and Rayleigh scattering. Thomson scattering is the low-energy limit of Compton scattering and Rayleigh scattering applies to transparent materials, so Compton scattering is the primary focus.

3.2.4.1 Photoelectric Effect

In the photoelectric effect, incident photons are absorbed by a material and electrons are emitted. While it was first observed by Heinrich Hertz in 1887 [24], the emission mechanism was explained almost 20 years later by Albert Einstein in 1905. A photon interacts with an absorber atom in which the photon completely disappears. In its place, an energetic photoelectron is ejected from one of the bound shells of the atom. For gamma rays, this is probably a K-shell electron. The photoelectric effect is most pronounced for photon energies that are close to the binding energies of shell electrons (Figure 2.3 shows the binding energy for shell electrons across a range of atomic numbers). It can be shown that the photoelectric effect only applies to bound electrons.

The energy of the photoelectron is dependent on the energy of the photon, not the intensity of the illumination, as would be predicted by a more classical analysis. The photoelectron energy is given by

$$E_e = h\nu - E_b. \tag{3.54}$$

The energy E_b is the binding energy for the electron in its original shell location. When gamma ray energy is higher than about 200 keV, most of the photon energy can be carried away by the photoelectron. As shown in Figure 3.14, the exit of the photoelectron from its shell location results in a rearrangement of shell electrons in the absorber atom. When another shell electron fills the photoelectron vacancy, a characteristic X-ray photon is released. The photon is usually absorbed close to the original creation point. In some cases, an Auger electron may substitute for the characteristic X-ray, carrying away the excitation energy.

The photoelectric effect can be the predominant mode of interaction for low energy gamma rays and X-rays. While there is no single analytic expression that provides the probability of photoelectric absorption of a photon over all values of photon energy and absorber material, a rough approximation for the interaction probability is given by [25]

$$\tau \propto \frac{Z^n}{(h\nu)^3}. \tag{3.55}$$

In Eq. (3.55) the exponent n varies between 4 and 5. Clearly, the probability and thus the cross-section for the photoelectric effect decreases rapidly with photon energy. However, the extreme dependence on Z in Eq. (3.55) is the primary reason that high-Z materials like lead are good absorbers of gamma rays and other high energy photons.

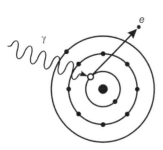

(a) Photoelectron emission

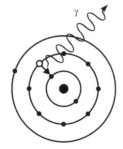

(b) Fluorescence

Figure 3.14 Photoelectric effect.

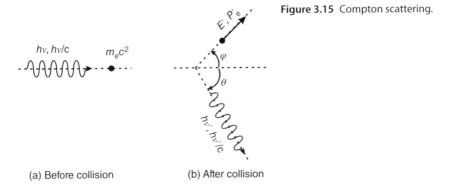

Figure 3.15 Compton scattering.

(a) Before collision (b) After collision

3.2.4.2 Compton Scattering

Compton scattering occurs when a photon imparts some of its energy to an unbound electron. The photon is not absorbed entirely, as with the photoelectric effect. Instead, the now lower energy (longer wavelength) photon is scattered to conserve momentum with the electron.

Figure 3.15 provides the basis for an analysis based on conservation of momentum and energy. We assume a photon moving in the x direction with initial energy $h\nu$ and final energy $h\nu'$ and an electron that is initially at rest. Recognizing the need to use the relativistic energy–momentum relation for the post-collision electron, conservation of energy leads to

$$h\nu + m_e c^2 = E' + h\nu' = \sqrt{(P'_e c)^2 + (m_e c^2)^2} + h\nu'. \tag{3.56}$$

When considering conservation of momentum, it is easier to work with the vector components of momentum. Using $\overrightarrow{P_\gamma}$ and $\overrightarrow{P'_\gamma}$ to represent the pre- and post-collision values for the photon momentum we have

$$\overrightarrow{P_\gamma} = \overrightarrow{P'_\gamma} + \overrightarrow{P'_e} \Rightarrow \overrightarrow{P'_e} = \overrightarrow{P_\gamma} - \overrightarrow{P'_\gamma}. \tag{3.57}$$

We need the value for the scalar $P'_e c^2$ in Eq. (3.56) so we use Eq. (3.57) and the definition of the dot product:

$$P'^2_e = (\overrightarrow{P_\gamma} - \overrightarrow{P'_\gamma}) \cdot (\overrightarrow{P_\lambda} - \overrightarrow{P'_\gamma}) = P^2_\gamma + P'^2_\gamma - 2P_\gamma P'_\lambda \cos\theta. \tag{3.58}$$

Multiply by c^2 and make substitutions for the pre- and post-collision photon energies:

$$P'^2_e c^2 = (h\nu)^2 + (h\nu')^2 - 2(h\nu)(h\nu') \cos\theta. \tag{3.59}$$

Substitution of the Eq. (3.59) result in Eq. (3.56) and using $\lambda = c/\nu$ and $\lambda' = c/\nu'$ yields Compton's derived result:

$$\Delta\lambda = \lambda - \lambda' = \frac{h}{m_e c}(1 - \cos\theta). \tag{3.60}$$

We see from Eq. (3.60) that the Compton shift in wavelength is not dependent on photon frequency. The quantity $h/m_e c$ is called the Compton wavelength and is a quantum mechanical property of the electron. It is equivalent to the wavelength of a photon with the same energy as the rest-mass energy of the electron.

When the photon energy is much less than the rest mass of the electron ($hv \ll m_e c^2$), the scattering will be elastic. This is the low-energy limit of Compton scattering called Thomson scattering. The electron oscillates in response to the electromagnetic electric field of the photon, radiating at the same frequency as the photon. The emitted radiation scatters the incident photon.

3.2.4.3 Pair Production

Photons with energy greater than twice the rest mass of the electron can interact with a nucleus or other nearby mass to produce electron–positron particle pairs. When a photon splits into a pair of particles, momentum and energy as well as particle properties like charge and spin are conserved. Electrons have a rest mass of 0.511 MeV, so photons with energy greater than 1.02 MeV can produce a particle pair.

We now examine the kinematics of pair production using conservation laws. A simplified diagram for pair production is shown in Figure 3.16. A photon traveling in the x-direction splits into a positron/electron pair emitted at equal angles to the x-axis each with momentum of magnitude $|\mathbf{p}_e| = |\mathbf{p}_{e-}| = |\mathbf{p}_{e+}|$. We first consider whether pair production can occur in a vacuum and ignore the influence of the nucleus with mass M in Figure 3.16. By conserving momentum, we find

$$E_\gamma = hv = 2|\mathbf{p}_e|c \cos\theta = 2\gamma m_e |v_e| c \cos\theta \tag{3.61}$$

so that

$$E_\gamma < 2m_e c^2. \tag{3.62}$$

Conserving energy, we find that the photon must supply at least enough energy for the rest mass of the positron and electron:

$$E_\gamma = 2m_e c^2. \tag{3.63}$$

We conclude from Eq. (3.61) and (3.62) that not all of the photon momentum can be imparted to the electron and positron. A *spectator mass* is needed to absorb some of the photon momentum. If a spectator mass M is added to the reaction, the energy and momentum can be conserved because the spectator mass absorbs the recoil from the pair production.

We now turn to some properties of Minkowski space to gain some better insight into pair production. In special relativity, the four-momentum $\mathbf{P}$ is the generalization

Figure 3.16 Pair production.

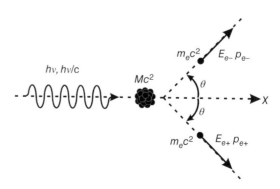

of the three-dimensional momentum in Euclidian space. For a particle with momentum $\mathbf{p} = (p_x, p_y, p_z)$ and energy E, the four-momentum is defined as

$$\mathbf{P} = (P^0, P^1, P^2, P^4) = (E/c, p_x, p_y, p_x) = (E/c, \mathbf{p}). \tag{3.64}$$

The above definition results from replacing Euclidian space coordinates (x, y, z) with a metric that includes a measure for the interval of time between two events, for example (ct, x, y, z), defining a four-dimensional space–time called Minkowski space.

The Minkowski inner product $\langle \mathbf{P}, \mathbf{P} \rangle$ is a Lorentz invariant quantity proportional to the square of the rest mass L:

$$\langle \mathbf{P}, \mathbf{P} \rangle = |\mathbf{P}|^2 = (mc)^2 = \mathbf{P}\eta\mathbf{P}^\mathrm{T} = \left(\frac{E}{c}\right)^2 - |\mathbf{p}|^2 \tag{3.65}$$

where η is the Minkowski metric with signature $(+ - - -)$

$$\eta = \begin{pmatrix} 1 & 0 & 0 & 0 \\ 0 & -1 & 0 & 0 \\ 0 & 0 & -1 & 0 \\ 0 & 0 & 0 & -1 \end{pmatrix}. \tag{3.66}$$

The Lorentz invariance of the Minkowski inner product means its value is the same for the initial and final states of the pair production reaction. We can use this property to explore pair production a little further. When the rest mass M of the nearby nucleus in Figure 3.16 is added to the interaction, we find that

$$E_\gamma = h\nu = 2m_e c^2 \left(1 + \frac{m_e}{M}\right) > 2m_e c^2. \tag{3.67}$$

Note that for a large mass M, like a nucleus the threshold photon energy for pair production is approximately twice the electron rest mass as expected, but if the mass M is an electron, the threshold photon energy is multiplied by a factor of two.

3.2.4.4 Photonuclear Reactions

An energetic photon can be absorbed by a nucleus causing a nucleon to be ejected. This type of reaction is called *photodisintegration*. The process is analogous to the photoelectric effect except that it occurs with the nucleus instead of shell electrons. Like the photoelectric effect, photodisintegration can happen only when the photon energy is above a threshold value. The kinetic energy of the ejected nucleon is equal to the energy of the photon minus the binding energy of the nucleon in the nucleus. Typical binding energies range around several MeV. Photonuclear reactions use shorthand expressions like those in Table 3.5 to describe the reaction. For example, a photonuclear reaction that results in an expelled neutron is written (γ,n). For an expelled proton, the expressing is (γ,p). Other reactions are also possible, where multiple nucleons are expelled $(\gamma,2n)$, $(\gamma,2p)$, (γ, np), etc. As with neutron interaction with nuclei, the resulting nucleus from a photonuclear reaction may not be stable and may decay to a more stable state. Even the more common photonuclear reactions are rather unlikely with cross sections measuring in the milli-barns (mb).

3.3 Charge Trapping in Silicon Dioxide

The previous sections provided the background for transient effects caused by passing particles and photons. For the most part, the focus was on their interaction in bulk materials. There are significant transient effects caused by the passage of particles or radiation through the bulk areas of semiconductor devices. These are classified as single-event effects (SEE) and are treated in detail in Chapter 4. This section as well as the following sections will concentrate on some of the more persistent effects caused by ionizing radiation in semiconductor materials and more specifically gate oxides, field oxides, and other oxide layers used in integrated circuits. As we shall see, the dielectrics used in semiconductor devices not only trap charge but introduce interfaces between dissimilar materials where additional processes occur.

Transient radiation can result in lasting effects on integrated circuit structures if continued over a period of time due to the differences in carrier transport between various material layers and the presence of electric fields that can separate electron–hole (e–h) pairs after their creation. These are the exact conditions for the most commonly used device in integrated circuits, the MOS transistor. Charge transport of electrons in SiO_2, the most common MOS gate material, is rapid when compared to hole transport. The electrons that do not recombine escape the oxide, but holes get left behind. Over time and continued exposure to radiation, the holes build up, creating a fixed oxide charge that can affect transistor and circuit operation. When there are electric fields present, the migration of charge toward interfaces between materials can cause lasting changes in device function.

Silicon dioxide and materials containing silicon dioxide are used everywhere in semiconductor processing. Silicon dioxide is used for gate dielectrics in MOS devices, isolation between devices in the same semiconductor substrate, isolation between metallization layers, and as passivation at the top of the semiconductor integrated circuit. It can be formed in many ways, including thermal oxidation of silicon and deposition, and is often combined with other materials like silicon nitride, phosphorous, and boron. There can be a wide variation in the response to passing particles and radiation, depending on the growth conditions for silicon dioxide and its derivative materials. For the purposes of this section, we shall consider only thermally grown oxides, as these represent the best-case scenario in terms of charge trapping in the bulk and at grown oxide–silicon interfaces.

3.3.1 Charge Generation/Recombination

Consider the simplified diagram of a high-energy particle hitting the gate of a MOS device, as shown in Figure 3.17a. A typical path for an ionizing event is indicated by the arrow, and an ionized region or trail along the event path is indicated by the dotted halo. A slightly slanted top-to-bottom path is chosen for illustration, but such events are likely to occur from all directions. We also ignore for the moment dielectric and interconnect layers that cover the MOS device in the actual integrated circuit. As the event progresses through the structure of the MOS device, each region receives an amount of deposited energy dictated by the stopping power for the event and proportional to the path length of the event in that material.

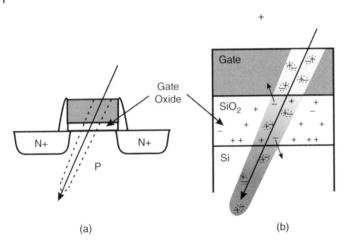

Figure 3.17 Basic mechanism for trapped positive charge in SiO_2: (a) Cross-sectional view of an NMOS transistor showing path of ionizing particle or photon; (b) close-up view of NMOS gate oxide.

Figure 3.17b shows a close-up of the MOS structure with the deposited energy in each of electron–hole pairs along the event path. Many of these e–h pairs will recombine shortly after generation, but some will get separated, especially if there is an electric field present, and recombine in other regions of the MOS device. As the gate and bulk Si region of the MOS device are conductive, both holes and electrons have good mobility in these areas. The net charge from these regions will show up as a current at the MOS device terminal connected to the collection region. For the gate oxide SiO_2 region, electrons also have a high mobility and can escape the oxide quickly after the event and transport through a conductive region, as indicated by the small arrows in Figure 3.17b. Holes, on the other hand, have a very low mobility in SiO_2, so they get left behind and contribute to a gradual increase in oxide charge with every passing ionizing event.

3.3.1.1 Geminate and Columnar Models

It is possible to estimate the deposited energy to the MOS gate oxide from a given event in terms of e–h pair generation using values from Table 3.1 and taking into account the geometry of the MOS structure. For example, 10 MeV protons have a stopping power around 50 MeV-cm^2/g in silicon dioxide (Figure 3.22). Using the value of 17 eV/pair for SiO_2 from Table 3.1 along with the density for SiO_2 (somewhere between 2.196 g/cm^3 for amorphous SiO_2 and 2.648 g/cm^3 for α-quartz) yields a line density around 7×10^6 pairs/cm. A certain amount of the generated pairs will recombine, but recombination will depend on the density of generated e–h pairs in the ionization trail or column of the event, as well as any applied electric fields. The amount of recombination is important because it determines how many carriers will be trapped in the oxide and potentially transport to the oxide interfaces. The measure of this leftover charge is called the *net fractional yield* and can range widely, depending on the radiation source and applied electric field, as shown in Figure 5.1 and Figure 3.19.

The current understanding of ionization and recombination has its origins in studies following the discovery of ionizing radiation in the 1890s. While this early work was focused on the ionization of gases, the two recombination models that were developed, columnar and geminate, as well as numerical methods have been applied to the current problem of recombination in SiO₂ with good results. In the columnar model, recombination is dependent on the collective e–h charge density, along the ionization path, so this model works well when the e–h pairs are close together [27]. The geminate model assumes that the Coulomb field between the generated e–h pair dominates, so that pair recombination is the most likely outcome. The geminate model is accurate for far apart e–h pairs [27].

3.3.1.2 Geminate Recombination

In general terms, geminate or initial recombination takes place when an electron produced by ionization recombines with its parent ion. For SiO₂, geminate recombination takes place when the e–h pair recombines after generation. This type of recombination is most likely when the density of ionization is low so that the Coulomb interaction between the e–h pair is unscreened by other nearby generated charges. This problem was analyzed by Lars Onsager in 1938 [28] by treating the electron motion as diffusional. Onsager assumed that the electron-recombination kinetics can be described by one Brownian particle in the presence of a Coulomb field, as shown in Figure 3.18a. After generation, the electron and hole reach thermal equilibrium with their surroundings at a distance r from each other. If the electron is not energetic enough to escape the Coulomb field, it will be captured and recombine with the hole. By equating the thermalized energy of the electron with the corresponding potential energy of the Coulomb field, we find the Onsager radius:

$$r_c = \frac{e^2}{4\pi\varepsilon kT}. \tag{3.68}$$

The radius r_c provided by Eq. (3.68) defines a maximum distance between the thermalized hole and electron in the absence of an applied electric field where recombination is the likely outcome.

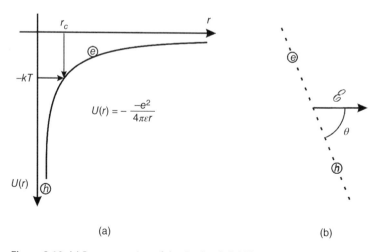

(a)   (b)

Figure 3.18 (a) Representation of the Coulomb field between an electron and hole showing the Onsager radius; (b) the applied electric field is at an angle θ with the line intersecting the electron and hole.

One of the assumptions in the Onsager treatment is that the electron is represented by a probability function $p(\mathbf{r}, t)$. We can then require conservation of probability

$$\frac{\partial p(\mathbf{r}, t)}{\partial t} + \nabla \cdot \mathbf{j} = 0 \tag{3.69}$$

and write the probability current $\mathbf{j}$ as the sum of a diffusion term (where D is the diffusivity) and a drift term due to the influence of the applied electric field $\mathbf{E}$:

$$\mathbf{j} = -D\nabla p(\mathbf{r}, t) + \mu \mathbf{E} p(\mathbf{r}, t). \tag{3.70}$$

Using Eqs. (3.68) and (3.70) along with the Einstein relation, we arrive at the Smoluchowski equation:

$$\frac{\partial p(\mathbf{r}, t)}{\partial t} = D \left\{ \nabla^2 p(\mathbf{r}, t) - \left(\frac{e}{cT} \right) \nabla \cdot [\mathbf{E} p(\mathbf{r}, t)] \right\}. \tag{3.71}$$

Onsager solved the Smoluchowski equation analytically showing the probability for escape of the e−h pair is given by

$$\begin{aligned}
\varphi(r, \theta, \beta) &= e^{-\beta r(1+\cos\theta)} \int_{r_c/r}^{\infty} J_0 \left\{ 2[-\beta r(1+\cos\theta)s]^{\frac{1}{2}} \right\} e^{-s} ds \\
&= e^{-(r_c/r)-\beta r(1+\cos\theta)} \sum_{m,n=0}^{\infty} \beta^{m+n} \frac{(1+\cos\theta)^{m+n} r_c^m r^n}{m!(m+n)!}
\end{aligned} \tag{3.72}$$

where θ is the angle between the direction of the applied electric field and a line connecting the electron and hole as shown in Figure 3.18b, $\beta = (q/2kT)E$, r_c is as defined in Eq. (3.66) and J_0 is the ordinary Bessel function of order zero. Note that the dependence on the electric field in Eq. (3.72) is in the parameter β. It is worth noting that at small initial separations r between the electron and hole, the relative effect of the external electric field reaches its maximum:

$$e^{r_c/r} \varphi(r, \theta, \beta) \to J_0 \left(4\left(-\frac{\beta r_c}{2} \right)^{1/2} \cos\frac{\theta}{2} \right). \tag{3.73}$$

The angular dependence of Eq. (3.72) is largely due to the initial condition of one hole and one electron. If the distribution of electrons around the holes is assumed to be isotropic, the angular dependence of $\phi(r,\theta,\beta)$ can be removed by integration so that

$$\varphi(r, \beta) = \frac{1}{\pi} \int_0^{\pi} e^{-\beta r(1+\cos\theta)} I_0 \left\{ 2[\beta r(1+\cos\theta)s]^{\frac{1}{2}} \right\} e^{-s} ds. \tag{3.74}$$

This simplified function has been solved numerically by Ausman [29] and is shown in Figure 3.19 along with measured values for two types of radiation sources in thermally grown SiO_2. The 12 MeV electron fractional yield data are from measurements by Boesch and McGarrity on MOS capacitors at 80 K [30]. At this temperature, the holes are effectively frozen and electrons either recombine or are swept out of the oxide quickly (in about a picosecond). Measurements of flatband voltage shift can then be used to estimate the charge in the oxide escaping recombination. The ^{60}Co data in Figure 3.19 are from measurements by Oldham and McGarrity on MOS capacitors at 77 K [31]. The ^{60}Co and electron fractional yield data overlap because the stopping powers for ^{60}Co ($\sim$1.2 MeV γ photon) and 12 MeV electrons are about the same.

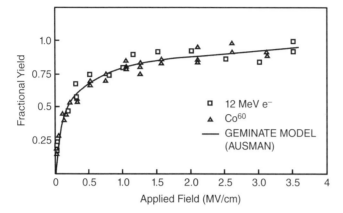

Figure 3.19 Fractional yield as a function of applied electric field for 12 MeV electrons, Co^{60} gamma rays and geminate model calculations. Source: Oldham 1984, Figure 3, p. 1238. © 1984 IEEE. Reprinted, with permission, from [26].

An important parameter in Ausman's model is r_t, the thermalization distance, the most probable distance from the original particle or radiation track where the hole or electron achieves thermal equilibrium with its surroundings. The distance was thought to be in the range of 5–10 nm. Ausman's model data in Figure 3.19 is for a temperature of 80 K and good agreement is obtained by using a thermalization distance r_t of 8 nm. The distance λ between e−h pairs along the radiation trajectory will on average be the inverse of the line density N_0:

$$\lambda = \frac{1}{N_0} = \frac{w}{S\rho}. \tag{3.75}$$

For stopping powers in the 1 MeV-cm^2/g range, represented by the data in Figure 3.19, we can use the stopping power definition from Eq. (3.4) along with the w value from Table 3.1 and the density of SiO_2 to find that λ is about 6.5 μm. This value of λ is much larger than the thermalization distance r_c, justifying the assumptions used to formulate the Onsager and Ausman models.

3.3.1.3 Columnar Recombination

The columnar model is the result of work by George Jaffe [32] following earlier work by Bragg [33] and Langevin. French physicist Paul Langevin was first to realize that ionization from alpha particles formed extremely dense columns around the particle track [34, 35] and proposed the recombination was governed by

$$\frac{\partial n_\pm}{\partial t} = -\alpha_R n_+ n_- \tag{3.76}$$

where n_+ and n_- are the densities of positive and negative charge, respectively, and α_R is a recombination coefficient. Langevin also derived an analytical solution for the value of α_R

$$\alpha_R = \frac{q}{\varepsilon\varepsilon_0}(\mu_+ + \mu_-). \tag{3.77}$$

In Eq. (3.77) μ_+ and μ_- correspond to the mobilities of the positive and negative charges.

Jaffe expanded on Langevin's theory by incorporating a diffusion term as in geminate recombination, to account for thermal motion and drift due to an external electric field. Figure 3.20 shows the time evolution of a section of an ionization track from a time right after the passage of a particle. In Figure 3.20a, the distribution of electrons and holes at time t_0 after the passage of ionizing radiation are at thermal equilibrium with their surroundings and overlap completely. At a later time t_1, as in Figure 3.20b, the electron and hole distributions have each diffused outward from their respective centers and but also drift apart under the influence of the applied electric field, E. Figure 3.20c shows the distributions at a time t_2 later than time t_1. Columnar recombination assumes a much higher density of deposited charge carriers than geminate recombination. Each charge may undergo geminate recombination, but there are many other charges nearby that contribute to recombination. Jaffe described the time evolution of the columns of charge depicted in Figure 3.20 with the following (combined) pair of differential equations [32]:

$$\frac{\partial n_\pm}{\partial t} = D_\pm \nabla^2 n_\pm \mp \mu_\pm E_x \frac{\partial n_\pm}{\partial x} - \alpha_R n_+ n_- \tag{3.78}$$

where the particle is traveling along the z-axis and E_x is the electric field in the x-direction (normal to the direction of particle travel).

Jaffe's solution to Eq. (3.78) was not exact. He ignored the recombination term in Eq. (3.78), finding the more-simple diffusion-only solution first. He then reintroduced recombination by allowing the total number of charges N to vary over time and arrived at the following approximate expression for amount of charge escaping recombination

$$Y = \left[1 + \left(\frac{\pi}{2} \right)^{1/2} \left(\frac{N_0 e}{4\pi \varepsilon \varepsilon_0 bE \sin \theta} \right) \right]^{-1}. \tag{3.79}$$

Equation (3.79) utilizes the total electric field E as well as the previously introduced charge pair line density N_0 and angle θ (Figure 3.18b). The effective column radius b in

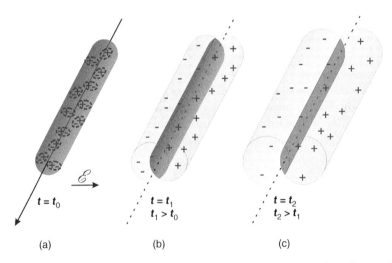

$t = t_0$ $t = t_1$ $t = t_2$
              $t_1 > t_0$ $t_2 > t_1$

(a)   (b)   (c)

Figure 3.20 Time evolution of a section of an ionization track under the influence of an electric field.

Eq. (3.79) comes from the initial charge distribution used by Jaffe:

$$n(r,0) = \frac{N_0}{\pi b^2} e^{\left(\frac{-r^2}{b^2}\right)}. \tag{3.80}$$

The column radius is difficult to determine experimentally, but it can be used as a fitting parameter in Jaffe's approximation as all other parameters are known, as shown in Figure 3.21.

Jaffe's approximate solution was aimed at modeling recombination in ionization experiments using gases. His solution did not agree well with experiments using liquefied gases [37, 38] nor did it agree with the experimental results using α-particles incident on SiO_2 films by Oldham and McGarrity [39, 40]. In the early 1980s, Oldham published a technical report [36] that provided Jaffe's derivation of Eq. (3.79) and outlined a computer program that solved Eq. (3.78) directly using a finite difference algorithm. The output of this program, along with fractional yield data for 2 MeV α-particles incident at a 45° angle to a SiO_2 film are also shown in Figure 3.21.

3.3.1.4 Numerical Methods

The geminate and columnar models represent the two limiting cases in describing recombination, as there is no analytic model that describes recombination for e–h pair densities that fall between the two models. A figure of merit can be applied to understand whether either model can be used with good accuracy. For example, a low ionizing particle like an electron with an energy of 10 MeV has a stopping power in SiO_2 of about 2.0 MeV/g cm^2 corresponding to a distance between e–h pairs of about 32 nm. This separation is an order of magnitude larger than the thermalization distance so the recombination is largely geminate. On the other hand, 10 MeV α-particle has a stopping power of 400 MeV/g cm^2 corresponds to a 0.16 nm distance between e–h pairs. This

Figure 3.21 Fractional yield versus applied field for columnar recombination according to Jaffe's approximate solution. Experimental and calculated results using various values for the column radius, b, are shown. Source: Oldham 1982, Figure 13, p. 25 [36].

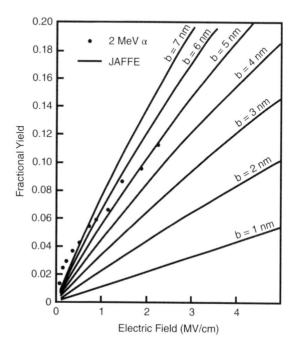

separation is at least an order of magnitude smaller than the thermalization distance so recombination will be columnar. In practice, there are no definite dividing lines between geminate and columnar recombination. Lightly ionizing particles may not have a uniform e–h pair generation, but rather, a more clustered generation along the particle trajectory [41]. Figure 3.22 shows electron, proton, and alpha particle stopping powers in SiO_2. The dividing lines between geminate and columnar recombination are shown to illustrate the large gap between the application of these models and a range in particle energies where the recombination is not well modeled.

Numerical approaches to determining charge yield eliminate the assumptions used in the prior models. For example, the initial charge has been shown [42, 43] to be quite different from the initial Gaussian distributions used by Jaffe and Oldham. The uniform distribution of charge along the ionization track assumed in both columnar and geminate models is also a simplification. Brown [44] provides a description of energy deposition regions for SiO_2 that call for a more complex distribution of charge. Such a charge distribution can be simulated using a Monte Carlo approach, for example, as was done by Murat et al. [45, 46]. The simulations recreate possible distributions of e–h pairs produced by electrons, protons, heavy ions and X-rays, which are then used to estimate the fractional yield. This approach can take into account ionization from all generations of secondary electrons up to their thermalization as well as ionization from the fragments of nuclear interactions and can bridge the perceived transition between geminate and columnar recombination, as shown in Figure 3.23.

3.3.2 Hole Trapping and Transport

Charge generated by radiation that does not recombine will transport through silicon dioxide, although electrons and holes have quite different mobilities in this dielectric. At low fields the room temperature mobility of electrons is about $20 \, cm^2/V \, s$ in SiO_2, a value that doubles when the temperature is lowered to 150 K [47]. Electron velocity saturates

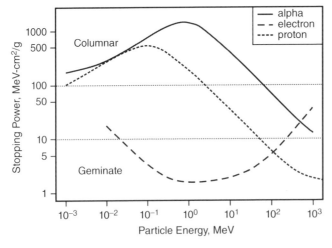

Figure 3.22 Region of application for columnar and geminate recombination in silicon dioxide for electrons, protons, and alpha particles. Source: Adapted from Berger et al. 2005 [5], http://physics.nist .gov/Star.

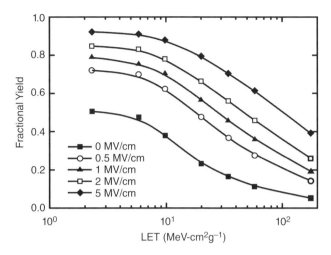

Figure 3.23 Dependence of fractional yield in SiO_2 on ion LET with electric field as a parameter. For each line, the two left-most points are for 200 keV electrons; all other points are for protons (left to right: 100, 50, 20, 5, and 1 MeV). Source: Murat et al. 2006, Figure 6, p. 4. © 2006 IEEE. Reprinted, with permission, from [46].

at approximately 10^7 cm/s for operating fields 10^5 V/cm or greater, so for typical oxide thicknesses of around 100 nm operating at a gate bias of several volts the electrons will transport out of the oxide in a matter of a few picoseconds. Holes, on the other hand, have a very low mobility in silicon dioxide that is dependent on temperature and the applied electric field typically ranging from 10^{-4} to 10^{-11} cm^2/ V s.

The basic transport picture for both holes and electrons in a MOS gate region is shown in the band diagram of Figure 3.24. Ionizing radiation generates an e–h pair, which is separated by the electric field in the gate region. Electrons are swept out of the gate very

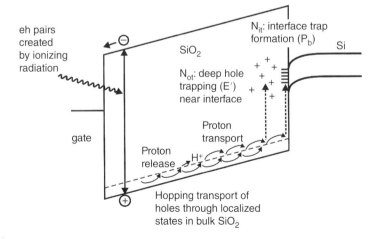

Figure 3.24 Band diagram showing the transport of generated charge in a MOS gate after the passage of ionizing radiation. Source: Schwank et al. 2008, figure 1, p. 1834. © 2008 IEEE. Reprinted, with permission, from [48].

quickly, but holes are trapped. Under a positive gate bias the holes will transport toward the SiO_2/Si interface by hopping from one localized trap to another. When the holes approach the SiO_2/Si interface closely enough, they can be captured by border traps (Section 3.3.3.2 Border Traps). As the holes transport to the interface under positive bias they release hydrogen (H+), which, in turn, also transports to the interface where it affects the interface state density (Section 3.3.3.3 Hydrogen).

Even the best thermally grown SiO_2 will have microscopic imperfections in stoichiometry throughout the material and at silicon/silicon dioxide interfaces. Unlike crystalline materials where there is a predictable order, amorphous materials like SiO_2 have only a short range order as shown in Figure 3.25. Its order is measured by the stoichiometry and coordination of bonds between silicon and oxygen. Such materials still have a band structure and a band gap, however. Instead of crystalline defects like vacancies and interstitials, defects in amorphous materials are described as coordination defects like dangling or unsatisfied bonds.

An example of correctly bonded SiO_2 along with some of the more common coordination defects is shown in Figure 3.26. The perfect bonding (Figure 3.26a) has every silicon atom bonded to four oxygen atoms and every oxygen atom bonded to two silicon atoms. When silicon bonds to another silicon atom, as in Figure 3.26b, a neutral oxygen vacancy is formed. The Si–Si bond is very weak, however, and can be broken by the capture of a hole, forming the well-known E′ center or tri-coordinated silicon (Figure 3.26c). The E′ center is so named because the uncompensated spin of the electron in the dangling bond produces a signal in electron paramagnetic resonance (EPR) measurements. Two other possible defects where bonds are not formed and left "dangling" as with the E′ center include the nonbridging oxygen (Figure 3.26d) and twofold coordinated silicon (Figure 3.26e). An extra oxygen atom can result in a peroxy bridge, where oxygen bonds to oxygen, as in Figure 3.26f. The defects in Figure 3.26 may present several charge states due to electron or hole trapping.

3.3.2.1 *E′ Centers*

The E′ center is one of the most studied defects in silica [49–55]. It is a paramagnetic defect that was first observed in 1956 in irradiated α quartz and named $E_1{'}$ center [49]. The defect was also observed in silica [51–53] and designated the $E_\gamma{'}$ center. EPR studies

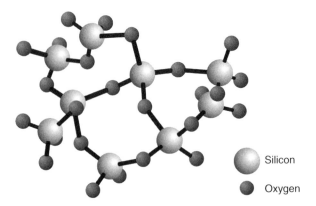

○ Silicon

● Oxygen

Figure 3.25 Fragment of the SiO_2 structure.

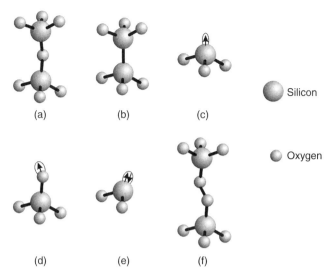

Figure 3.26 Defects in SiO_2: (a) Example of correct bonding; (b) oxygen vacancy; (c) tricoordinated silicon or E′ center; (d) nonbridging oxygen; (e) twofold coordinated silicon; and (f) peroxy bridge.

provided evidence that the E′ center comprises an unpaired electron highly localized in a silicon hybrid orbital orientated toward an oxygen position [56]. In one model of the E′ center, an oxygen vacancy (Figure 3.27a) traps a hole to form the E′ center, as in Figure 3.27b. An early model [57] relaxed the positively charged silicon atom away from the unpaired electron into a planar configuration (Figure 3.27c), but a better fit to the EPR data was obtained by relaxing the silicon into a puckered configuration (Figure 3.27d) [58]. In this configuration, the silicon atom is back-bonded to a nearby oxygen atom. A variant E′ center discernable from EPR measurements, designated E_2' in α quartz and E_β' in silica [54, 58], is shown in Figure 3.27e. A proton is trapped in the oxygen vacancy and the unpaired spin is relaxed away from the vacancy and points

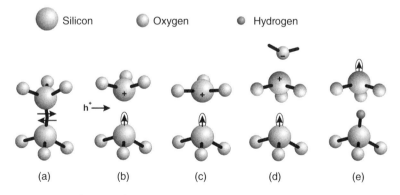

Figure 3.27 E′ centers: (a) oxygen vacancy $O \equiv Si–Si \equiv O$; (b) traps a hole forming the E′ center $O \equiv Si^* + Si \equiv O$ (unrelaxed); (c) relaxed or planar structure of E′ center [57]; (d) refined model [58] where the planar silicon relaxes into a "puckered" configuration and back-bonds to another oxygen atom; and (e) E_β' center in silica [54] (E_2' center in α-quartz [58]) where a proton is trapped in the oxygen vacancy and the unpaired spin is relaxed away from the vacancy.

into a void or a space where there are no nearby oxygen atoms. Many of the oxygen vacancies remain in the dimer configuration of Figure 3.27b. Nicklaw et al. [59] found that in amorphous bulk SiO_2, vacancies at the vast majority of O sites (~90%) do not relax to Figure 3.27c, d, but simply form Si–Si dimers in both the neutral and positively charged states (this is also true for suboxide Si–Si dimers found near the Si–SiO_2 interface).

As expected, oxide growth factors like temperature and pressure have a large influence on oxide defect density, and thus the radiation tolerance and response of the oxide. Hydrogen is often used to passivate or anneal oxygen vacancies and other defects in silicon dioxide and has a profound effect on bulk and interface trap densities, but the effect can be compromised by ionizing radiation and the subsequent transport of holes and release and transport of H+.

3.3.2.2 Continuous-Time Random-Walk (CTRW)

There has been extensive study of hole transport in SiO_2, but the best overall description is provided by the continuous-time random-walk (CTRW) developed by Montroll, Weiss, Sher, and others [60–63]. The CTRW formalism was later applied to hole transport in silicon dioxide by McLean, Hughes, and others. Typical experiments involve irradiating MOS capacitor samples and monitoring the electron and hole currents over time under different applied electric fields and temperatures. Key characteristics of CTRW emerge from such experiments; transient photocurrents decay very slowly, and experimental data from the various temperatures and applied electric fields can be superimposed when the transport time is normalized. The results from one such experiment [64] are shown in Figure 3.28. The shift in flatband voltage is normalized to the initial shift post irradiation shift noted for each sample. The time scale is also normalized to the time at which half of the initial flatband voltage shift recovers. In this way, measurements taken at the various applied field strengths line up to the CTRW model indicated by the solid line.

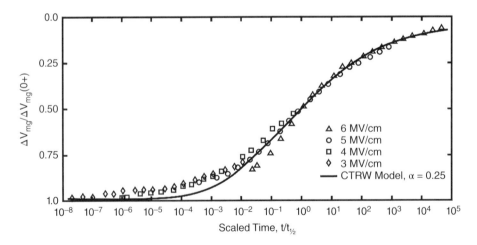

Figure 3.28 Normalized midgap voltage shift recovery curves shifted to coincide at half-recovery point. Solid curve is CTRW model for $\alpha = 0.25$. Source: Adapted from McLean et al. 1987 [64], Figure 18, p. 40.

The CTRW model is a more complicated version of the discrete-time random-walk (DTRW) model. In each model, particles take steps in random directions with the position of the particle at a given time predicted by the sum of the steps up to that time. In the DTRW model, the time steps are at regular intervals. If the distance traveled during each step taken is a fixed length or has a length defined by a bounded probability distribution, the final position distribution will be Gaussian, as predicted by the central limit theorem. In CTRW the time steps are also random variables, so the model is a random walk in both time and space. Typically, the spatial steps are assumed to be drawn from a distribution that has finite variance but the steps in time are from a distribution with infinite mean and variance.

One consequence of the CTRW model is that some particles will get "stuck" in a very long time step and stop moving. This aspect is best illustrated by comparing two plots shown in Figure 3.29. An initial distribution of particles is subjected to an electric field. In Figure 3.29a, transport by discrete time random walk is shown for several points in time with the starting point at time $t = t_0$. The x-axis is the distance traveled by the particles. The initial Gaussian distribution of particles transports in response to an applied field but also broadens due to the random (but Gaussian) aspect in the distance traveled by each particle. In Figure 3.29b, transport according to the CTRW model is illustrated. When the time of transport and the distance of transport are both randomized, a large number of particles get stuck in long time steps near the origin. With no upper bound to the time of transport, the distribution of particles diffuses and moves much more slowly.

The specific mechanism behind hole transport in SiO_2 appears to be small polaron hopping between localized, shallow trap states having a random spatial distribution, but having an average separation of about 1 nm [64]. Polarons are quasi-particles that form when a charged particle moves through a material. The electrons and nuclei in the atoms comprising the material shift positions slightly in response to the charged particle. This

Figure 3.29 Comparison of (a) discrete-time random-walk; and (b) continuous-time random-walk (CTRW) transport.

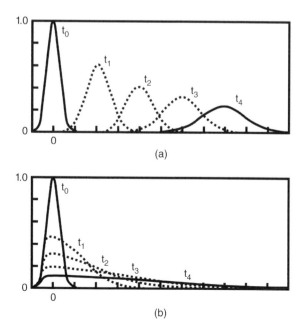

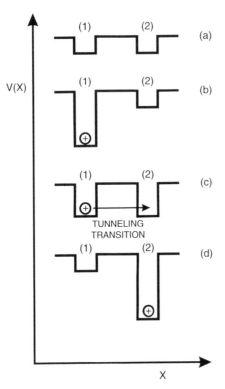

Figure 3.30 Illustration of hole transport via polaron hopping between two nearby localized trap sites, (1) and (2): (a) both sites unoccupied; (b) hole localized on site (1); (c) thermal fluctuations bring trap energies into coincidence and hole tunnels from site (1) to (2); (d) final state with hole localized on site (2). Source: McLean et al. 1987 [64], Figure 2, p. 16.

creates a region of polarization surrounding the particle, and the ensemble is called a polaron. In the case of SiO_2 a small polaron is created when a hole is trapped at an oxygen vacancy as in Figure 3.27b, for example. Hole trapping and transport by polaron hopping are shown schematically in Figure 3.30 [64]. Two nearby localized trap sites, (1) and (2), are initially unoccupied in (a). In (b) a hole is localized at trap site (1). The hole along with the local polarization results in a stable lower overall energy level at the trap site. Over time thermal fluctuations may bring trap sites (1) and (2) into equilibrium at which point the hole can tunnel from site (1) to (2) as shown in (c). In (d) the final state is shown with the hole localized to trap site (2).

3.3.3 The Silicon/Silicon Dioxide Interface

The discussion thus far has applied to the properties of bulk SiO_2, but what happens at the $Si-SiO_2$ interface? The answers are still evolving but have their beginnings during the quest for reliable production of MOS devices and the use of solid-state electronics in the space environment. During the 1960s, there was a lot of research aimed at getting MOS structures to behave in a reproducible way, and this required understanding the components that influence the silicon surface potential. Snow et al. [65] used capacitance-voltage (C-V) measurements to establish that mobile alkali ions like Na+ and K+ from processing chemicals were responsible for much of the instability and reproducibility issues. The contribution from the lattice orientation was evident from oxidation and C-V measurements reported by Balk in 1965 [66] from a series of experiments carried out on (111), (110), and (100) oriented wafers. In 1967, Deal et al. [67]

published a study that showed the surface state charge, Q_{ss}, is an intrinsic property of the Si–SiO$_2$ system. The authors provided insight into the influence of surface orientation, oxidation temperature and condition (wet or dry), pull rates, and post-oxidation anneals.

The "space race" in the 1960s also promoted better understanding of interfacial trapping of charge. With the discovery of the van Allen radiation belts in 1958 and the failure of *Telstar 1* in 1962, survivability of electronics in space became a focus. Studies at the time showed the importance of semiconductor surface effects (see, e.g. [68]). In 1967 Snow et al. [69] reported the effects of ionizing radiation on oxidized silicon surfaces as well as planar devices, including MOS, bipolar junction transistor (BJT), and JFET structures.

Over the next 15 years, researchers established that four general types of charges are associated with the Si–SiO$_2$ system and agreed to the following naming convention in 1980 [70]:

Fixed-oxide charge	Q_f, N_f
Mobile ionic charge	Q_m, N_m
Interface-trapped charge	Q_{it}, N_{it}
Oxide-trapped charge	Q_{ot}, N_{ot}.

Measurements of net effective charge, Q, are per unit area at the Si–SiO$_2$ interface (C/cm^2) and N denotes the net number of charges per unite area at the Si–SiO$_2$ interface (number/cm^2). The mobile ionic charge, Q_m, is not really a problem now that high purity chemicals and processes are used in modern foundries, so it will not be discussed. The oxide-trapped charge, Q_{ot}, includes trapped holes in the oxide as discussed in Section 3.3.1.2, which are localized at oxide defects (E centers) but also includes trapped electrons. The interface-trapped charge, Q_{it}, is located exactly at the Si–SiO$_2$ interface, can be positive or negative, and is due to structural defects, as described in the next section. The fixed-oxide charge, Q_f, is positive and located within 25 Å of the interface. Studies since this nomenclature was accepted indicate that some of the E centers in this region can exchange charge with the substrate. In 1992, Fleetwood [71] suggested these near-interfacial oxide traps be called border traps. Section 3.3.3.2 will provide a brief introduction and review of the more pertinent findings. Some good starting points for more comprehensive coverage can be found in [27, 72, 73] and [74].

3.3.3.1 Interface Traps

We now take a more comprehensive look at the Si–SiO$_2$ interface. In the bulk oxide region, it is likely that the composition is closely stoichiometric, but containing some coordination defects discussed earlier and illustrated in Figure 3.26. At the interface between silicon and silicon dioxide the situation is a little bit less than stoichiometric with excess Si and some Si bonds are uncoordinated. Figure 3.31 is provided in order to visualize defects at or near the Si–SiO$_2$ interface. These include the previously discussed E′ centers as well as uncoordinated or dangling bonds at the interface, which are associated with atoms electrically connected to the silicon substrate.

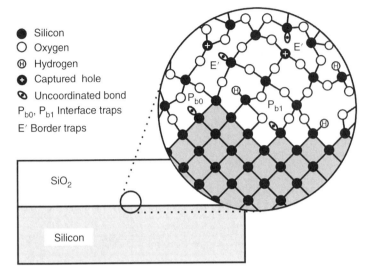

- ● Silicon
- ○ Oxygen
- Ⓗ Hydrogen
- ⊕ Captured hole
- ✿ Uncoordinated bond
- P_{b0}, P_{b1} Interface traps
- E′ Border traps

Figure 3.31 Structural model of the (100) Si–SiO$_2$ interface.

Poindexter [75] used electron spin resonance (ESR) and C-V measurements to identify and correlate interface states in thermally oxidized (100) and (111) silicon wafers subjected to various processing treatments. The ESR signal of interest, called P_b, appears to be the only signal that is connected to the interface state density. There are two components of this signal for (100) Si, P_{b0} corresponds to a trivalent silicon atom back bonded to three other silicon atoms while P_{b1} corresponds to a trivalent silicon atom back bonded to two silicon atoms and one oxygen atom. These states exist because there simply isn't enough room at the interface to fully coordinate the transition from Si to SiO$_2$. They are positioned within one or two atomic bond lengths from the silicon lattice (approx. 0.5 nm). After oxidation, most of the silicon at the interface is bonded to oxygen but there still remains an interface trap density of about 10^{12} cm^{-2} eV^{-1}. MOS processes will introduce hydrogen to passivate the remaining dangling bonds and this reduces the interface trap density to around 10^{10} cm^{-2} eV^{-1}, a level that is useful for MOS fabrication.

The interface trap density is related to the number of dangling silicon bonds as can be seen in the Si–SiO$_2$ interface state densities reported by [76] and others. The (111) surface has a higher density of Si bonds than the (100) surface resulting in a higher interface trap density. Electrons and holes in the silicon conduction and valence bands can readily make quantum mechanical transitions into and out of the interface traps. The interface traps have an amphoteric nature as they are donor-like states in the lower half of the bandgap and acceptor-like states in the upper half of the bandgap. The donor-like states are positively charged when empty and electrically neutral when occupied by an electron. The acceptor-like states are electrically neutral when empty and negatively charged when occupied by an electron. Interface traps can communicate with the silicon substrate very rapidly. Their geometry with two or three back bonds to the silicon substrate and a dangling bond pointing into the SiO$_2$ provides a likely avenue for tunneling electrons to affect the trap's charge.

3.3.3.2 Border Traps

In 1992, Fleetwood [71] pointed out the inadequacies of the Deal convention [70] in describing near interfacial oxide traps that communicate with the Si substrate. At the time, there were many names for these traps including "slow" interface (or surface) states, anomalous positive charge, and rechargeable E′ centers (a more comprehensive list can be found in [73, 74]). Fleetwood advocated calling these border traps to help standardize the naming convention. He reasoned that border traps would be located within 3 nm of the Si substrate or the gate, based on the electron tunneling probability, and oxides thinner than 6 nm would have no "bulk-like" traps.

Figure 3.32 shows Fleetwood's updated nomenclature. The right-most diagrams distinguish between the defect location and the defect electrical response. Border traps have a wide time scale over which they trap or release charges. The fastest border traps are those closest to the interface and are indistinguishable from interface traps using standard subthreshold current-voltage (I-V) or C-V techniques [77–79].

Radiation annealing studies [80–82] indicated some odd behavior when the post anneal bias across the oxide is switched. Trapped charge in irradiated MOS oxides that was thought to have been removed by positive-bias annealing could be partially restored during negative-bias annealing (annealing temperatures are typically around 100 °C). In 1989, Lelis et al. [83] provided a model for this behavior. Following irradiation holes will transport via polaron hopping toward the Si–SiO$_2$ interface when there is a positive bias on the gate. The lattice deformation that surrounds the hole causes additional strain on the already weak bond at oxygen vacancies in the region. If the bond breaks, the hole is trapped on one of the Si atoms, leaving it with a positive charge, while the other Si atom remains neutral with a dangling orbital and an unpaired electron. This is shown schematically in Figure 3.33 as the transition from state A (oxygen vacancy) to state B (E′ center). Note that the lattice around the positively charged Si atom in state B relaxes into a more planar state (also see Figure 3.27c).

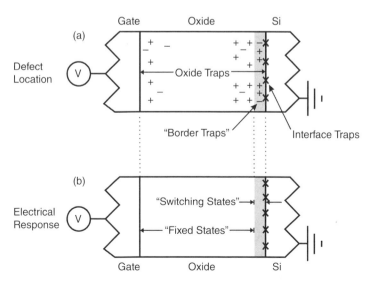

Figure 3.32 Update to the Deal convention suggested by Fleetwood. Source: Fleetwood 1996, Figure 2, p. 780. © 1996 IEEE. Reprinted, with permission, from [77].

When a positive bias is applied to the gate, electrons tunnel from the substrate and join unpaired electrons in nearby E′ centers, resulting in the configuration shown in Figure 3.33 state C. This configuration accomplishes the removal of the unpaired spin and also neutralizes the trapped hole. While the original oxygen vacancy bond is not reformed, there is an attractive electrostatic force between the two ends of the dipole formed by the two Si atoms. If the ends of the dipole get close enough, the bond will reform as shown in Figure 3.33 as the transition between state C and state A. Otherwise, when a negative bias is applied to the gate, one of the electrons from the negatively charged Si atoms tunnels back to the substrate. The result is a return to Figure 3.33 state B. Depending on the gate bias, electrons can tunnel either way, continuing the cycle between state B and state C, until the bond is reformed. In fact, repeated cycles of positive and negative bias anneals indicate some bonds do reform, as the amount of switched charge removed decreases over time [83].

3.3.3.3 Hydrogen

By the mid-1960s it was well known that incorporating hydrogen during the fabrication process resulted in a large decrease in interface traps. Hydrogen will passivate the P_b centers at the Si–SiO$_2$ interface, producing a higher-quality interface and leading to better MOS device characteristics. Hydrogen also tends to bond to other defects in bulk SiO$_2$. Sources of hydrogen during wafer fabrication include thermal oxidation steps, chemical vapor deposition (CVD), and stability bakes/anneals. Hydrogen will interact strongly with impurities as well as defects in the Si crystal and one of the strongest bonds is with silicon dangling bonds. The Si—H bond at these locations can have a strength up to

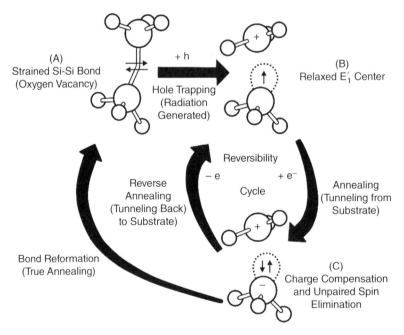

Figure 3.33 A model for hole trapping at interfacial E' centers based on the characteristics of switched bias annealing experiments. Source: Lelis et al. 1989, Figure 1, p. 1808. © 1989 IEEE. Reprinted, with permission, from [83].

3.6 eV [84, 85], thus forming a deep trap for hydrogen. Removing hydrogen from such a trap would face a lower barrier, however, as only about 2.5 eV is needed to place the hydrogen in an interstitial location [86].

Exposure to ionizing radiation not only increases the oxide-trapped charge, N_{ot}, but leads to breaking of Si—H and Si—OH bonds and an increase in interface traps [87] over time. The change in threshold voltage over time for n-channel MOS devices shown in Figure 3.34 [73] illustrates the effects from irradiation. There is an immediate change in threshold voltage following irradiation due to the trapped charge in the oxide. Many of the holes will recombine, but with a positive bias on the gates some of the holes are swept to the Si–SiO$_2$ interface over time. Here they either recombine or occupy border traps as discussed in the previous section. There is also a longer-term build-up of interface traps that is primarily due to the removal of the hydrogen passivation at the interface.

In 1980, McLean [89] formulated an empirical two-stage model for hole and hydrogen transport following ionizing radiation. This model helped to explain experimental results from his co-workers [90–93]. In the first stage, holes transporting through the oxide release hydrogen, most likely the ion H$^+$. In the second stage, a positive bias causes the holes as well as the hydrogen to transport to the Si–SiO$_2$ interface. When the hydrogen ions reach the interface they react to break Si—H bonds forming H$_2$ and a dangling Si bond (indicated by Si$^+$):

$$SiH + H^+ \rightarrow Si^+ + H_2. \tag{3.81}$$

Figure 3.35 illustrates one of the experiments supporting the McLean two-stage process [94]. Hardened oxides were exposed to a dose of 0.6 Mrad and then subjected to various voltage bias (4 MV/cm) scenarios. Curve A represents a positive bias on the gate throughout the experiment while curve E represents a negative bias throughout

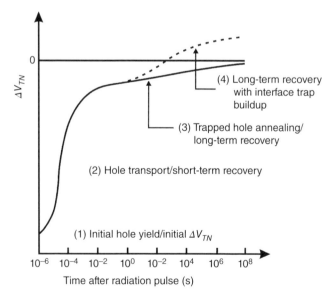

Figure 3.34 Time-dependent threshold voltage recovery of n-channel MOSFET. Source: Oldham and McLean 2003, Figure 3, p. 484. © 2003 IEEE. Reprinted, with permission, from [88].

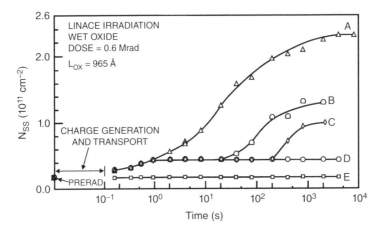

Figure 3.35 Bias dependency of the interface state buildup after irradiation. Source: Oldham et al. 1989 [94], Figure 7, p. 12.

the experiment (control sample). The remaining curves are from samples that were initially biased positive for one second, but then negative biased until switched back to positive bias at 20s (B), 200 s (C), and 2000s (D).

Rashkeev et al. [95] provided first-principles calculations showing that H^+ is the only stable charge state at the Si–SiO_2 interface. They simulated the Si–SiO_2 interface using a supercell approach and found that H^+ arriving at the interface releases a net 1.1 eV upon completion of the reaction in Eq. (3.81). It takes about 0.2 eV to get an H^+ ion close enough to the Si—H bonded hydrogen (about 1.6 Å), however. This means there is a 1.3 eV barrier to forming the Si—H bond (the reverse of Eq. (3.81)). For an initially neutral dangling bond, the activation energy would include the ionization energy of the P_b center, which is about half of the Si bandgap (0.55 eV) for a total barrier of 1.85 eV.

In a study of molecular hydrogen, Tuttle et al. [96] modeled the possible reactions of hydrogen molecules with common defects found in SiO_2. They found that hydrogen molecules are unlikely to dissociate at oxygen vacancies at room temperature before irradiation. After irradiation, molecular hydrogen is cracked at vacancies that have trapped a hole. They calculated the barrier height for this reaction at about 0.4 eV. The most likely locations correspond to the puckered condition (Figure 3.27d) because there is a large open region between the Si atoms. Only about 20% of the oxygen vacancies relax to the puckered condition, the other sites were found to have a higher barrier of 1.4 eV to H_2 dissociation.

3.3.3.4 ELDRS
The enhanced low dose rate sensitivity (ELDRS) effect in bipolar devices and circuits has been studied extensively following its discovery in the early 1990s [97]. Typical dose rates used to simulate total ionizing dose (TID) are in the 50 to 300 rad(Si)/s range but low dose rates (0.01 rad(Si)/s or lower) have been found to cause accelerated failure or parametric shift in bipolar circuits compared to testing done with higher dose rates. The lower dose rates are typical of space applications so the discovery of potential failures not

predicted by high-dose-rate testing caused an industrywide reevaluation of space-rated components and a search for cost-effective hardness assurance test methods.

ELDRS can be seen in BJTs base oxides when irradiated under low electric fields. The failure mechanism is the buildup of interface traps in the base oxide regions, which degrades the transistor current gain. The dose rate dependency can be dramatic, as seen in Figure 5.2. There are several models that have been developed to account for ELDRS in bipolar circuits including:

Space charge. At high dose rates, holes accumulate in the oxide under the low electric field conditions and create a space charge that inhibits hole and proton transport to the Si–SiO$_2$ interface [98–101].

Charge yield. Competition between the probability for recombination and the probability for trapping of radiation-induced carriers leads to dose rate dependency. The model considers the probability of electron trapping, which is higher at low electric fields [102].

Bimolecular reactions. At high dose rates, bimolecular reactions like recombination, hydrogen recapture and hydrogen dimerization dominate over the radiolysis reaction. At low dose rates the radiolysis reaction dominates, leading to maximum effect from the radiation [103].

Time and temperature dependency. At high dose rates, holes recombine before protons can be released. At low dose rates, holes can release protons with higher probability before recombination or deep hole trapping. This is especially true for oxides with high concentrations of weakly bonded hydrogen [104].

Components and ICs tested at low dose rates were seen to have dependencies to wafer processing such as final passivation [105–107] and also packaging including thermal stress [108] and sources of hydrogen [109, 110]. Pease et al. [111] showed that the amount of hydrogen present influences the maximum-low-dose-rate degradation as well as where the transition from low-dose-rate enhancement to the high-dose-rate regime occurs.

3.4 Bulk Damage

Damage to the semiconductor lattice can occur during irradiation by heavy particles. Figure 3.36 provides an example of the damage to a crystal lattice by neutrons. A neutron strikes a lattice atom and displaces it from its lattice position. For energetic collisions there can be a damaged region where several atoms have been displaced from their lattice positions. The recoil atom leaves an open spot in the lattice called a vacancy and usually ends up between lattice points somewhere as an interstitial atom. Often the track of the recoil atom will also produce damage to the lattice. It takes about 15 eV to displace a Si atom from its lattice position [112].

Both vacancy and interstitial defects are not stable at room temperature, so they will anneal over time. Often the annealing is in the form of complexes with other defects or impurities. For example, if a vacancy is located next to a donor atom in n-type material a donor-vacancy defect trapping complex is formed with a discrete energy level in the bandgap. The complex is initially positively charged due to the donor atom and can capture an electron from the conduction band. If the Fermi energy is greater than the trap

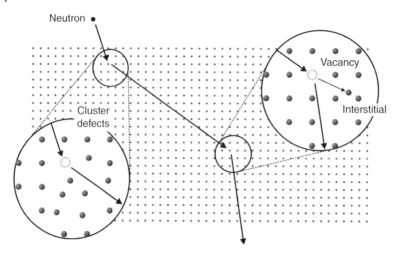

Figure 3.36 Crystal damage caused by neutron irradiation.

energy, the complex can capture an additional electron changing its polarity to negative. Acceptor-vacancy defect trapping complexes are likewise formed in p-type material. Divacancy complexes are formed with two vacancies end up side by side. This defect complex can have five charge states from double negative to double positive. The two positive states can occur due to the stress caused on adjacent Si bonds, which can rupture and release one or two electrons to the conduction band. Some silicon can have oxygen as an impurity, for example silicon grown using the Czochralski (CZ) process. Oxygen atoms can be accommodated in silicon vacancies and if such an oxygen atom is situated next to another vacancy, an oxygen-vacancy defect can form. These defects can capture an electron from the conduction band if the Fermi energy is greater than the energy of the oxygen-vacancy.

Any particle can cause damage through knock-on interactions with target atoms. Displacement damage and the resulting impurity-vacancy complexes create recombination centers that remove carriers. This can affect resistivity, mobility, and space charge regions in semiconductor devices. Buehler [113] compared experimental to calculated resistivity for n-type and p-type for neutron fluences (reactor neutrons with $E > 10\,keV$) from 10^{13} to $10^{16}/cm^2$. He found that the experimental data fit the resistivity-fluence relations:

$$\rho_p = \rho_{p0}e^{\phi/k_p} \tag{3.82}$$

$$\rho_n = \rho_{n0}e^{\phi/k_n} \tag{3.83}$$

where φ is the neutron fluence and $k_{p,n}$ are the p-type and n-type damage constants. The damage constants are determined by assuming the irradiation causes changes only in carrier concentration and not in mobility. Buehler found that

$$k_p = K_p p_0^{0.77} \tag{3.84}$$

$$k_n = K_n n_0^{0.77} \tag{3.85}$$

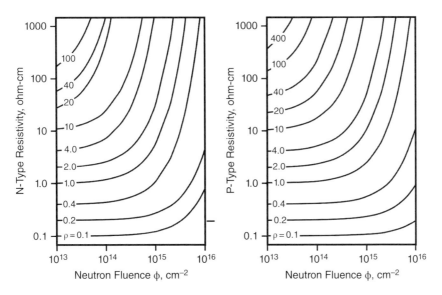

Figure 3.37 Resistivity for uniformly doped n-type and p-type silicon versus neutron fluence.

with the coefficient values of $K_p = 387$ and $K_n = 444$. Eqs. (3.84) and (3.85) indicate the strong dependence of the damage constant on the pre-irradiation carrier concentration. The plots in Figure 3.37 were generated using Eqs. (3.82)–(3.85) along with Irvin's curves [114] and Buehler's coefficient values for n- and p-type silicon.

3.5 Summary and Closing Comments

This chapter provided a working knowledge of the basic effects caused by radiation in materials like silicon and silicon dioxide. Section 3.2 reviewed the basic effects caused by heavy charged particles, electrons, neutrons, and photons. The idea of stopping power was introduced and explained in terms of the various interactions that particles have with matter. Stopping power, a characteristic of the material and not the radiation source, is an important concept that follows from the interaction of radiation and matter. Some effects, such as bremsstrahlung, occur only with light relativistic particles like electrons and muons while other effects are unique to photons. Section 3.3 provided the basic ideas behind charge trapping and transport in silicon dioxide starting with the columnar and geminate models for recombination after an ionizing event. Hole trapping and transport were discussed in relation to SiO_2 defects and the CTRW model, respectively.

The $Si-SiO_2$ interface microstructure was described in terms of interface and border traps with the former identified as P_b centers (Si dandling bonds) and the latter as E centers in close communication with the substrate (silicon). The role of hydrogen in controlling properties on the $Si-SiO_2$ interface was described and The McLean two-stage model introduced to explain interface trap build-up after radiation exposure. The dose rate effect called ELDRS was described, along with several explanations for the low-dose-rate enhancement and a final section on bulk damage rounded out the chapter.

Problems

3.1 Relativistic Q_{max} Calculation
Repeat Eqs. (3.6) through (3.10) but include relativistic effects.

3.2 Electromagnetic Radiation
Examine the colinear and perpendicular components of radiated power in Eq. (3.25). Confirm that the colinear component is proportional to γ^6 and m^{-6} and that the perpendicular component is proportional to γ^3 and m^{-4}.

3.3 Larmar Formula
Start with Eq. (3.31) and show the steps that result in the Lienard result, Eq. (3.32).

3.4 Critical Energy
There are two definitions for critical energy, E_c:
EGS4 definition: Critical energy is the energy at which losses by ionization are equal to losses by radiation (EGS4 is a computer program that uses Monte Carlo simulations to model EM showers).
The Rossi definition: Critical energy is the energy at which the ionization loss per radiation length equals the electron's own energy.
Show that the two definitions for critical energy are the same.

3.5 Elastic Scattering
a) Show that for Eq. (3.53) when E_2/E_1 is at a minimum the maximum energy is transferred.
b) What substances are good for slowing neutrons?

3.6 Photoelectric Effect
Show that the photoelectric effect applies only to bound electrons.

3.7 Thomson Scattering
The low energy limit of Compton scattering is called Thomson scattering. In this case, the particle kinetic energy and photon frequency do not change from scattering. Derive Eq. (3.60), but assume that $h\nu \ll mc^2$.

3.8 Pair Production
Derive Eq. (3.65) by adding a "spectator mass" M to the calculation.

3.9 Oxide Traps
Discuss possible explanations for the departure from linear behavior seen in the MOSFET threshold voltage shift after radiation for thin gate oxides as seen in Figure 1.14 (from Chapter 1 reference [27]).

References

1 Bohr, N. (1913). On the theory of the decrease of velocity of moving electrified particles on passing through matter. *Philosophical Magazine* 25: 10–31.

2 Bloch, F. (1933). Bremsvermögen von Atomen mit mehreren Elektronen. *Zeitschrift für Physik* 81 (5–6): 363–376.

3 Bethe, H.A. (1932). Bremsformel für Elektronen relativistischer Geschwindigkeit. *Zeitschrift für Physik* 76: 293–299.

4 Fano, U. (1963). Penetration of protons, alpha particles, and mesons. *Annual Review of Nuclear and Particle Science* 13: 1–66.

5 Berger, M.J., Coursey, J.S., Zucker, M.A., and Chang, J. (2005). ESTAR, PSTAR, and ASTAR: Computer Programs for Calculating Stopping-Power and Range Tables for Electrons, Protons, and Helium Ions (version 1.2.3), [Online] Available: http://physics.nist.gov/Star [2018, August 21], National Institute of Standards and Technology, Gaithersburg, MD.

6 Diver, D.A. (2001). *A Plasma Formulary for Physics, Technology, and Astrophysics*, 46–48. Berlin: Wiley-VCH Verlag.

7 Jackson, J.D. (1962). *Classical Electrodynamics*, 464–504. New York: Wiley.

8 Leo, W.R. (1987). *Techniques for Nuclear and Particle Physics Experiments, a How-to Approach*, 35. Berlin: Springer-Vertag.

9 Seltzer, S.M. (1988). Cross sections for bremsstrahlung production and electron impact ionization. In: *Monte Carlo Transport of Electrons and Photons* (eds. T.M. Jenkins, W.R. Nelson and A. Rindi). New York: Plenum Press.

10 Joseph, J. and Rohrlich, F. (1958). Pair production and bremsstrahlung in the field of free and bound electrons. *Reviews of Modern Physics* 30 (2): 354–368.

11 Koch, H.W. and Motz, J.W. (1959). Bremsstrahlung cross-section formulas and related data. *Reviews of Modern Physics* 31 (4): 920–955.

12 Pratt, R.H. and Feng, I.J. (1985). Electron-atom bremsstrahlung. In: *Atomic Inner-Shell Physics* (ed. B. Crasemann), 533. New York: Plenum.

13 Messel, H. and Crawford, D.F. (1970). *Electron-photon Distribution function tables for Lead, Copper and Air Absorbers*. Headington Hill Hall, Oxford: Pergamon Press.

14 Tanabashi, M., Hagiwara, K., Hikasa, K. et al. (Particle Data Group) (2018). Review of particle physics. *Physical Review D* 98 (030001): 447.

15 Tsai, Y.S. (1974). Pair production and bremsstrahlung of charged leptons. *Reviews of Modern Physics* 46 (4): 815–851.

16 Rossi, B. (1952). *High Energy Particles*. Englewood Cliffs, NJ: Prentice-Hall, Inc.

17 Chadwick, J. (1933). Bakerian lecture: the neutron. *Proceedings of the Royal Society A: Mathematical, Physical and Engineering Sciences* 142 (846): 1–25.

18 Chadwick, J. (1932). Possible existence of a neutron. *Nature* 129 (3252): 312.

19 Rutherford, E. (1920). Bakerian lecture: nuclear constitution of atoms. *Proceedings of the Royal Society A: Mathematical, Physical and Engineering Sciences* 97 (686): 374–400.

20 Bothe, W. and Becker, H. (1930). Künstliche Erregung von Kern-γ-Strahlen. *Zeitschrift für Physik* 66 (5–6): 289–306.

21 Becker, H. and Bothe, W. (1932). Die in Bor und Beryllium erregten γ-Strahlen. *Zeitschrift für Physik* 76 (7–8): 421–438.

22 Joliot-Curie, I. and Joliot, F. (1932). Émission de protons de grande vitesse par les substances hydrogénées sous l'influence des rayons γ très pénétrants. *Comptes Rendus* 194: 273.

23 Chadwick, M. B., Young, P. G., and Hetrick, D. (2018). ENDF/B-VIII.0 evaluation, MAT # 1425 MF3 MT1, Los Alamos National Laboratory and Oak Ridge National Laboratory, http://www-nds.iaea.org.

24 Hertz, H. (1887). Ueber den Einfluss des ultravioletten Lichtes auf die electrische Entladung. *Annalen der Physik* 267 (8): 983–1000.

25 Davisson, C.M. (1965). Interaction of γ-RADIATION with Matter. In: *Alpha-, Beta- and gamma-ray Spectroscopy*, vol. 1 (ed. K. Siegbahn), 37–78. Amsterdam: North-Holland Publishing Company.

26 Oldham, T.R. (1984). Analysis of damage in MOS devices for several radiation environments. *IEEE Transactions on Nuclear Science* 31 (6): 1236–1241.

27 McLean, F.B., Boesch, H.E. Jr.,, and Oldham, T.R. (1989). Electron-hole generation, transport, and trapping in SiO$_2$. In: *Ionizing Radiation Effects on MOS Devices and Circuits* (eds. T.P. Ma and P.V. Dressendorfer), 87–192. New York: Wiley.

28 Onsager, L. (1938). Initial recombination of ions. *Physics Review* 54 (8): 554–557.

29 Ausman, G. A. (1986). Field Dependence of Geminate Recombination in a Dielectric Medium, Harry Diamond Laboratories Report No. 2097, Adelphi, MD.

30 Boesch, H.E. Jr., and McGarrity, J.M. (1976). Charge yield and dose effects in MOS capacitors at 80 K. *IEEE Transactions on Nuclear Science* 23: 1520–1525.

31 Oldham, T.R. and McGarrity, J.M. (1983). Comparison of Co60 response and 10KeV X-ray response in MOS capacitors. *IEEE Transactions on Nuclear Science* 30: 4377–4381.

32 Jaffe, G. (1913). Zur Theorie der Ionisation in Kolonnen. *Annalen der Physik* 42: 303–344.

33 Bragg, W.H. and Kleeman, R.D. (1906). On the recombination of ions in air and other gases. *Philosophical Magazine* 12: 273.

34 Langevin, M.P. (1903). Recombinaison et mobilites des ions dans les gaz. *Annales de Chimie Physique* 28: 433–530.

35 Langevin, M.P. (1903). L'ionization des gaz. *Annales de Chimie Physique* 28: 289–384.

36 Oldham, T. R., (1992). Charge Generation and Recombination in Silicon Dioxide from Heavy Charged Particles, Harry Diamond Laboratories Technical Report ADA114713, April 1992.

37 Gerritsen, A.N. (1948). Ionization by alpha-particles in liquids at low temperatures: 1. Measurements in liquid nitrogen and liquid hydrogen. *Physica* 14 (6): 381–406.

38 Gerritsen, A.N. (1948). Ionization by alpha-particles in liquids at low temperatures: 2. Measurements in liquid helium and liquid argon. *Physica* 14 (6): 407–424.

39 Oldham, T.R. and McGarrity, J.M. (1981). Ionization in SiO2 by heavy charged particles. *IEEE Transactions on Nuclear Science* 28 (6): 3975–3980.

40 Oldham, T.R. (1985). Recombination along the tracks of heavy charged particles in SiO2 films. *Journal of Applied Physics* 57 (8): 2695–2702.

41 Bradford, J.N. (1986). Clusters in ionization tracks of electrons in silicon dioxide. *IEEE Transactions on Nuclear Science* 33 (6): 1271–1275.

42 Fageeha, O., Howard, J., and Block, R. (1994). Distribution of radial energy deposition around the track of energetic particles in silicon. *Journal of Applied Physics* 75 (5): 2317–2321.

43 Akkerman, A. and Barak, J. (2002). Ion-track structure and its effects in small size volumes of silicon. *IEEE Transactions on Nuclear Science* 49 (6): 3022–3031.

44 Brown, D.B. and Dozier, C.M. (1981). Electron-hole recombination in irradiated SiO_2 from a microdosimetry viewpoint. *IEEE Transactions on Nuclear Science* 28 (6): 4142–4144.

45 Murat, M., Akkerman, A., and Barak, J. (2004). Spatial distribution of electron-hole pairs induced by electrons and protons in SiO_2. *IEEE Transactions on Nuclear Science* 51 (6): 3211–3218.

46 Murat, M., Akkerman, A., and Barak, J. (2006). Charge yield and related phenomena induced by ionizing radiation in SiO_2 layers. *IEEE Transactions on Nuclear Science* 53 (4): 1973–1980.

47 Hughes, R.C. (1973). Charge-carrier transport phenomena in amorphous SiO_2: direct measurement of the drift mobility and lifetime. *Physical Review Letters* 30 (26): 1333–1336.

48 Schwank, J.R., Shaneyfelt, M.R., Fleetwood, D.M. et al. (2008). Radiation effects in MOS oxides. *IEEE Transactions on Nuclear Science* 55 (4): 1833–1853.

49 Weeks, R.A. (1956). Paramagnetic resonance of lattice defects in irradiated quartz. *Journal of Applied Physics* 27 (11): 1376–1381.

50 Weeks, R.A. and Nelson, C.M. (1960). Irradiation effects and short-range order in fused silica and quartz. *Journal of Applied Physics* 31 (9): 1555–1558.

51 Weeks, R.A. and Nelson, C.M. (1960). Trapped electrons in irradiated quartz and silica: II, electron spin resonance. *Journal of the American Ceramic Society* 43 (8): 399–404.

52 Griscom, D.L., Friebele, E.J., and Sigel, G.H. (1974). Observation and analysis of the primary ^{29}Si hyperfine structure of the E' center in non-crystalline SiO_2. *Solid State Communications* 15 (3): 479–483.

53 Griscom, D.L. (1978). Defects in amorphous insulators. *Journal of Non-Crystalline Solids* 31 (1–2): 241–266.

54 Griscom, D.L. (1984). Characterization of three E'-center variants in X- and γ-irradiated high purity a-SiO_2. *Nuclear Instruments and Methods in Physics Research Section B* 1 (2–3): 481–488.

55 Griscom, D.L. and Friebele, E.J. (1986). Fundamental radiation-induced defect centers in synthetic fused silicas: atomic chlorine, delocalized E' centers, and a triplet state. *Physical Review B* 34 (11): 7524–7533.

56 Silsbee, R.H. (1961). Electron spin resonance in neutron-irradiated quartz. *Journal of Applied Physics* 32 (8): 1459–1462.

57 Feigl, F.Y., Fowler, W.B., and Yip, K.L. (1974). Oxygen vacancy model for the E_1' center in SiO_2. *Solid State Communications* 14 (3): 225–229.

58 Rudra, J.K., Fowler, W.B., and Feigl, F.J. (1985). Model for the E_2' center in alpha quartz. *Physical Review Letters* 55 (23): 2614–2617.

59 Nicklaw, C.J., Lu, Z.-Y., Fleetwood, D.M. et al. (2002). The structure, properties, and dynamics of oxygen vacancies in amorphous SiO_2. *IEEE Transactions on Nuclear Science* 49 (6): 2667–2673.

60 Montroll, E.W. and Weiss, G.H. (1965). Random walks on lattices II. *Journal of Mathematical Physics* 6 (2): 167–181.

61 Scher, H. and Lax, M. (1973). Stochastic transport in a disordered solid, I – Theory. *Physical Review B* 7 (10): 4491–4502.

62 Scher, H. and Lax, M. (1973). Stochasitc transport in a disordered solid, II – impurity conduction. *Physical Review B* 7 (10): 4502–4519.

63 Scher, H. and Montroll, E.W. (1975). Anomalous transit-time dispersion in amorphous solids. *Physical Review B* 12 (6): 2455–2477.

64 McLean, F. B., Boesch, H. E. Jr.,, and McGarrity, J. M. (1987). Dispersive Hole Transport in SiO_2, Report No. HDL-TR-2117, U. S. Army Laboratory Command, Harry Diamond Laboratories, Adelphi, MD.

65 Snow, E.H., Grove, A.S., Deal, B.E., and Sah, C.T. (1965). Ion transport phenomena in insulating films. *Journal of Applied Physics* 36 (5): 1664–1673.

66 Balk, P., Burkhardt, P.J., and Gregor, L.V. (1965). Orientation dependence of built-in surface charge on thermally oxidized silicon. *Proceedings of the IEEE* 53 (12): 2133–2134.

67 Deal, B.E., Sklar, M., Grove, A.S., and Snow, E.H. (1967). Characteristics of the surface-state charge (Qss) of thermally oxidized silicon. *Journal of the Electrochemical Society* 114 (3): 266–274.

68 Brown, W.L. (1953). *n*-type surface conductivity on *p*-type germanium. *Physics Review* 91 (3): 518–527.

69 Snow, E.H., Grove, A.S., and Fitzgerald, D.J. (1967). Effects of ionizing radiation on oxidized silicon surfaces and planar devices. *Proceedings of the IEEE* 55 (7): 1168–1185.

70 Deal, B.E. (1980). Standardized terminology for oxide charges associated with thermally oxidized silicon. *IEEE Transactions on Electron Devices* 27 (3): 606–608.

71 Fleetwood, D.M. (1992). "Border Traps" in MOS devices. *IEEE Transactions on Nuclear Science* 39 (2): 269–271.

72 Nicollian, E.H. and Brews, J.R. (1982). *MOS (Metal Oxide Semiconductor) Physics and Technology*. New York: Wiley.

73 Oldham, T.R. (1999). *Ionizing Radiation Effects in MOS Oxides*. Singapore: World Scientific Publishing.

74 Fleetwood, D.M., Pantelides, S.T., and Schrimpf, R.D. (eds.) (2009). *Defects in Microelectronic Materials and Devices*. Boca Raton, Florida: CRC Press.

75 Poindexter, E.H., Caplan, P.J., Deal, B.E., and Razouk, R.R. (1981). Interface states and electron spin resonance centers in thermally oxidized (111) and (100) silicon wafers. *Journal of Applied Physics* 52 (2): 879–884.

76 White, M.H. and Cricchi, J.R. (1972). Characterization of thin-oxide MNOS memory transistors. *IEEE Transactions on Electron Devices* 19 (12): 1280–1288.

77 Fleetwood, D.M. (1996). Fast and slow border traps in MOS devices. *IEEE Transactions on Nuclear Science* 43 (3): 779–786.

78 Fleetwood, D.M., Winokur, P.S., Reber, R.A. Jr., et al. (1993). Effects of oxide traps, interface traps, and border traps on MOS devices. *Journal of Applied Physics* 73 (10): 5058–5074.

79 Fleetwood, D.M., Shaneyfelt, M.R., Warren, W.L. et al. (1995). Border traps: issues for MOS radiation response and long-term reliability. *Microelectronics Reliability* 35 (3): 403–428.

80 Schwank, J.R., Winokur, P.S., McWhorter, P.J. et al. (1984). Physical mechanisms contributing to device "Rebound". *IEEE Transactions on Nuclear Science* 31 (6): 1434–1438.

81 Dozier, C.M., Brown, D.B., Throckmorton, J.L., and Ma, D.I. (1985). Defect production in SiO2 by X-Ray and Co-60 radiations. *IEEE Transactions on Nuclear Science* 32 (6): 4363–4368.

82 Lelis, A.J., Boesch, H.E. Jr.,, Oldham, T.R., and McLean, R.B. (1988). Reversibility of trapped hole annealing. *IEEE Transactions on Nuclear Science* 35 (6): 1186–1191.

83 Lelis, A.J., Oldham, T.R., Boesch, H.E. Jr.,, and McLean, F.B. (1989). The nature of the trapped hole annealing process. *IEEE Transactions on Nuclear Science* 36 (6): 1808–1815.

84 Van de Walle, C.G. (1994). Energies of various configurations of hydrogen in silicon. *Physical Review B* 49 (7): 4579–4585.

85 Tuttle, B. and Adams, J. (1998). Energetics of hydrogen in amorphouse silicon: An ab initio study. *Physical Review B* 57 (20): 12859–12868.

86 Van de Walle, C.G. and Tuttle, B.R. (2000). Microscopic theory of hydrogen in silicon devices. *IEEE Transactions on Electron Devices* 47 (10): 1779–1786.

87 Schwank, J.R., Fleetwood, D.M., Winokur, P.S. et al. (1987). The role of hydrogen in radiation-induced defect formation ini polysilicon gate MOS devices. *IEEE Transactions on Nuclear Science* 34 (6): 1152–1158.

88 Oldham, T.R. (2003). Total ionizing dose effects in MOS oxides and devices. *IEEE Transactions on Nuclear Science* 50 (3): 483–499.

89 McLean, F.B. (1980). A framework for understanding radiation-induced interface states in SiO_2 MOS structures. *IEEE Transactions on Nuclear Science* 27 (6): 1651–1657.

90 Winokur, P.S., McGarrity, J.M., and Boesch, H.E. Jr., (1976). Dependence of interface-state buildup on hole generation and transport in irradiated MOS capacitors. *IEEE Transactions on Nuclear Science* 23 (6): 1580–1585.

91 Winokur, P.S., Boesch, H.E. Jr.,, and McLean, F.B. (1977). Field- and time-dependent radiation effects at the SiO_2/Si interface of hardened MOS capacitors. *IEEE Transactions on Nuclear Science* 24 (6): 2113–2118.

92 Pantelides, S.T. (ed.) (1978). *Physics of SiO_2 and its interfaces*, 428. New York: Pergamon Press.

93 Winokur, P.S. and Boesch, H.E. Jr., (1980). Interface-state generation in radiation-hard oxides. *IEEE Transactions on Nuclear Science* 27 (6): 1647–1650.

94 Oldham, T. R., McLean, F. B., Boesch, H. E. Jr.,, and McGarrity, J. M. (1989). An Overview of Radiation-Induced Interface Traps in MOS Structures, Report No. HDL-TR-2163, U.S. Army Laboratory Command, Harry Diamond Laboratories, Adelphi, MD.

95 Rashkeev, S.N., Fleetwood, D.M., Schrimpf, R.D., and Pantelides, S.T. (2001). Defect generation by hydrogen at the Si–SiO_2 interface. *Physical Review Letters* 87 (16): 165506.

96 Tuttle, B.R., Hughart, D.R., Schrimpf, R.D. et al. (2010). Defect interactions of H2 in SiO$_2$: implications for ELDRS and latent Interface trap buildup. *IEEE Transactions on Nuclear Science* 57 (6): 3046–3053.

97 Enlow, E.W., Pease, R.L., Combs, W. et al. (1991). Response of advanced bipolar processes to ionizing radiation. *IEEE Transactions on Nuclear Science* 39 (4): 1342–1351.

98 Fleetwood, D.M., Kosier, S.L., Nowlin, R.N. et al. (1994). Physical mechanisms contributing to enhanced bipolar gain degradation at low dose rates. *IEEE Transactions on Nuclear Science* 41 (6): 1871–1883.

99 Fleetwood, D.M., Riewe, L.C., and Schwank, J.R. (1996). Radiation effects at low electric fields in thermal, SIMOX, and bipolar-base oxides. *IEEE Transactions on Nuclear Science* 43 (6): 2537–2546.

100 Witczak, S.C., Lacoe, R.C., Mayer, D.C. et al. (1998). Space charge limited degradation of bipolar oxides at low electric fields. *IEEE Transactions on Nuclear Science* 45 (6): 2339–2351.

101 Rashkeev, S.N., Cirba, C.R., Fleetwood, D.M. et al. (2002). Physical model for enhanced interface-trap formation at low dose rates. *IEEE Transactions on Nuclear Science* 49 (6): 2650–2655.

102 Boch, J., Saigne, F., Schrimpf, R.D. et al. (2006). Physical model for the low-dose-rate effect in bipolar devices. *IEEE Transactions on Nuclear Science* 53 (6): 3655–3660.

103 Hjalmarson, H.P., Pease, R.L., Witczak, S.C. et al. (2003). Mechanisms for radiation dose-rate sensitivity of bipolar transistors. *IEEE Transactions on Nuclear Science* 50 (6): 1901–1909.

104 Fleetwood, D.M., Schrimpf, R.D., Pantelides, S.T. et al. (2008). Electron capture, hydrogen release, and enhanced gain degradation in linear bipolar devices. *IEEE Transactions on Nuclear Science* 55 (6): 2986–2991.

105 Pease, R.L., Platteter, D.G., Dunham, G.W. et al. (2004). Characterization of enhanced low dose rate sensitivity (ELDRS) effects using gated lateral PNP transistor structures. *IEEE Transactions on Nuclear Science* 51 (6): 3773–3780.

106 Shaneyfelt, M.R., Pease, R.L., Schwank, J.R., and al, e. (2002). Impact of passivation layers on enhanced low-dose rate sensitivity and thermal-stress effects in linear bipolar Ics. *IEEE Transactions on Nuclear Science* 49 (6): 3171–3179.

107 Seiler, J.E., Platteter, D.G., Dunham, G.W. et al. (2004). Effects of passivation on the enhanced low dose rate sensitivity of National LM124 operational amplifiers. *Proc. IEEE Radiation Effects Data Workshop Record*: 42.

108 Shaneyfelt, M.R., Witczak, J.R., Schwank, J.R. et al. (2000). Thermal-stress effects on enhanced low dose rate sensitivity in linear bipolar ICs. *IEEE Transactions on Nuclear Science* 47 (6): 2539–2545.

109 Chen, X.J., Barnaby, H., Vermeire, B. et al. (2007). Mechanisms of enhanced radiation-induced degradation due to excess molecular hydrogen in bipolar oxides. *IEEE Transactions on Nuclear Science* 54 (6): 1913–1919.

110 Pease, R.L., Platteter, D.G., Dunham, G.W. et al. (2007). The effects of hydrogen in hermetically sealed packages on the total dose and dose rate response of bipolar linear circuits. *IEEE Transactions on Nuclear Science* 54 (6): 2168–2173.

111 Pease, R.L., Adell, P.C., Rax, B.G. et al. (2008). The effects of hydrogen on the enhanced low dose rate sensitivity (ELDRS) of bipolar linear circuits. *IEEE Transactions on Nuclear Science* 55 (6): 3169–3173.

112 Messenger, G.C. and Ash, M.S. (1986). *The Effects of Radiation on Electronic Systems*, 162. New York: Van Nostrand Reinhold.

113 Buehler, M.G. (1968). Design curves for predicting fast-neutron-induced resistivity changes in silicon. *Proceedings of the IEEE* 56 (10): 1741–1743.

114 Irvin, J.C. (1962). Resistivity of bulk silicon and of diffused layers in silicon. *The Bell System Technical Journal* 41 (2): 387–410.

4

Radiation-Induced Single Events

4.1 Introduction – Single-Events Effects (SEE)

This chapter discusses both nondestructive and destructive single-event effects (SEEs) [1–75]. Nondestructive events can change the state of a semiconductor device, or circuit but do not lead to a functional failure. Nondestructive events can be classified as the following (Figure 4.1):

- Single-event upsets (SEUs) [1]
- Multiple-bit upsets (MBUs)
- Single-event functional interrupts (SEFIs) [2]
- Single-event transients (SETs) [4, 5]
- Single-event disturb (SED) [4, 5]

Destructive SEEs can change the state of a semiconductor device, or circuit, or system and does lead to a functional failure. Destructive events can be classified as the following (Figure 4.2):

- Single-event snapback (SESB) [1, 10–14]
- Single-event latchup (SEL) [76–83]
- Single-event gate rupture (SEGR) [6–9]
- Single-event burnout (SEB) [10–15]

4.1.1 Single-Event Upsets (SEU)

SEU is an upset of a device, circuit, or system where the state is changed in a nondestructive process [1]. The most commonly discussed SEUs are associated with memory circuits. In more recent times, SEUs are a concern in latch networks in logic circuitry. SEUs can occur in dynamic read access memory (DRAM), and static read access memory (SRAM) circuitry (Figure 4.3).

4.1.2 Multiple-Bit Upset (MBU)

MBUs is an upset of a device, circuit, or system where the state is changed in a nondestructive process where more than one element is involved. A single-particle can influence more than one device or circuit (Figure 4.4). The most commonly discussed MBUs are associated with memory circuits. MBUs can occur in DRAM, and static read access memory (SRAM) circuitry.

Integrated Circuit Design for Radiation Environments, First Edition.
Stephen J. Gaul, Nicolaas van Vonno, Steven H. Voldman and Wesley H. Morris.
© 2020 John Wiley & Sons Ltd. Published 2020 by John Wiley & Sons Ltd.

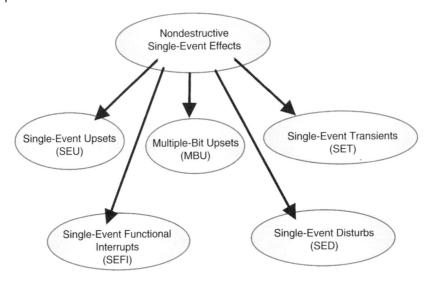

Figure 4.1 Nondestructive single-event effects.

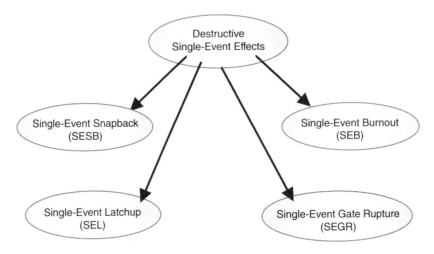

Figure 4.2 Destructive single-event effects.

4.1.3 Single-Event Transients (SET)

SETs is an upset of a device, circuit, or system where the state is changed due to an impulse response of a given amplitude and duration in a nondestructive process (Figure 4.5). SET can occur in analog and mixed signal circuitry [4, 5].

4.1.4 Single-Event Functional Interrupts (SEFIs)

An SEFI is an upset of a device, circuit, or system where there is a corruption of a data path leading to a loss of operation, where the loss of operation is a nondestructive

Figure 4.3 Single-event upset.

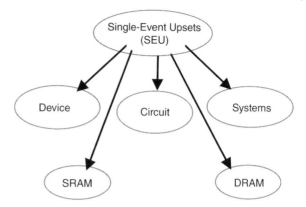

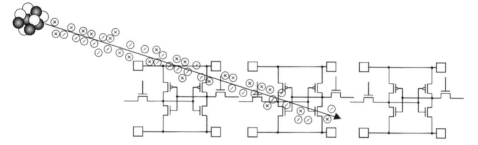

Figure 4.4 Multiple-bit upset (MBU).

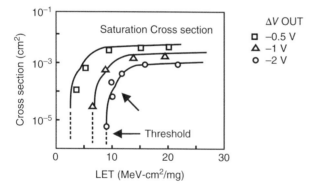

Figure 4.5 Single-event transient.

process [2]. SEFIs can occur in complex circuitry, such as state machines and control systems.

4.1.5 Single-Event Disturb (SED)

SEDs is an upset of a bit where the information stored in the bit is changed in a non-destructive process. SEDs can occur in combinational logic and latches within logic devices.

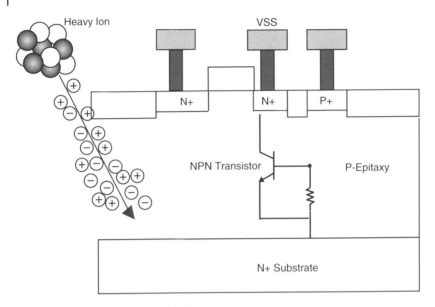

Figure 4.6 Single-event snapback (SESB).

4.1.6 Single-Event Snapback (SESB)

SESB is snapback of a device, where the device undergoes a transition from a high-voltage/low-current state to a low-voltage/high-current state where the transition was triggered by a single-event particle [1, 10–14]. SESB can be destructive. SESB can occur in a bipolar junction transistor (BJT), a heterojunction bipolar transistor (HBT), or an n-channel metal-oxide semiconductor field-effect transistor (MOSFET) device (Figure 4.6). Single-event snapback, similar to SEL but not requiring the PNPN structure, can be induced in n-channel MOS transistors switching large currents, when an ion hits near the drain junction and causes avalanche multiplication of the charge carriers. The transistor then opens and stays opened. Hard errors are irreversible.

4.1.7 Single-Event Latchup (SEL)

SEL is a latchup associated with an inherent or parasitic pnpn structure where the device undergoes a transition from a high-voltage/low-current state to a low-voltage/high-current state, where the transition was triggered by a single-event particle [76–83]. SEL can be destructive if the structure approaches thermal breakdown. SEL can occur in a parasitic BJT formed from a p-channel MOSFET device and a n-channel MOSFET device (Figure 4.7). SEL can occur intracircuit or intercircuit.

4.1.8 Single-Event Burnout (SEB)

SEB is associated with a device, circuit, or system that undergoes destructive burnout triggered by a single-event particle [10–15]. SEB is destructive due to the high current conditions. SEB can occur in a parasitic BJT or an n-channel power MOSFET device [10–15].

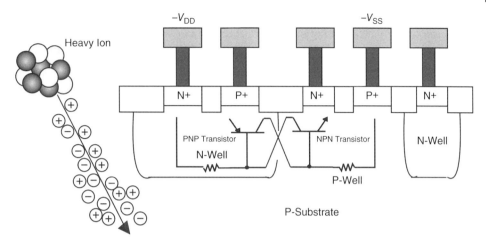

Figure 4.7 Single-event latchup.

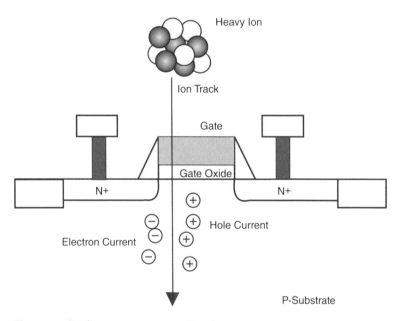

Figure 4.8 Single-event gate rupture (SEGR).

4.1.9 Single-Event Gate Rupture (SEGR)

SEGR is failure of a device, circuit, or system associated with a gate dielectric rupture due to high field conditions, where the rupture was triggered by a single-event particle [6–9]. SEGR can occur in a parasitic BJT, a power MOSFET (e.g. LDMOS, or DeMOS), and other integrated circuits (Figure 4.8).

4.1.10 Single-Event Hard Errors (SHE)

Single-event hard errors (SHEs) a hard error where the transition was triggered by a single-event particle [6–16]. SHE is an upset of a device, circuit, or system where the state is changed in a destructive process. The most commonly discussed SHEs are associated with memory circuits. In more recent times, SHEs are a concern in latch networks in logic circuitry. SHEs can occur in DRAM, and static read access memory (SRAM) circuitry.

4.2 Single-Event Upset (SEU)

SEUs exist in semiconductor memory devices. Figure 4.9 highlights the type of memory that SEUs occur. SEU can occur in both CMOS and bipolar memory cells (Figure 4.9). CMOS SEU occur in DRAM and SRAM semiconductor cells [31]. Bipolar technology soft errors occur in SRAM cells [25–30].

4.2.1 SEU – Memory

SEUs exist in semiconductor memory devices. SEU can occur in both CMOS and bipolar memory cells. CMOS SEU occurs in DRAM and SRAM semiconductor cells. Bipolar technology soft errors occur in SRAM cells.

4.2.2 SEU in CMOS Memory

CMOS SEU occurs in DRAM semiconductor cells [17, 19–23]. Soft errors from alpha particles in DRAMs were first discovered by Intel by May and Woods, published at the International Reliability Physics Symposium in 1979 [17]. This was a result of the ceramic substrate from ceramics mined from a source with high uranium and thorium concentrations.

4.2.3 SEU in Bipolar Memory

SEUs can occur in bipolar SRAM cells [25–30]. At one time, there was a belief that bipolar SRAM cells were immune to SEUs because it was a static RAM. Early bipolar SRAM cells had a critical charge of 500–2000 fC. With the scaling of bipolar SRAMs, the cell size and critical charge were reduced. There was no evidence of bipolar SRAM failures in this period.

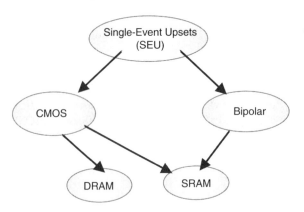

Figure 4.9 Single-event upsets in technology types.

In late 1982, it was discovered in IBM Burlington, Vermont, that bipolar SRAMs were not immune to alpha particle events [25–30]. A bipolar SRAM whose critical charge was 100 fC was placed under an alpha particle source, leading to a significant number of soft error fails per minute. This discovery led to the beginning of the evaluation of bipolar SRAM chips to SEUs. To address this issue, experimental work, circuit analysis, and alpha particle soft error rate (SER) simulators were initiated for bipolar memory [25–30]. Critical charge analysis was initiated by R. Flaker and S. Voldman in IBM. Alpha particle experimental work was completed in IBM Burlington [25–30]. Alpha particle simulation development was initiated by G. Sai-Halasz, and S. Voldman [19, 25–29].

By 1985, it was discovered that system-level failures in IBM computers were occurring due to terrestrial cosmic rays. Systems at higher altitude had a higher frequency of failure than sea-level systems. Ten machines in Denver, Colorado, had a failure rate that was four times the expected failure rate. At this time, the IBM computer memory in the high-speed cache was bipolar SRAM chips. This discovery lead to the SEU accelerated testing at particle accelerators.

4.2.4 SEU in CMOS SRAM

The focus on SER on bipolar SRAMs led to the development of CMOS SRAM SER simulation and testing [31]. A CMOS 6-D SRAM SER simulation was developed in IBM by S. Voldman in 1985 to compare the failure rates of bipolar SRAMs, to CMOS SRAMs [31].

4.2.5 SEU in Future Technology – FINFETs

SEUs will occur in future CMOS technologies as they are scaled from bulk planar transistors to three-dimensional (3-D) CMOS FinFET devices. FinFETs are being introduced in sub-20 nm technologies.

4.3 SEU – Particle Sources

Different particle sources can lead to SEUs. The sources of radiation come from radioactive decay and cosmic rays [18]. Alpha particles are from radioactive decay known as the uranium series and thorium series.

Cosmic rays can be grouped into classes of primary cosmic rays, which subdivide into solar particles and galactic particles [18, 42]. These can further classified as cascade cosmic rays and sea-level cosmic rays. Cascade cosmic rays are the intermediate flux within the atmosphere. Sea-level cosmic rays are the final terrestrial flux of particles.

The incident primary particle will lead to a cascade of particles. These include pions, muons, protons, neutrons, electrons, and positrons, as well as gamma radiation [42]. The electromagnetic or "soft component" can include pions, electrons, positrons, and gamma radiation. Meson or "hard component" can include pions and muons. The nucleonic component can include both high energy nucleons, and disintegration product nucleons. The disintegration products can include neutrons and protons.

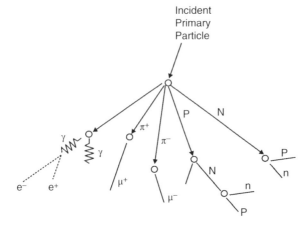

Incident
Primary
Particle

Figure 4.10 Single-event upsets particle source.

These sources can include the following (Figure 4.10):

- Alpha particles
- Muons
- Pions [41]
- Electrons
- Positrons
- Neutrons
- Protons
- Heavy ions

These particles sources can be from natural radioactive decay to cosmic ray events. Alpha particles are from radioactive decay, whereas muons, neutrons, and protons are from cosmic ray sources. Heavy ions are also found in space environments.

4.3.1 SEU Source – Alpha Particles

When the alpha particle traverses silicon, energy is absorbed by silicon, leading to the generation of electron−hole pairs (EHPs) along the alpha particle track [17, 19–23]. In semiconductor materials, the amount of energy to generate an EHP is typically three to four times the bandgap of the material. In silicon, one EHP is formed for every 3.5 eVeV. As the alpha particle loses energy, the EHP generation increases. At the beginning of its track, it generates approximately on the order of 5000 EHPs, which increases to 15 000 EHPs at the end of its track. As a result, the end of the alpha particle track has a generation approximately three times that of the beginning of the alpha track. The range of the alpha particle in silicon can be expressed as [25–30]:

$$R(E) = 1.53E1.67$$

where $R(E)$ is the range in microns, and E is the energy in MeV. The energy range of alpha particle in the uranium series (Figure 4.11) and thorium series (Figure 4.12) are less than 8.8 MeV. An 8.8 MeV alpha particle can traverse approximately 80 μm in a silicon wafer.

The amount of charge can be calculated from the energy of the alpha particle. Since it is known that a single-electron equal 1.6×10^{-19} fC, it can be shown that an alpha particle

Figure 4.11 Uranium series.

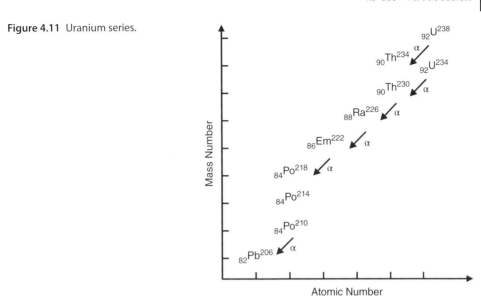

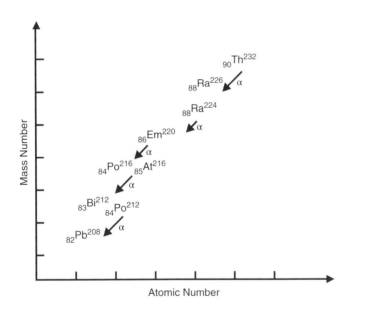

Figure 4.12 Thorium series.

can generate on the order of 400 fC of charge. From this charge, some EHPs recombine and some are collected at the metallurgical junction. The current time constant is a function of the charge collected by the electric field in the metallurgical junction, and the time it to diffuse to the metallurgical junctions.

As the EHPs are generated, the randomly diffuse from the particle track in the silicon through diffusion. EHPs can recombine, or become collected at metallurgical junctions. Figure 4.13 shows an example of an alpha particle track penetrating the silicon devices, highlighting the EHP generation.

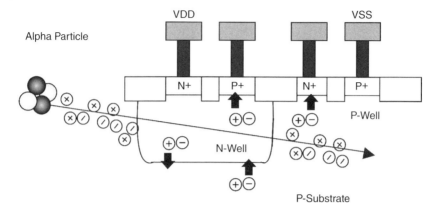

Figure 4.13 Electron hole pair charge transfer from alpha particle events.

The time response and the total charge collected is a function of alpha particle characteristics as well as the circuit layout that it traverses. Alpha particles are emitted in random angles of incident relative to the silicon surface. As a result, the collection process of the carriers is a function of the emission energy, initial point of the emission, and the angle of incident. For vertical angles of incident, the carrier generation occurs deep in the substrate region. EHPs generated deep in the substrate have a higher probability of recombining or diffusing radially outward. In heavily doped substrates, the EHPs recombine prior to reaching the silicon surface devices. In lightly doped substrates, the charge generation influences the electrostatic potentials, forming an effect known as the "funneling effect." In the funneling effect, the electric field generated by the EHP, distorts the n-well to substrate depletion region, leading to a larger number of carriers collected by the electric drift field. The funneling effect is observed as a fast collection process followed by a slower diffusion

Alpha particles can initiate SEUs in memory cells [19–23]. Figure 4.14 illustrates a circuit schematic highlighting the charge transfer processes with EHP generation.

4.3.2 SEU Source – Pions and Muons

Cosmic ray pions are present at sea level. Pions have a mass of 135 MeV, and a lifetime of approximately 26 ns. The pion flux rate is 450 pions/cm²-year. Pions undergo "pion capture" where negative-charged pions interact with a nucleus, entering an orbital. The pion interacts with a nucleon, leading to nuclear fission (Figure 4.15). The nuclear fission event of pion capture leads to a release of 22 nC [42].

Cosmic-ray muons are present at sea level. Similar to the pion, the muon undergoes "muon-capture" at a rate of 510 cm³/year. Muons can introduced soft errors in semiconductor chips [42].

Pion production occurs from neutrons generated by cosmic rays in the upper atmosphere. Pions have rest masses about 140 MeV, and they interact strongly with the nucleons in the atomic nuclei. If a pion is created within a heavy nucleus and immediately absorbed, the rest mass of about 140 MeV would be available for the continued breakup of the target nucleus and eventually result in a dramatic increase in the localized energy deposition. However, if the pion escapes from the target nucleus with momentum p, that

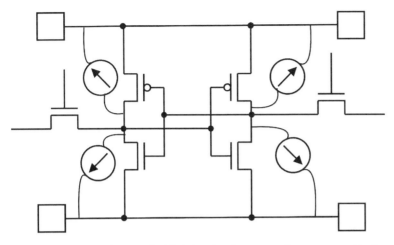

Figure 4.14 Circuit schematic highlighting the charge transfer processes with electron–hole pair generation.

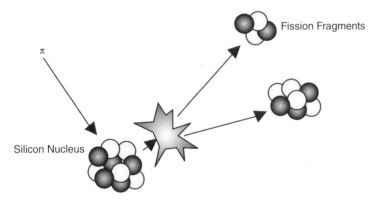

Figure 4.15 Pion capture.

momentum of the target nucleus and the resulting localized energy deposition will also be increased, but by a much smaller amount.

4.3.3 SEU – Neutrons

Neutrons can lead to SEUs in the atmosphere, or at sea level. At sea level, the neutron flux has a wide energy spectrum from low energy (e.g. 0.01 eV) to 1000 MeV. There are three peaks of neutrons in the spectrum, a low-energy thermal neutron peak population (e.g. 0.01–0.2 eV), spallation neutron population (e.g. 0.1–5 MeV), and high-energy neutron population (e.g. 10–1000 MeV) [42]. Figure 4.16 illustrates a neutron spallation event.

4.3.4 SEU Source – Protons

Protons can generate SEUs in space and terrestrial. Cosmic ray protons lose energy as they enter the earth's atmosphere. The cosmic ray protons lose energy because of their

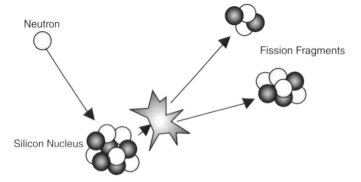

Figure 4.16 Neutrons.

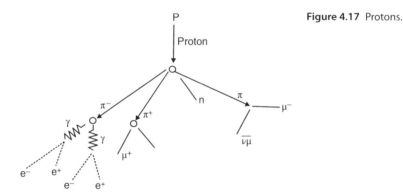

Figure 4.17 Protons.

charged nature. Protons that enter the atmosphere generate a number of species that lead to single-upset events (Figure 4.17).

Protons typically do not cause an upset through direct ionization, but through direct interaction with the nucleus of atoms, known as nuclear spallation. Spallation is a nuclear reaction in which fragments are generated from the target nucleus. Nuclear fragments are the result of the nuclei ion recoil. The spallation nuclei ions can lead to SEU [42].

4.3.5 SEU – Heavy Ions

Heavy ions occur in galactic and solar cosmic rays that can lead to SEUs in the atmosphere. Heavy ions can be incident primary particles, fission fragments, and disintegration product nucleons [42]. Figure 4.18 shows a heavy ion interacting with a semiconductor device.

4.4 Single-Event Gate Rupture (SEGR)

SEGR is a unique mechanism that introduces an ion track and failure mechanism. SEGR is one type of condition that occurs from SEB. SEB was observed by Waskiewicz, Groninger, Strahan, and Long from heavy ions [10]. In 1987, Fischer noted

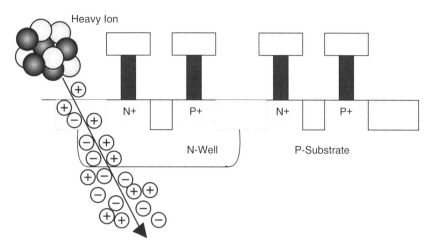

Figure 4.18 Heavy ions.

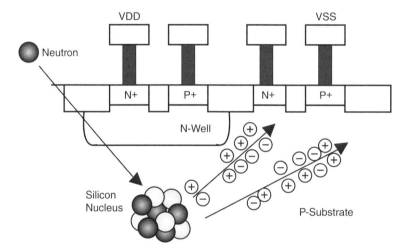

Figure 4.19 Single-event gate rupture (SEGR).

that heavy-ion induced gate rupture can occur in power MOSFETs [13]. Figure 4.19 illustrates an example of SEGR.

4.4.1 Definition SEGR

SEGR is failure of a device, circuit, or system associated with a gate dielectric rupture due to high field conditions latchup where the rupture was triggered by a single-event particle [6–15]. SEGR can occur in a parasitic BJT, a power MOSFET (e.g. LDMOS or DeMOS), and other integrated circuits.

4.4.2 SEGR Source – Ion Track

In the SEGR event, an ion track is formed associated with a charged particle. EHPs are formed in the silicon and the dielectric region.

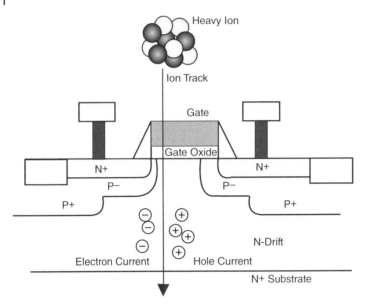

Figure 4.20 Power transistor SEGR.

4.4.3 SEGR Source – Failure Mechanism

SEGR is the formation of a conductive path through the gate dielectric of a MOSFET, or power transistor [6–9]. Destructive damage occurs in the gate dielectric region. In most SEUs, the damage is nondestructive. But, in this case, a destructive damage occurs.

4.4.4 SEGR – Modeling and Simulation

SEGR can be modeled and simulated to quantify the probability of failure. Johnson, Hohl, Schrimpf, and Galloway simulated single-event burnout in n-channel power devices [11]. Simulation can be performed experimentally using particle accelerators, or with semiconductor device simulation.

4.4.5 Power Transistors and SEGR

SEGR occurs in MOSFET power transistors (Figure 4.20) [6–15]. Power transistors are typically significantly larger than standard memory cells or logic circuits. As a result, there is a high probability of striking a power transistor, which must be modeled and simulated to quantify the probability of failure.

4.4.5.1 Lateral Power Transistors SEGR

SEGR can occur in lateral power transistors [6–9]. Lateral power transistor SEGR typically occurs from a vertical impinging particle event. The single-event particle penetrates the gate dielectric region forming a conduction path.

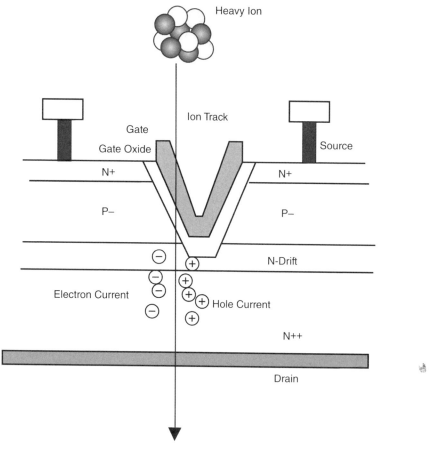

Figure 4.21 VMOS SEGR.

4.4.5.2 Vertical MOS (VMOS) SEGR

SEGR occurs in vertical MOS (VMOS) power transistors [16] (Figure 4.21). In a vertical MOS (VMOS) transistor, the structure has a trench region subtending into the substrate region. The VMOS gate is a vertical structure. A single-event particle can be a low-angle or lateral particle that establishes a conduction path through the vertical gate structure. The SEGR can also occur from a nuclear spallation event that initiates fission fragments in the substrate. J. Titus and C.F. Wheatley showed that SEGR and SEB can occur in vertical power MOSFETs [16].

4.4.5.3 Advanced Technologies – Planar MOSFET SEGR

SEGR will occur in future technologies. As the devices are scaled, the gate oxide and the area of the power MOSFET will be reduced. Yet, SEGR will still occur in sub-20 nm technologies.

4.4.5.3.1 *Advanced Technology – FinFET SEGR* SEGR will occur in sub-20 nm technologies. In these advanced technologies, the structures may be FinFET devices. FinFETs will be a plurality of MOSFET fingers. Due to the 3-D nature and plurality of the MOSFET

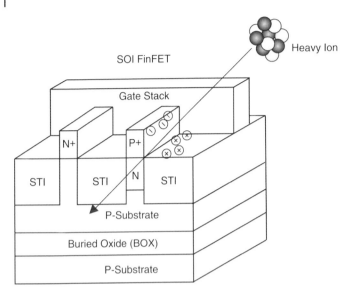

Figure 4.22 FinFET SEGR.

fingers, SEGR can occur as a particle penetrates through the MOSFET gate structure. Particle events that impinge parallel to the semiconductor substrate, can provide a conduction path in the gate dielectric. In a FinFET device, the incident particle, or spallation events can traverse many FinFET fingers that are in parallel with the surface of the wafer, or a low-angle particle (Figure 4.22).

4.5 Single-Event Transients (SETs)

SETs occurs in semiconductor chip circuitry in both digital and analog circuits. Single event from radiation particles may cause one or more voltage pulses that can impact timing of a circuit, and propagate through a circuit.

SET are transient events that may not upset the logic state of the digital circuit [4]. An SET event is not a SEU unless the state of the circuit is changed (e.g. logic 1 to logic 0). An SET may not lead to a change that results in an incorrect value of a latch circuit.

4.5.1 SET Definition

SETs is an upset of a device, circuit, or system where the state is changed due to an impulse response of a given amplitude and duration in a nondestructive process. SET can occur in analog and mixed signal circuitry [4].

4.5.2 SET Source

Single-event transients (SET) can occur in both CMOS and bipolar memory cells and logic circuits. An ionizing particle can introduce a transients in a memory cell or logic circuit leading to change of state, or a delay. The source can be an alpha particle, cosmic ray particle, or heavy ion.

4.5.3   SET Source Failure Mechanisms

SET can occur from both external transients or internal transients initiated by a ionizing particle, or a spallation event. Single-event transients can occur in analog and digital circuitry [4]. SET can occur in both CMOS and bipolar memory cells and logic circuits, causing failures. These failures can be soft failures or hard failures.

4.5.4   SET in Integrated Circuits

SET can cause system-level impacts on both digital and analog circuitry. In the following sections, example of digital and analog SET impacts are discussed.

4.5.4.1   Digital Circuitry
A transients on an internal signal can also occur in digital CMOS circuitry. For example, a transient signal can occur in a ring oscillator, or string of inverters. The SET can lead to erroneous data. A transients that disturbs a latch can lead to a sensing error or delay.

4.5.4.2   Continuous Time Analog Circuitry
A transients on an external signal can occur in an analog comparator circuit. In a comparator circuit, a differential signal induced by the particle event can lead to false triggering of the comparator circuit. This may affect subsequent circuits if not well filtered in the design. In a comparator circuit, the SET can change the pulse amplitude. The SET cross section (cm^2) is a function of the linear energy transfer (LET) (MeV-cm^2/mg).

4.5.5   Prediction and Hardening

Prediction of SET can be evaluated in circuit simulation or experimentally. Both can be used to provide prediction.

For circuit simulation, a current pulse can be applied to a digital or analog circuit. Simulation can provide insight into the sensitivity of the nodes and the timing delay. Simulation can also evaluate the propagation of the SET to subsequent circuit stages.

For experimental verification, particle accelerators can be used to evaluate the SET cross section (cm^2) as a function of the LET (MeV-cm^2/mg). Various particle sources can be used for experimental verification and sensitivity to source type and energies.

4.6   Single-Event Latchup (SEL)

SEL is a latchup associated with an inherent or parasitic pnpn structure, where the device undergoes a transition from a high-voltage/low-current state to a low-voltage/high-current state, where the transition was triggered by a single-event particle [76–83]. SEL can be destructive if the structure approaches thermal breakdown. SEL can occur in a parasitic BJTs formed from a p-channel MOSFET device and an n-channel MOSFET device. SEL can occur intracircuit, or intercircuit (Figure 4.23).

Alpha particles can also initiate CMOS latchup if the charge collected within the CMOS pnpn structure is enough to switch the parasitic pnpn state [76]. EHPs within the regenerative feedback loop can lead to switching. EHPs can be generated in the emitter,

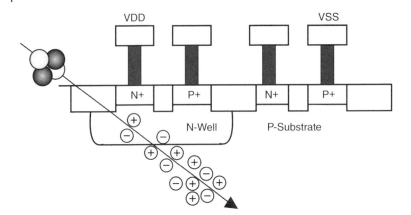

Figure 4.23 Single-event latchup (SEL).

base, and collector regions of the pnpn; this includes the p– substrate, the n-well, the p+ source/drain diffusion, and the n+ source/drain diffusion.

For the case of an EHP generated in the substrate, EHPs that diffuse to the n-well-to-substrate metallurgical junction, lead to the electron being collected in the n-well electrical node, and leaving the hole behind in the substrate. From a circuit perspective, the electron is pulling the n-well "low" (e.g. toward logical 0), and the hole is raising the p– substrate "high" (toward logical 1). In a second case, EHPs that diffuse to the n+ diffusion-to-substrate metallurgical junction, leads to electron collected by the pnpn cathode, and the hole remain in the local substrate region. From a circuit perspective, the electron is pulling the cathode "low" (e.g. toward logical 0), and the hole is raising the p– substrate "high" (toward logical 1) [76].

For the case of an EHP generated in the n-well, EHPs that diffuse to the n-well-to-substrate metallurgical junction, lead to the hole being collected in the n-well electrical node, and leaving the electron behind in the n-well. From a circuit or electrical potential perspective, the electron is pulling the n-well "low" (e.g. toward logical 0), and the hole is raising the p– substrate "high" (toward logical 1). For the case of an EHP generated in the n-well, EHPs can also diffuse to the p+ diffusion-n-well metallurgical junction; in this case, the hole is collected in the pnpn anode, and leaving the electron behind in the n-well. This case pulls the anode potential "high" and pulls the n-well potential "low." As a result, the electrical potential is perturbed in potentially all physical regions of the pnpn from the alpha particle event. For latchup to occur, the charge collected has to be enough to change the state of the pnpn structure.

4.6.1 SEL Definition

For SEL, an ionized particle or a spallation event generates enough charge to trigger a pnpn structure into a high current state.

4.6.2 SEL Source

The source of SEL can be from a cosmic ray, or a terrestrial source. SEL can be initiated by neutrons, protons, and heavy ions. In terrestrial applications, the source for SEL can be

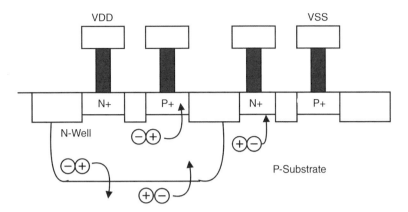

Figure 4.24 Cross section highlighting minority carrier transport.

radioactive decay particles such as alpha particles [76–83]. Figure 4.24 shows examples of the charge transport that can lead to latchup.

4.6.3 SEL Time Response

Using alpha particles as an example, the procedure to quantify the response can be understood. To understand the response of the circuit to alpha particle events, circuit simulation can be performed. To simplify the response of a circuit to an alpha particle, the waveform can be modeled as a two-state response. The first form is a rapid response associated with the charge collected in the metallurgical junction and the "funnel." Estimates of this time can be obtained analytically, or semiconductor device simulation. The alpha waveform can consist of a rapid pulse, followed by a decay characteristic associated with the diffusion time of the EHPs. R.R. Troutman simplified the problem by assuming the worst-case alpha particle is a Dirac delta function; in this fashion, analytical circuit analysis using Laplace transforms can be used for evaluation of the circuit response.

To evaluate the statistical probability that latchup can occur from an alpha particle source, Monte Carlo alpha particle simulators can be used for analysis. Monte Carlo alpha particle simulators can provide the computation of the probability that a given structure collects a given amount of charge for randomly emitted alpha emissions. This technique has been successfully employed for the evaluation of alpha particle SER for DRAMs and SRAM cells. These simulators undergo the following steps:

- Generate a sample layout of the physical design. Create a doping map, a isolation map, a region map, and a depletion field map.
- Choose the number of alpha emissions. Initiate one at a time.
- Provide a random choice of which alpha particle from the uranium or thorium discrete energy spectrum. Note that this can be also replaced with a lead spectrum of alpha energies.
- Choose a location of the alpha emission (in space).
- Choose a random radial angle.
- Choose a random vertical angle of incident.
- Assume all traversed materials are equivalent to silicon.

- Define the alpha track (based on the above information).
- Represent an EHP along the track, where the EHP density is associated with the generation rate as it slows down.
- Choose the number of representative EHPs (one thousand to one million EHPs are defined along the track).
- Allow the alpha track be the initial location of the EHPs.
- Follow each EHP in a random walk process consisting of a random directional step (3-D). After each step, determine if the carrier recombined, absorbed, or collected at a metallurgical junction. Follow each EHP packet along its random path
- Sum all the collected EHPs at the junctions or electrical nodes of interest.
- Repeat the process for the next alpha particle.
- At the last particle, the statistics of the probability distribution function (pdf) and the cumulative distribution function (cdf) are obtained.

Given the probability of charge collected is less than the charge to initiate latchup, no latchup will occur. Given the probability of the charge collected exceeds the charge needed to initiate latchup, CMOS latchup will occur. Hence, from the cumulative distribution, knowing the needed charge to flip the specific circuit, one can determine the probability of a latchup event from the alpha particle source. Note that this methodology can be applied to other SEE to evaluate SEL, where the method would be modified to the specific source event interaction and the LET.

4.6.4 SEL Maximum Charge Collection Evaluation in a Parallelepiped Region

To evaluate the maximum charge collection of an alpha particle through a structure that can initiate latchup, a simple analytical expression can be defined. Assume an analytical relationship of the range of a particle can be expressed as follows [26, 31]:

$$R(E) = aE^b$$

where $R(E)$ is the range of the alpha particle, E is the energy, and two constant parameters, a and b. Alpha particles can traverse a region, in silicon initiating EHPs along the track. We can idealize a semiconductor device as a parallelepiped region where within the parallelepiped region, the EHP generation occurs. This region can be a metallurgical junction, a diffusion, a well, or any region of interest. The region can be represented as a parallelepiped of width, W, length, L, and depth, t. If we assume that the alpha particle traverses the parallelepiped, the largest path through the parallelepiped is main diagonal. Since the maximum generation of EHP occurs at the end, we can assume the alpha particle track ends at the corner of the parallelepiped. The maximum alpha particle track in a parallelepiped region is then

$$R = (W^2 + L^2 + t^2)^{1/2}$$

For a particle whose energy is sufficient to traverse the region, and which ends at the corner of the parallelpiped, we can equate with the range-energy equation:

$$R = (W^2 + L^2 + t^2)^{1/2} = R(E) = aE^b$$

Solving for the alpha particle energy of this event:

$$E = \left\{ \frac{1}{a}(W^2 + L^2 + t^2)^{1/2} \right\}^{1/b}$$

Hence, the total energy absorbed in the medium to generate EHPs is the complete energy of the alpha particle. The number of EHPs, N_{EHP}, can be evaluated assuming it takes ε energy to create one EHP. Then, the number of EHPs in the parallelepiped is

$$N_{EHP} = n' = p' = \frac{E}{\varepsilon} = \frac{1}{\varepsilon}\left\{\frac{1}{a}(W^2 + L^2 + t^2)^{1/2}\right\}^{1/b}$$

The total charge of a given carrier (e.g., either the electron or the hole) can be calculated from the product of the charge per carrier, q, and the number of carriers, N_{EHP}.

$$Q = qN_{EHP} = qn' = qp' = \frac{qE}{\varepsilon} = \frac{q}{\varepsilon}\left\{\frac{1}{a}(W^2 + L^2 + t^2)^{1/2}\right\}^{1/b}$$

From this expression, the amount of charge generation (e.g. EHP) within the parallelepiped can be calculated.

$$R(E) = 1.53E^{1.67}$$

In silicon, for calculation, the coefficient $a = 1.53$, $b = 1.67$, $\varepsilon = 3.5\,\text{eV/EHP}$, and $1\,\text{eV} = 1.6 \times 10^{-19}\,\text{fC}$. From alpha particle emission, the highest energy alpha particle emission from the uranium and thorium series is 8.78 MeV. This is equivalent to a worst-case track length of approximately 80 µm and approximately 400 fC of electrons and 400 fC of holes. These are the maximum distance and charge levels generated by the worst-case alpha particle emission. As an example, given an n-well region whose length is 80 µm, it is possible that an 8.78 MeV alpha particle will traverse this region, collecting all of the electrons generated in the n-well-to-substrate metallurgical junction. In this case, this will lead to a 400 fC charge impulse to the circuit pulling the n-well toward ground potential. Additionally, this will induce a 400 fC charge impulse to the substrate potential toward V_{DD}. Given larger structures, this is the maximum charge collection that can be obtained. For smaller circuit elements, the alpha particle range exceeds the maximum parallelpiped distance, and the circuit will collect less charge.

In this discussion, there are a few assumptions. First, it is assumed all the charge is collected along the track. This is only true if there is no recombination in the region. Second, it assumes this represents one physical node. Third, in this simple worst-case analysis, it assumes all the collection is being collected at a given physical node.

In the pnpn structure, the charge collection can occur in the metallurgical junctions, or physical regions. The metallurgical junctions of concern are the p+/n-well, the n-well/p- substrate, and n+/p- substrate junctions. Electrons in the p+ diffusion can diffuse to the p+/n-well junction. Holes generated in the n-well can also diffuse to the p+/n-well metallurgical junction. Holes in the n-well can also diffuse to the n-well/p- substrate junction, and electrons in the p- substrate can diffuse to the n-well/substrate junction or the n- diffusion. In order for these alpha particle events to accumulate charge in the physical regions of the p+ diffusion, the n+ diffusion and the n-well, the angle of incident relative to the surface needs to be small. Low-angle incident alpha particles must be emitted through the inter-level dielectric (ILD) films and have energy enough to influence the semiconductor device. Because of the issue of the angle of incidence, although these alpha events can occur, the probability is low.

This chord length analysis can also be applied to other events using the LET and the chord length through a collection volume. In this process, the charge in the volume would be associated with the "collected charge" and compared to the "critical charge" to initiate SEL.

4.6.5 A SEL Design Practice

A CMOS SEL latchup design practice can be summarized as follows:

- Alpha particle induced latchup can be evaluated using hand calculations of the amount of charge collected along an alpha particle track, knowing the physical layout of interest.
- Worst-case analysis can be completed by assuming (i) 100% charge collection along the track (e.g. collected in a critical junction); (ii) the complete track fits within a given circuit layout; and (iii) collection time is the collection time across a depletion region (e.g. n-well junction) or a Dirac delta function.
- Alpha particle induced latchup can be evaluated using semiconductor simulators to evaluate the collected charge using semiconductor device simulators.
- Accurate alpha particle analysis that evaluates the statistics of a latchup event can be obtained using Monte Carlo alpha particle random walk simulators.

The time response, and the total charge collected, is a function of alpha particle characteristics as well as the circuit layout that it traverses. Alpha particles are emitted in random angles of incident relative to the silicon surface. As a result, the collection process of the carriers is a function of the emission energy, initial point of the emission, and the angle of incident. For vertical angles of incident, the carrier generation occurs deep in the substrate region. EHPs generated deep in the substrate have a higher probability of recombining or diffusing radially outward. In heavily doped substrates, the EHPs recombine prior to reaching the silicon surface devices. In lightly doped substrates, the charge generation influences the electrostatic potentials, forming the funneling effect. In the funneling effect, the electric field generated by the EHP distorts the n-well to substrate depletion region, leading to a larger number of carriers collected by the electric drift field. The funneling effect is observed as a fast collection process followed by a slower diffusion tail.

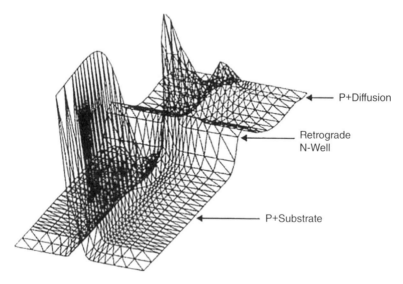

Figure 4.25 Semiconductor device simulation time sequence of an alpha particle through a CMOS process (e.g. profile is a retrograde n-well and p++ substrate).

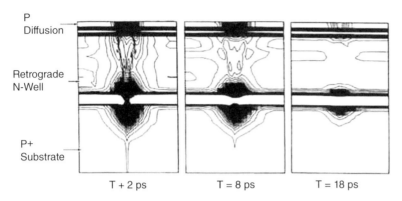

P Diffusion

Retrograde N-Well

P+ Substrate

T + 2 ps   T = 8 ps   T = 18 ps

Figure 4.26 Semiconductor device simulation of the electric potentials as an alpha particle penetrates through a CMOS device.

4.6.6 SEL Semiconductor Device Simulation

SEL can be evaluated using semiconductor device simulation. The simulator requires a particle source for the simulation. Alpha particle or other ionized particles can be initiated in a semiconductor device simulator. Figure 4.25 shows a sequence of the potential fields as an alpha particle penetrates vertically through a modern CMOS process. The structure consists of a p+ diffusion, a retrograde n-well, and a p+ substrate. This simulation results were achieved using a 3-D mesh in a 3-D device simulator (FIELDAY).

Figure 4.26 shows the electric potentials as an alpha particle penetrates through a CMOS profile.

4.7   Summary and Closing Comments

In this chapter, radiation-induced SEUs were discussed. The topics of SEUs, MBUs, SEFIs, SETs, SED, SESB,SEL, SEGR, and SEB were highlighted. The chapter focused on SEU in DRAM and SRAM memory.

In Chapter 5, radiation testing will be discussed. Radiation test for neutrons, protons, prompt gamma, and other particle sources will be highlighted. The chapter will also discuss sources, as well simulation and fidelity.

Problems

4.1   List all possible SEU phenomena that are nondestructive.

4.2   List all possible SEU phenomena that are destructive.

4.3   Explain SEU of a DRAM cell. Show the charge transfer that leads to SEU failure.

4.4   Explain SEU of a 6-D CMOS SRAM cell. Show the charge transfers that lead to a SEU failure.

4.5 Explain SEU of a 6-D CMOS SRAM cell on a circuit schematic.

4.6 Explain SEU of a bipolar SRAM cell. Show the charge transfers that lead to a SEU failure.

4.7 Explain SEU of a bipolar SRAM cell on a circuit schematic.

4.8 Explain the difference between charge transfers of a CMOS SRAM versus a bipolar SRAM.

4.9 In a bipolar SRAM, the charge transfers are doubled. Explain this issue.

4.10 Explain the difference between common mode charge collection versus differential mode charge collection in an SRAM cell. What leads to soft errors in an SRAM cell?

References

1 Dodd, P.E., Shaneyfelt, M.R., Schwank, J.R., and Hash, G.L. (2000). Single-event upset and snapback in silicon-on-insulator devices. In: *Proceedings of the IEEE Nuclear and Space Radiation Effects Conference (NSREC)*, 1–4. IEEE.
2 Koga, R., Penzin, S.H., Crawford, K.B., and Crain, W.R. (1997). Single-event functional interrupt (SEFI) sensitivity in micro-circuit. In: *Proceedings of the Radiation and its Effects on Components and Systems (RADECS)*, 311–318. IEEE.
3 Struesson, F. (2003) *Single-event effects (SEE) mechanisms and effects*, Short Course, Radiation and Its Effects on Components and Systems (RADECS).
4 Ferlet-Cavrois, V., Messengill, L.W., and Gouker, P. (2013). Single-event transients in digital CMOS – a review. *IEEE Transactions on Nuclear Science* 60 (3): 1767–1790.
5 Buchner, S., and McMorrow, D. (2005) *Single-event transient in linear integrated circuits*, Short Course, Proceedings of the Nuclear and Space Radiation Effects Conference (NSREC).
6 Allenspach, M., Brews, J.R., Galloway, K.F., and Johnson, G.H. (1996). SEGR: a unique failure mode for power MOSFETS in spacecraft. *Microelectronics Reliability* 36 (11/12): 1871–1874.
7 Diehl, S.E. (1984). A new class of single-event soft errors. *IEEE Transactions on Nuclear Science* 31 (6): 1145–1148.
8 Mouret, I., Allenspach, M., Schrimpf, R.D. et al. (1994). Temperature and angular dependence of substrate response in SEGR. *IEEE Transactions on Nuclear Science* 41 (6): 2216–2221.
9 Mouret, I., Calvet, M.C., Calvel, P. et al. (1995). Experimental evidence of the temperature and angular dependence in SEGR. In: *Proceedings of the Third European Conference on Radiation and its Effects on Components and Systems (RADECS)*, 313–321. IEEE.

10 Waskiewicz, A.E., Groninger, J.W., Strahan, V.H., and Long, D.M. (1986). Burnout of power MOS transistors with heavy ions of Californium-252. *IEEE Transactions on Nuclear Science* 33 (6): 1710–1713.

11 Johnson, G.H., Hohl, J.H., Schrimpf, R.D., and Galloway, K.D. (1993). Simulating single-event burnout in n-channel power MOSFETs. *IEEE Transactions on Electron Devices* 40 (2): 1001–1008.

12 Johnson, G.H., Schrimpf, R.D., Galloway, K.D., and Koga, R. (1992). Temperature dependence on single-event burnout in n-channel power MOSFETs. *IEEE Transactions on Nuclear Science* 39 (6): 1605–1612.

13 Fischer, T.A. (1987). Heavy-ion induced gate-rupture in power MOSFETs. *IEEE Transactions on Nuclear Science* 34 (6): 1786–1791.

14 Titus, J.L., Johnson, G.H., Schrimpf, R.D., and Galloway, K.D. (1991). Single-event burnout of power bipolar junction transistors. *IEEE Transactions on Nuclear Science* 38 (6): 1315–1322.

15 Swift, G.M., Padgett, D.J., and Johnston, A.H. (1994). A new class of single-event hard errors. *IEEE Transactions on Nuclear Science* 41 (6): 2043–2048.

16 Titus, J. and Wheatley, C.F. (1996). Experimental studies of a single-event gate rupture and burnout in vertical power MOSFETs. *IEEE Transactions on Nuclear Science* 43 (2): 533–545.

17 May, T.C. and Woods, M.H. (1979). Alpha-particle-induced soft errors in dynamic memories. *IEEE Transactions on Electron Devices* 26: 2.

18 Ziegler, J.F. and Lanford, W.A. (1979). The effect of cosmic rays on computer memories. *Science* 206: 776–788.

19 Sai-Halasz, G.A. and Wordeman, M.R. (1980). Monte Carlo modeling of the transport of ionizing radiation created carriers in integrated circuits. *IEEE Electron Device Letters* 1 (10): 211–213.

20 Hsieh, C.H., Murley, P.C., and O'Brien, R.R. (1981). A field-funneling effect on the collection of alpha particle generated carriers in silicon devices. *IEEE Electron Device Letters* 20 (4): 103–105.

21 Hu, C. (1982). Alpha particle-induced field and enhanced collection of carriers. *IEEE Electron Device Letters* 3 (2): 31–34.

22 Sai-Halasz, G.A., Wordeman, M.R., and Dennard, R.H. (1982). Alpha-particle-induced soft error rate in VLSI circuits. *IEEE Transactions on Electron Devices* 29: 725–731.

23 Sai-Halasz, G.A., Wordeman, M.R., and Dennard, R.H. (1982). Alpha particle induced soft error rate in VLSI circuits. *IEEE Journal of Solid-State Circuits* 17 (2): 355–361.

24 Diehl, S.E. (1982). Error analysis and prevention of cosmic ion-induced soft error in CMOS RAMs. *IEEE Transactions on Nuclear Science* 29: 2032–2039.

25 Sai-Halasz, G.A. and Tang, D.D. (1983). Soft errors rates in static bipolar RAMs. In: *International Electron Device Meeting (IEDM) Technical Digest, Session 20.7 1983*, 344–347. IEEE.

26 Voldman, S. and Patrick, L. (1983) Alpha particle induced single-event upsets in bipolar static emitter coupled logic (ECL) cells. *IEEE Nuclear and Space Radiation Effect Conference (NSREC)*, Boulder Colorado, 1983.

27 Voldman, S. and Patrick, L. (1984). Alpha particle induced single-event upsets in bipolar static emitter coupled logic (ECL) cells. Special issue on 1984 annual conference on nuclear and space radiation effects. *IEEE Transactions on Nuclear Science* 31 (6): 1196–1200.

28 Sai-Halasz, G.A. and Tang, D.D. (1983). Soft error rates in static bipolar RAMs. In: *Proceedings of the International Electron Device Meeting (IEDM)*, vol. 29, 344–347.

29 Voldman, S. and Patrick, L. (1984). Alpha particle induced single-event upsets in bipolar static ECL cells. *IEEE Transactions on Nuclear Science* 31 (6): 1196–1200.

30 Voldman, S., Patrick, L., and Wong, D. (1985). Alpha particle effects on bipolar emitter-coupled logic static arrays. In: *Proceedings of the International Solid State Circuit Conference (ISSCC), Vol. XXVII, 1985*, 262–263. IEEE.

31 Voldman, S., Corson, P., Patrick, L. et al. (1987). CMOS SRAM alpha particle modeling and experimental results. In: *Proceedings of the International Electron Device Meeting (IEDM) Technical Digest, Session 20.7 1987*, 518–521. IEEE.

32 Sai-Halasz, G.A. (1983). Cosmic ray induced soft error rate in VLSI circuits. *IEEE Electron Device Letters* 4 (6): 172–174.

33 Barth, J.E., Drake, C.E, Fifield, J.A. et al. (1992) Dynamic RAM with on-chip ECC and optimized bit and word redundancy, U.S. Patent No. 5,134,616, issued July 28th, 1992.

34 Zoutendyk, J. (1987). Single-event upset (SEU) in a DRAM with on-chip error correction code. *IEEE Transaction on Nuclear Science* 34 (6): 1310–1316.

35 NASA (1989). Internal correction of errors in a DRAM. In: *NASA Technical Briefs*, 30–31. NASA Jet Propulsion Laboratory.

36 Diehl, S.E., Vinson, J., Shafer, B.D., and Mnich, T.M. (1983). Considerations of single-event immune VLSI logic. *IEEE Transactions on Nuclear Science* 30: 4501–4507.

37 Chappell, B., Schuster, S.E., and Sai-Halasz, G.A. (1985). Stability of SER analysis of static RAM cells. *IEEE Transactions on Electron Devices* 32 (2): 463–470.

38 Idei, Y., Hanna, N., Nambu, H., and Sakurai, Y. (1991). Soft error rate characteristics in bipolar memory cells with small critical charge. *IEEE Transactions on Nuclear Science* 38 (6): 2465–2471.

39 McCall, D., Kroesen, P., Jones, F. et al. (1991). Alpha particle immune ECL latches. In: *Proceedings of the Bipolar Circuits and Technology Meeting (BCTM)*, 124–129.

40 Dooley, J.G., (1994) SEU-immune latch for gate array, standard cell, and other ASIC applications, U.S. Patent No. 5,311,070, issued May 10th, 1994.

41 Gelderloos, C.J., Peterson, R.J., Nelson, M.E., and Ziegler, J.F. (1997). Pion induced soft upsets in 16-Mbit DRAM chips. *IEEE Transactions on Nuclear Science* 44 (6): 2237–2242.

42 Ziegler, J.F. and Puchner, H. (2004). *SER-History, Trends, and Challenges.* Cypress Semiconductor Company.

43 Kobayashi, H., Shiraishik, K., and Tsuchiya, H. (2002). Soft error in SRAM devices induced by high energy neutrons, thermal neutrons, and alpha particles. In: *Proceedings of the International Electron Device Meeting (IEDM)*, 337–340.

44 Hazuchi, P., Karnik, T., and Walstra, S. (2003). Neutron soft error measurements in a 90-nm CMOS process and scaling trends in SRAM 0.25-μm to 90-nm generation. In: *Proceedings of the International Electron Device Meeting (IEDM)*, 523–526.

45 Binder, D., Smith, E.C., and Holman, A.B. (1975). Satellite anomalies from galactic cosmic rays. *IEEE Transactions on Nuclear Science* 22: 2675–2680.

46 Wyatt, R.C., McNulty, P.J., Toumbas, P. et al. (1979). Soft errors induced by energetic protons. *IEEE Transactions on Nuclear Science* 26: 4905–4910.

47 Guenzer, C.S., Wolicki, E.A., and Allas, R.G. (1979). Single-event upset of dynamic RAM's by neutrons and protons. *IEEE Transactions on Nuclear Science* 26: 5048–5086.

48 Baumann, R.C. (2005). Radiation-induced soft errors in advanced semiconductor technologies. *IEEE Transactions on Device and Materials Reliability* 5: 305–306.

49 Gordon, M.S., Goldhagen, P., Rodbell, K.P. et al. (2004). Measurement of the flux and energy spectrum of cosmic-ray induced neutrons on the ground. *IEEE Transactions on Nuclear Science* 51: 3427–3434.

50 Gover, J. E., Johnston, A. H., Halpin, J. and Rudie, N. J., *Radiation effects and systems hardening*, IEEE Nuclear and Space Radiation Effects Conference (NSREC) Short Course (1980).

51 Srour, J. R., Longmire, C. L., Raymond, J. P. and Allen, D. J., *Radiation effects and systems hardening*, IEEE Nuclear and Space Radiation Effects Conference (NSREC) Short Course (1982).

52 Gover, J. E., Rose, M. A., Tigner, J. E., Tasca et al., (1984) *Radiation effects and systems hardening*, IEEE Nuclear and Space Radiation Effects Conference (NSREC) Short Course.

53 Schwank, J.R., Shaneyfelt, M.R., Dodd, P.E. et al. (2009). Hardness assurance test guideline for qualifying devices for use in proton environments. *IEEE Transactions on Nuclear Science* 56: 3844–3940.

54 Schwank, J.R., Shaneyfelt, M.R., Baggio, J. et al. (2005). Effects of particle energy on proton-induced single-event latchup. *IEEE Transactions on Nuclear Science* 52: 2622–2629.

55 Enlow, E.W., Pease, R.L., Combs, W. et al. (1991). Response of advanced bipolar processes to ionizing radiation. *IEEE Transactions on Nuclear Science* 38: 1342–1351.

56 McClure, S., Pease, R.L., Will, W., and Perry, G. (1994). Dependence of total dose response of bipolar linear microcircuits on applied dose rate. *IEEE Transactions on Nuclear Science* 41: 2544–2549.

57 Witczak, S.C., Lacoe, R.C., Mayer, D.C. et al. (1998). Space charge limited degradation of bipolar oxides at low electric field. *IEEE Transactions on Nuclear Science* 45: 2339–2351.

58 Pickel, J.C. and Blandford, J.T. Jr., (1978). Cosmic ray induced errors in MOS memory cells. *IEEE Transactions on Nuclear Science* 25: 1161–1171.

59 Reed, R.A., Kinnison, J., Pickel, J.C. et al. (2003). Single-event effects ground testing and on-orbit rate prediction methods: the past, present, and future. *IEEE Transactions on Nuclear Science* 50: 622–634.

60 Schwank, J. R., Shaneyfelt, M. R. and Dodd, P. E., (2008) Radiation hardness assurance testing of microelectronic devices and integrated circuits: radiation environments, physical mechanisms and foundations for hardness assurance, *Sandia National Laboratories document* SAND-2008-6851P, 2101–2118.

61 Dodd, P.E. and Massengill, L.W. (2003). Basic mechanisms and modeling of single-event upset in digital microelectronics. *IEEE Transactions on Nuclear Science* 50: 583–602.

62 Sexton, F.W. (2003). Destructive single-event effects in semiconductor devices and ICs. *IEEE Transactions on Nuclear Science* 50: 603–621.

63 Massengill, L.W. (1996). Cosmic and terrestrial single-event radiation effects in dynamic random access memories. *IEEE Transactions on Nuclear Science* 43: 576–593.

64 Anton, T., Seichik, J., Joyner, K. et al. (1996). Direct measurement for SOI and bulk diodes of single-event upset charge collection from energetic ions and alpha particles. In: *Symposium on VLSI Technology*, 98–99.

65 Koga, R., Crain, W.R., Hansel, S.J. et al. (1995). Ion induced charge collection and SEU sensitivity of emitter coupled logic (ECL). *IEEE Transactions on Nuclear Science* 42: 1823–1828.

66 Reeds, R.A., Carts, M.A., Marshall, P.W. et al. (1997). Heavy ion and proton induced single-event multiple upsets. *IEEE Transactions on Nuclear Science* 44: 2224–2249.

67 Normand, E., Oberg, D.L., Wert, J.L. et al. (1994). Single-event upset and charge collection measurements using high energy protons and neutrons. *IEEE Transactions on Nuclear Science* 41: 2203–2209.

68 Miroshkin, V.V. and Tverskoy, M.G. (1998). A simple approach to SEU cross section evaluation of semiconductor memories. *IEEE Transactions on Nuclear Science* 45: 2884–2890.

69 Pickel, J.C. (1996). Single-event effects rate prediction. *IEEE Transactions on Nuclear Science* 43: 483–495.

70 Granlund, T., Granborn, B., and Olsson, N. (2003). Soft error rate increase for new generations of SRAM. *IEEE Transactions on Nuclear Science* 50: 2065–2080.

71 Tang, H.H. (1996). Nuclear physics of cosmic rays interaction with semiconductor materials. *IBM Journal of Research and Development* 40: 91–108.

72 Dirk, J.D., Nelson, M.E., Ziegler, J.F. et al. (2003). Terrestial thermal neutrons. *IEEE Transactions on Nuclear Science* 50: 2060–2064.

73 Baumann, R., Hossain, T., Smith, E. et al. (1995). Boron as a primary source of radiation in high density DRAMs. In: *Proceedings of the International Reliability Physics Symposium (IRPS)*, 297–302.

74 Baumann, R., Hossain, E., Murata, S., and Kitagawa, H. (1995). Boron compounds as a dominant source of alpha particles in semiconductor devices. In: *Proceedings of the International Reliability Physics Symposium (IRPS)*, 297–302. IEEE.

75 Baumann, R. and Smith, E.B. (2000). Neutron-induced boron fission as a major source of soft errors in deep submicron SRAM devices. In: *Proceedings of the International Reliability Physics Symposium (IRPS)*, 152–157. IEEE.

76 Voldman, S. (2007). *Latchup*. Wiley.

77 Morris, W. (2003). CMOS Latchup. In: *Proceedings of the International Reliability Physics Symposium (IRPS), May 2003*, 86–92.

78 Voldman, S., Gebreselasie, E., Liu, X.F. et al. (2005). The influence of high resistivity substrates on CMOS latchup robustness. In: *Proceedings of the Electrical Overstress/Electrostatic Discharge (EOS/ESD) Symposium, 2005*, 90–99. IEEE.

79 Voldman, S. (2005). A review of CMOS latchup and electrostatic discharge (ESD) in bipolar complimentary MOSFET (BiCMOS) silicon germanium technologies: part I-ESD. *Microelectronics and Reliability* 45: 323–340.

80 Voldman, S. (2005). A review of CMOS latchup and electrostatic discharge (ESD) in bipolar complimentary MOSFET (BiCMOS) silicon germanium technologies: part II-Latchup. *Microelectronics and Reliability* 45: 437–455.

81 Voldman, S. (2005). Latchup and the domino effect. In: *Proceedings of the International Reliability Physics Symposium (IRPS), 2005*, 145–156. IEEE.

82 Voldman, S. (2005). Cable discharge event and CMOS latchup. In: *Proceedings of the Taiwan Electrostatic Discharge Conference (T-ESDC)*, 23–28. IEEE.

83 Cottrell, P., Warley, S., Voldman, S. et al. (1988). N-well design for trench DRAM arrays. In: *Proceedings of International Electron Device Meeting (IEDM), December 1988*, 584–587. IEEE.

5

Radiation Testing

5.1 Introduction

This chapter provides a brief introduction to the various radiation environments that may be encountered by electronic systems and components. We discuss ionizing radiation, single-event effects (SEE) and proton effects, which are the three predominant natural space environments. We also briefly discuss man-made environments found in nuclear weapons scenarios. We further discuss radiation testing in the laboratory for these environments and the simulation fidelity issues arising from these methodologies.

5.1.1 Radiation Units and Measurements

We will lead off with a short discussion of the units involved in radiation effects. As in many other disciplines, this field has a somewhat convoluted system of units, with a good mixture of SI and non-SI units and of older and newer units, and we will restrict our comments to those in general use in radiation effects in electronics. Also, as in many disciplines, a good understanding of the units involved and of their relationships and derivation is essential to a clear understanding. We will discuss five categories of interest:

1. Radioactivity: the Curie, Becquerel, and Rutherford
2. Ionizing radiation exposure: the Röntgen
3. Absorbed ionizing radiation dose: the rad, Gray (Gy), and rem
4. Particle fluence and particle flux
5. Ionizing radiation dose equivalent: the Sievert

Activity is a measure of the radioactivity of a sample of a specific radioactive element or compound. A common older non-SI unit for activity is the Curie (Ci), defined as $1\,\text{Ci} = 3.7 \times 10^{10}$ decays per second, which is quite a large amount of activity. It can also be defined as the amount of radioactivity emitted each second by the decay of a gram of ^{226}Ra. A second, somewhat more modern activity unit is the Becquerel, a derived SI unit and a very small one: one nuclear decay per second, which is thus equivalent to $1\,\text{s}^{-1}$. The Becquerel was introduced in 1975 by the Conférence Générale des Poids et Mesures (General Conference on Weights and Measures), but the Curie and its derivatives the picocurie and microcurie are still in very widespread use in for example irradiator design and applications and in radiation physics. For completeness we should perhaps also mention the Rutherford, which equals 10^6 Bq or 10^6 decays/s but is not in general use.

Integrated Circuit Design for Radiation Environments, First Edition.
Stephen J. Gaul, Nicolaas van Vonno, Steven H. Voldman and Wesley H. Morris.

The main unit of *exposure* to ionizing radiation is the Röntgen, named after the discoverer of X-rays, Wilhelm Röntgen. The Röntgen is a legacy unit of radiation exposure dating back to the 1920s and is based on the ionization in air caused by X-rays and gamma rays. This unit is used in X-ray and gamma ray work, mainly in a medical context; both of these radiation types are photons. One Röntgen deposits 0.00877 Gy (0.877 rad or 87.7 ergs/g see below) of absorbed dose in dry air, or 0.0096 Gy (0.96 rad) in soft tissue. The Röntgen measures radiation exposure but its relationship to absorbed dose is not all that clearly defined. It lacks calibration to different materials, it is not accurate for alpha and beta radiation and it cannot be used for nonionizing radiation, such as, for example, neutrons. Its use in the field of radiation effects in electronics is hence minimal.

Absorbed dose is a dose quantity representing the energy imparted to a material per unit mass by ionizing radiation. The rad and the rem are the two main units used to measure radiation-absorbed dose. The historical non-Si unit of absorbed dose is the rad (radiation absorbed dose), which is specific to the material in which the energy is deposited in order to take stopping power differences into account. The linear stopping power is defined as the rate of energy loss per unit path length (dE/dx) of the charged particle. The mass stopping power is defined as the linear stopping power divided by the density of the absorbing medium:

$$\text{Stopping power} = \frac{1}{\rho}\frac{dE}{dx} \tag{5.1}$$

Dividing by the density of the absorbing material largely eliminates the dependence of the mass stopping power on mass density. Typical units for the linear and mass stopping powers are MeV/cm and MeV·cm^2/g, respectively, and ion beam SEE testing uses the mass stopping power expressed in MeV cm^2/mg, somewhat inaccurately called the linear energy transfer (LET).

A rad is equivalent to a deposited energy of 0.01 J/kg or 100 ergs/g in the absorbing material. The material of interest doing the absorbing must be specified; in integrated circuit work, we end up with the rad(Si) for the bulk material and the rad(SiO$_2$) for energy deposited in dielectric layers, although these two units differ only slightly and the rad(Si) dominates. In the SI system the unit of absorbed dose is the Gray, which is defined as 1 J/kg and which is hence 100 rad. As with the rad, the Gray is energy-specific, and the Gy(Si) is in common use in radiation effects. The Gray is considered the "correct" unit to use, but the rad still dominates in the US radiation effects community. Reluctance to the adoption of the Gray has even led to sporadic use of the informal *centigray*, which is 0.01 Gy, which of course (surprise!) is a rad. Since ionizing radiation damage in silicon devices is driven almost entirely by volume charging in the overlying silicon dioxide layers, a pretty good case can be made for using the rad (SiO$_2$) instead, and this is done by some leading research organizations.

The rem or "Röntgen equivalent [in] man" is an absorbed dose measurement that is based on the biological damage potential of different forms of radiation by using a quality (biological damage) factor Q. A rem is then defined as the product of the absorbed dose in rads multiplied by the quality factor Q, or

$$\text{rem} = \text{rad} \cdot Q \tag{5.2}$$

The quality factor Q varies with the type of radiation under consideration. For beta radiation, hard X-rays and gamma radiation, $Q = 1$; for alpha radiation, $Q = 20$; for neutron radiation, $Q = 5$, for soft X-rays of energy less than $10\,\text{keV}$, $Q = 15$ and for protons, $Q = 5$.

Particle flux is the number of particles per unit area per unit time, while fluence is the time integral of flux for a given irradiation time. As an example, in ion irradiation applications the flux is expressed as ions/cm^2 s and the fluence is simply in ions/cm^2. These units are important in SEE testing, in which samples are irradiated with energetic heavy ions or protons.

Dose equivalent is used mostly in health and safety settings, quantifying the health effects of low levels of ionizing radiation on the human body. The dose equivalent is related to the absorbed dose concept discussed above and also adds a measure of the biological effectiveness of the radiation, which is clearly a function of the specific type and energy of the radiation of interest. The SI unit in wide use is the Sievert (Sv). As an example, an alpha particle dose of one gray absorbed in tissue has a dose equivalent of 20 Sieverts, quantifying the enhanced biological damage done by the alpha particles. The dose equivalent of an absorbed gamma ray dose of 1 Gy is 1 Sv, and we note the wide range of relative damage potential for various types of radiation. This relationship can also be interpreted, as we saw above, as indicating that alpha particle radiation has a weighting factor of 20 as compared to gamma rays.

5.2 Radiation Testing and Sources

Testing integrated circuits and electronic systems in the actual end-use radiation environment would be ideal but is rarely convenient or economical, although limited on-orbit testing has been performed by missions such as the Combined Release and Radiation Effects Satellite (CRRES) [1] and Microelectronics and Photonics Test Bed (MPTB) [2] satellites. Because of the difficulty and high cost of on-orbit testing, many methods for "on the ground" simulation of space and other radiation environments have been developed. Some radiation sources such as gamma ray or X-ray irradiators as used for total ionizing dose (TID) testing are available commercially at manageable cost, so that corporate ownership and on-site usage are possible. Other sources, such as high-energy ion and proton beams, are only available at large research or medical facilities so that *beam time* must be scheduled and purchased by the testing organization and testing will be remote from that organization's facilities. Throughout our discussion, it will be important to recall that these tests are simulations of actual radiation environments and to understand the correlation between the simulated and actual use environments.

In this section, we discuss radiation-testing methodologies [3] for five commonly encountered radiation environments and some of the physics behind these tests. We also review the various types of radiation sources in common use for each one of these environments and show some typical examples of commercially available testing facilities or equipment. TID equipment typically uses either gamma rays or X-rays; while X-rays are generated electronically using an X-ray tube, gamma rays may be generated by a radioactive source or by high-energy accelerators. Ion beams used for simulating the effects of cosmic rays also use a particle accelerator to achieve the extremely high ion energy required for such testing. Neutron irradiation may be performed at nuclear facilities using a fast burst reactor (FBR) or may use a neutron generator to supply these particles. Proton testing is similar to heavy ion testing and requires a particle accelerator, while transient gamma testing requires a high-intensity pulsed gamma ray source.

5.2.1 Total Ionizing Dose (TID) Testing

The actual space environment consists almost entirely of protons, electrons, and heavy ions, but testing for this environment has historically been performed using gamma rays, which are energetic photons. This approach thus provides a simulation of the ionizing radiation effects of charged particles but does not address the displacement damage (DD) caused by these particles. The most common gamma ray source used in routine characterization and acceptance testing is ^{60}Co, which is a synthetic isotope of cobalt. Industry and government interest in simulation fidelity and combined environments is increasing, but testing experience and regulatory compliance with ^{60}Co goes back more than a half-century and correlation with actual environments has been well established. In Figure 5.1, we show the net positive charge yield in silicon dioxide for several species, including X-rays, protons, gamma rays, and alpha particles. The charge yield is a measure of the number of trapped, unrecombined holes left in the oxide, and is hence a measure of positive volume charging in that dielectric layer, which is the cause of most radiation damage in silicon integrated circuits.

So why use ^{60}Co at all? ^{60}Co gamma ray testing is a convenient, economical, and safe method, using highly penetrative photons that enable testing in any standard production

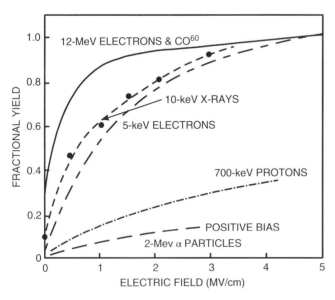

Figure 5.1 Net positive charge yield in silicon dioxide for 10 keV X-rays, 700 keV protons, ^{60}Co gamma rays, and 2 MeV alpha particles, plotting the fraction of unrecombined holes as a function of the electric field in the oxide. The charge yield determines the simulation fidelity for commonly used ^{60}Co and 10 keV X-ray irradiation sources. Source: McLean and Oldham 1987 [4], Figure 12, p. 28.

package without shielding issues. Photons do not activate fixturing or test articles, so no cooldown of samples is required, as is commonly encountered in proton and neutron testing, a critical attribute in revenue-critical lot acceptance testing work. Simulation fidelity issues must be understood by the experimenter in the context of the final application of the parts or system. In particular, it must be understood that energetic photon irradiation is being used to simulate particle radiation of electrons and protons. These last particles have both electric charge and mass, and each of these attributes will interact with a semiconductor device in different ways. The charge properties of the particle will drive ionization in insulating layers such as silicon dioxide (SiO_2), which will lead to net positive volume charging in these layers and to resulting changes in such device characteristics as the threshold voltage of a metal-oxide semiconductor (MOS) device. The massive properties of the particle will lead to an entirely different damage mechanism through displacement of semiconductor atoms from their positions in the single-crystal lattice, which will lead to a higher number of recombination centers, reduced minority carrier lifetime, and resulting changes in device characteristics such as the current gain (hFE) of a bipolar device. However, ^{60}Co simulation has been in use for some 60 years and has built up a strong set of correlations to actual particle irradiation, as well as being part of most radiation testing standards and procurement documents [5]. The TID specification for electronics is typically expressed as an end of life dose and is expressed in rad(Si) or krad(Si); a complete specification must state the dose rate as well, as we will see in subsequent discussions.

Alternatives to ^{60}Co sources have included electronically generated low-energy X-rays ranging perhaps between 10 keV and 60 keV, with dose rates that are generally much higher than ^{60}Co sources. This method was suitable for fast-turn wafer level testing and the Aracor Corporation built X-ray irradiators configured as a wafer probe station for many years. Low-energy X-ray irradiation further influences the simulation fidelity, as the charge yield is somewhat less (see Figure 5.1) than the corresponding ^{60}Co value. Another alternative is the use of a ^{137}Cs isotope source, which is a nuclear fission byproduct that emits gamma rays with a single energy peak at 662 keV; it has been used sporadically for gamma ray testing of electronics but is much more difficult to handle than ^{60}Co.

Next we explore the effects of the actual gamma ray dose rate. In 1991, researchers Ed Enlow and Ron Pease at Mission Research and Aerospace Corporation reported [6, 7] an unexpected increase in the parametric total dose degradation of an advanced bipolar technology at lower gamma ray dose rates; a rather counterintuitive result at the time, but quickly confirmed. This effect became known as enhanced low dose rate sensitivity (ELDRS) and has had a profound impact on bipolar integrated circuit hardening and hardness assurance. Prior to this time, TID qualification and acceptance testing had routinely been done per MIL-STD-883 Test Method 1019 test condition A [5], using a ^{60}Co source with a specified dose rate of 50–300 rad(Si)/s. The flaw in this approach was that the actual application dose rates (of proton and electron irradiation, to be sure) in space can be as low as a few millirad(Si)/s; *that's the real environment out there!* The arithmetic that follows is simple and inescapable: an acceptance test to 100 krad(Si) at 50–300 rad(Si)/s takes maybe an afternoon … and a test to 100 krad(Si) at 0.01 rad(Si)/s takes nearly four months! A simple rule of thumb for quick low-dose rate test duration approximation is 6 krad(Si)/wk. The 0.01 rad(Si)/s dose rate [5] has developed into a de facto low-dose rate standard, representing a middle ground between simulation fidelity

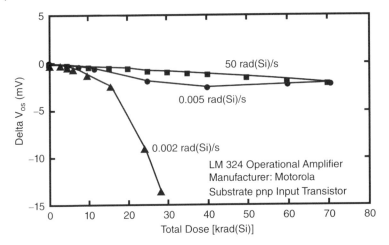

Figure 5.2 A classic set of curves: Where is the bottom? Input offset voltage degradation as a function of ^{60}Co gamma ray irradiation for a LM324 commercial operational amplifier at three dose rates, ranging from 50 to 0.002 rad(Si)/s. Source: Johnston et al. 1995, Figure 5, p. 1653. © 1995 IEEE. Reprinted, with permission, from [8].

and testing cost. The differences between high-dose rate and low-dose rate results can be dramatic. Figure 5.2 shows a classic set of TID response curves: the input offset voltage degradation as a function of ^{60}Co gamma ray irradiation for a commercial LM324 operational amplifier at three dose rates, ranging from 0.002 to 50 rad(Si)/s. The rapid degradation at 0.002 rad(Si)/s raises the question of just where the "bottom" of the low-dose rate response is in this part. Low-dose rate testing below the usual 0.010 rad(Si)/s has been proposed [9] sporadically, but the test times of such requirements would be entirely impractical and the idea has gotten little traction.

The space industry has historically performed TID acceptance testing on a lot-by-lot or wafer-by-wafer basis or on a characterization basis during product development at a high dose rate. Research results and the realization that a low-dose rate is the actual environment in space soon led to revised requirements with at least some low-dose rate data required for most programs. Due to the long duration of low-dose rate irradiations, the irradiator capacity requirements for supporting even a modest product line become substantial. One potential solution that has been advanced is the use of accelerated total dose testing as a means of bounding the low-dose rate response, and significant research has been carried out leading to innovative approaches such as high temperature irradiation at 100 °C [10] and switched-dose rate testing [11]. The drawbacks here are the requirements for detailed correlation studies before any approved production use of accelerated TID tests, which are rigorous enough to make accelerated testing potentially more expensive than actual low dose rate testing especially in the low-volume situations encountered in the space market. Also as a result of these advances the space market has shown what is perhaps decreasing interest in high dose rate tests for some applications. Note, though, that the coin has two sides: high-dose rate is (usually) best case for bipolars, but (also usually) worst case for complementary metal-oxide semiconductor (CMOS): there is no substitute for actual low-dose rate test data! Many current processes use both device types in BiCMOS flows supporting both MOS and bipolar devices, and both low- and high-dose rate testing are required on these parts [5]. Lot acceptance testing at low-dose rate is expensive in both cost and time, but has proven to be a strong value proposition due to customer preferences and a changing competitive situation.

The low-dose rate response of the bipolar junction transistor (BJT) is a true low-dose rate effect. MOS devices may show various time-dependent post-irradiation anneal [12] and rebound effects, also covered in MIL-STD-883, but these are not equivalent. A number of theoretical models for low-dose rate sensitivity (ELDRS) have been pursued by researchers in the field in an effort to resolve this phenomenon. One proposed model [13, 14], probably somewhat oversimplified here: high-dose-rate irradiation produces high local electric fields in the oxide, which assist in sweeping holes toward the interfaces. Charge yield is thus reduced at high dose rate due to increased local potential gradients leading to faster-moving positive carriers. At a low-dose rate, the electric field intensity diminishes and the holes move more slowly; an increased fraction gets trapped, enhancing the charge yield. This model correlates with extensive data indicating unbiased (all pins grounded) irradiation to be the worst-case condition at a low dose rate; applied bias sweeps the carriers out more rapidly. All oxides are different, though, and your results may vary; the authors caution against making assumptions, as in any other branch of engineering and science. Passivation composition [15, 16] (silicon nitride vs. silicon dioxide) and especially internal package ambient H_2 levels [17, 18] have been shown to play important roles as well. High-dose-rate testing may actually be interpreted as an excessively accelerated test, with limited application to the actual space environment but of great value in obtaining a quick assessment of radiation hardness assuming correlation data is available. Lot acceptance testing at low dose rate is much more expensive and schedule-intensive, but addressing the low dose rate issue in some form has proven to be unavoidable for nearly all space programs, and is hence equally unavoidable to the manufacturers of hardened semiconductors due to the competitive situation and strong customer preferences.

5.2.2 Total Ionizing Dose (TID) Sources

Moderate dose production TID testing of electronics is almost exclusively performed by the use of ^{60}Co as a source. ^{60}Co is a synthetic isotope of cobalt, which emits gamma rays with energy peaks at 1.17 and 1.33 MeV as a part of the gamma decay of the isotope. In gamma decay, a nucleus decays by the emission of an alpha or beta particle, and the remaining nucleus will be left in an excited state; it will then decay to a lower-energy state by emitting a gamma ray photon. There are other gamma ray sources such as ^{241}Am, in which alpha decay is also followed by gamma ray emission. The ^{241}Am gamma rays have an energy peak at 60 keV, which makes correlation difficult. ^{60}Co sources are available in a wide range of dose rates and irradiation facilities ranging from small research machines to large "panoramic" facilities using a concrete vault for shielding. The half-life of ^{60}Co is 5.27 years, which requires replacing the sources in these irradiators periodically in order to remain within the dose rate limits mandated by MIL-STD-883 Test Method 1019 [5]; this is the specification used by most US facilities and manufacturers, not just by military activities. The equivalent European Space Agency specification is ESA-ESCC-22900, which interestingly enough specifies a low dose rate of 0.10 rad (Si)/s.

The Gammacell 220^{TM} is a widely used commercial ^{60}Co irradiator initially manufactured by Atomic Energy of Canada Limited (now Nordion, Inc., Ottawa, Ontario, Canada). It is heavily shielded, enabling safe use in unshielded production environments. Figure 5.3 cross-sectional view of the irradiator with the sample chamber "drawer" in its closed and open positions. The irradiator consists of a 26 000 Ci nominal annular source of ^{60}Co "pencils" permanently enclosed within a lead shield and located at the bottom of the 6-in.-diameter cylindrical sample chamber of the irradiator. The stainless-steel holder contains 48 double-sealed source "pencils" set in an 8-in. annular diameter. The irradiator has a cylindrical elevator, and a drive mechanism is provided to move the elevator up or down in the irradiator chamber. For irradiation of electronic devices the elevator is provided with fixturing to enable electrical biasing of samples during irradiation, with the electrical connections introduced into the sample chamber through an access tube. The clearance between the chamber bore and the elevator diameter is tightly controlled to prevent excess radiation leakage. Fixtures of up to 6 in. in diameter and 8 in. high can be accommodated, and samples are routinely held and biased using standardized form factor PC boards. The Gammacell 220 product line was discontinued in 2008 but the irradiator had been produced in large numbers and remains a pervasive piece of equipment in many industries; an aftermarket service industry is well in place to perform recharging and calibration services. Figure 5.4 shows preproduction drawings from

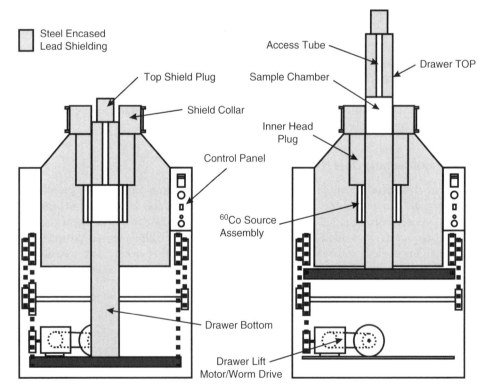

Figure 5.3 A Gammacell 220^{TM} ^{60}Co gamma ray irradiator built by Atomic Energy of Canada Limited (AECL, Ottawa, Ontario, Canada). The ^{60}Co gamma ray source surrounds the 6-in.-diameter irradiator chamber when the drawer is lowered to the closed position.

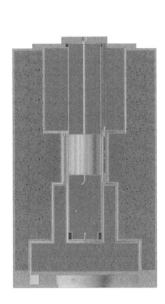 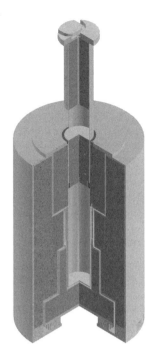

Figure 5.4 Preproduction drawings of the Hopewell Designs, Inc. Model GR420 high-dose-rate self-contained research irradiator. The design combines the capabilities and features of legacy systems but provides updates including improved dose uniformity, computer control, and the ability to service the source pencils on-site. Source: © 2018 Hopewell Designs, Inc. Reprinted with permission from Hopewell Designs, Inc., Alpharetta GA.

Hopewell Designs (Alpharetta, Georgia, USA) GR420 irradiator, designed as a replacement for legacy irradiators and available in late 2019. As seen in the cross-sectional and vertical section views, the irradiator is very similar to legacy designs but is capable of being resourced on-site.

At the low end of the dose-rate range we find compact, limited capacity R&D irradiators, such as the J.L. Shepherd and Associates Model 484, and high-capacity vault-type low-dose rate irradiators for production use. As we will see, these vault-type irradiators are indeed large facilities, typically housed in a concrete enclosure with inside dimensions of up to 15 × 15 ft. This very large capacity is an inescapable result of the arithmetic of low-dose-rate testing, as discussed previously. In Figure 5.5 we show a Hopewell Designs (Alpharetta, Georgia) model N40 low-scatter ^{60}Co irradiator modified for radiation effects in electronics work. This particular irradiator is in production use at Intersil, a Renesas Company. The base N40 provides a 360° "panoramic" gamma ray field and is predominantly used for calibration of radiation detection instruments and personnel dosimetry. The source can be readily replaced as its activity falls below the specified dose rate minimum. The 40 Ci "pop-up" ^{60}Co source is installed into a machined aluminum holder (colloquially known as the *rabbit*), which rides in a 2-in. stainless steel tube, which is, in turn, enclosed in a 12-in. diameter pipe that extends 10 ft below grade. The source holder is moved from the storage position (well below grade) to the expose position at

Figure 5.5 The fixturing section of a modified Hopewell Designs N40 vault-type low-dose-rate irradiator, with the 40 Ci ^{60}Co "pop-up" source at the center and test board holders surrounding the source in a spherical isodose arrangement, ensuring uniform exposure at all test positions. All test boards are contained in PbAl spectrum hardening filter boxes (black). The 16 test board holders are curved (obscured by the PbAl shields) to more closely approximate a sphere; they hold three boards each and are supported by movable arms to enable calibration. Source: Image reproduced with permission of Renesas Electronics Americas, Inc.

the top of the SS tube by compressed air, and is then held in place by a simple mechanical latch. The position of the source holder is monitored with optical and vacuum sensors to ensure the source is in the desired location. Safety interlocks in full regulatory compliance are provided for all system components that involve potential source exposure. The entire system is under computer control, including test scheduling, source depletion calculations, power supply voltages and supply sequencing. The Renesas modified N40 irradiator is equipped with fixturing for up to 64 large test boards, supported in an approximately isodose sphere arrangement around the pop-up source to insure equal gamma ray flux at all test positions. Secondary radiation (scintillation) from the interior and fixturing of the irradiator will generate lower-energy bremsstrahlung photons, and this component will cause dosimetry errors if not controlled. Shielding against this secondary radiation is mandated by Test Method 1019 [5], which specifies fixturing be enclosed in a lead-aluminum laminate (1 mm Pb/1 mm Al) spectrum hardening box.

5.2.3 Single-Event Effects (SEE) Testing

SEE were first predicted theoretically by Wallmark and Marcus in 1962 [19] and are defined as the results of the interaction of a single energetic ion with a semiconductor device, as opposed to the spatially uniform effects caused by multiple particles or photons as seen in seen in TID, neutron, and proton effects. The first actual satellite anomalies were reported by Binder et al. in 1975 [20], but this result was not immediately accepted as a real problem. Other early SEE work investigated alpha-particle-induced soft errors in dynamic read access memory (DRAM) samples [21] and cosmic ray induced errors in MOS memory cells [22]. This work was somewhat of a departure from SEE as observed in space; in it, the alpha particles originated from the natural decay of trace concentrations of uranium and thorium isotopes present in the ceramic materials then widely used in integrated circuit packaging. Later observations included satellite static random access memory (SRAM) upsets by cosmic rays and proton upsets under experimental conditions. Actual mission-related observations included soft in-flight errors on the Voyager and Pioneer missions and errors in the Hubble Telescope guidance system. The Hubble system used 1-kbit TTL bipolar static random access memory (SRAM) [23], which in the event turned out not to be a good choice for this application, and which was later replaced during Servicing Mission 1 flown on shuttle flight STS-61.

We have seen that SEE originate through the interaction of a single particle and an integrated circuit; this interaction can be through direct ionization (protons or heavy ions) but also through indirect ionization (neutrons). Direct ionization occurs when energetic charged particles such as electrons, protons, or alpha particles interact directly with atoms in the semiconductor lattice, resulting in electron–hole (e–h) pair generation. The charged particle interacts with multiple atoms, losing energy at each interaction. SEE in space are typically caused by energetic protons and heavy ions of galactic or solar origin. Indirect ionization is a more complex process in which neutral particles such as photons and neutrons interact with lattice atoms. These interactions release charged particles such as electrons, which then, in turn, interact with the material by direct ionization. Indirect ionization is characterized by this intermediate step creating charged particles. Protons that are trapped in the van Allen belts typically trigger indirect ionization effects. Also, atmospheric neutrons are an emerging area of concern [24] for surface (server farms) and avionics applications.

SEE is a classical stochastic process and results in random errors, both in time and in space. This randomicity is driven by the properties of cosmic rays; the name is really a misnomer as these are particles, not rays, but it has persisted in the popular literature and in science as well. Solar cosmic rays are anisotropic (directional) and their flux fluctuates with the 11-year solar activity cycle [25], while galactic cosmic rays are isotropic and are thought to originate at the galactic center, at massive black holes or in supernova explosions. Both are mostly protons and alpha particles, with less than 1% heavy ions. The galactic cosmic ray energy spectrum has a long "tail" at the high-energy end of the flux vs. energy spectrum. Research has demonstrated a low (but finite) flux of GeV and TeV particles as seen in Figure 2.31 where a flux of 10^{-22} m^{-2}-sr^{-1}-s^{-1}-GeV^{-1} corresponds to a few particles per square kilometer per day. This is a key issue for electronic systems in space; shielding of any practical thickness against these very penetrative relativistic particles at the higher energy levels is not effective, and mitigation must be done at the part or system level.

SEE may be conveniently categorized as nondestructive effects, destructive effects and single-event transients (SETs). Nondestructive effects are largely a digital issue and may be corrected by a rewrite or a reboot. Destructive effects imply permanent damage, either immediate or latent, and probable nonfunctionality. Transients in digital systems can result in soft errors if quantized by a bistable element, and transients in analog parts play a key role in power management applications. As we saw above, energy is lost by high-energy charged particles as they pass through the semiconductor lattice. The energy loss results in e–h pair generation, leaving a dense charge track (Figure 5.6). The effect scales: higher particle energy and mass lead to denser ionized tracks and more

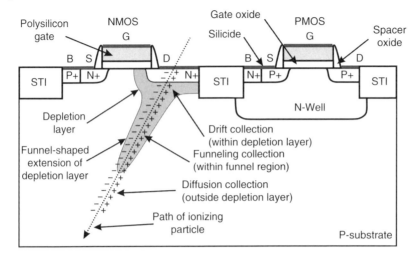

Figure 5.6 Charge collection in a CMOS semiconductor device structure, showing an ion strike on an n-channel drain, which is reverse biased with respect to the gate. Charge is collected in the depletion layer of the drain/body junction, in the "funnel" extension of the depletion layer, and in the bulk p−substrate region outside the depletion layer.

charge. In the presence of an electric field, which is the case in most operating devices, the electrons and holes are separated and collected. There are three collection mechanisms, as shown in Figure 5.6.

The collected charge may then transfer to sensitive circuit nodes, creating a current pulse and leading to node voltage changes, which may, in turn, cause circuit errors. The most common unit used in SEE work is the LET expressed in MeV cm^2/mg, which is another way of expressing $(1/\rho)\, dE/dx$, the energy loss per unit track length for a given material density. The material density must be taken into account to correct for differing stopping powers.

A common lament: *How can this be so difficult? Can't you just shield the electronics?* The low-energy component of the proton, electron, and heavy ion spectra of solar wind and galactic origin is easily and efficiently removed by commonly used aluminum spacecraft shielding, which is normally some 100 mils (0.001 in.) thick. As we saw in Figure 2.31, a small but nonzero part of the particle flux is very energetic and at the highest energy levels, shielding of any practical thickness will have little or no effect. To illustrate the selective removal of the lower-energy end of the spectrum shown in Figure 2.31, Figure 5.7 shows several iron (Fe) cosmic ray energy spectra behind various thicknesses of aluminum shielding, some practical (200 mils), some marginal (500 mils) and some quite impractical (5000 mils, or 5 in.). For the near-Earth charged particle environment, the Earth's magnetic field provides geomagnetic shielding, which is more intense at the equator and less so at the poles. The Earth's magnetic field is offset from its rotational axis, leading to an offset in the van Allen belts and resulting in enhanced proton and heavy ion flux in the South Atlantic Anomaly, off Brazil, and Argentina. Auroras ("northern lights" and "southern lights") are caused by high-energy charged particles interacting with the atmosphere, which restricts them to high latitudes, as the particle flux is higher there due to reduced geomagnetic shielding. The aurora light is emitted when the atoms

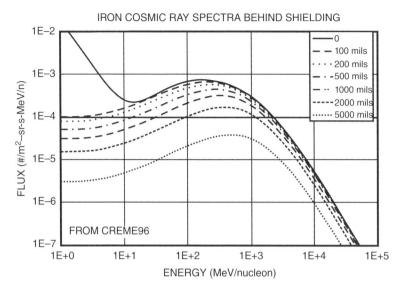

Figure 5.7 Shielding does you no good at all: Fe cosmic ray energy spectra behind various thicknesses of aluminum shielding, some practical (200 mils), some quite impractical (5000 mils, which equates to 5 in.). Based on the AP-8 environment model. Source: Adapted from Barth 1997, Figure 5.4.3.2, p. I-53. © 1997 IEEE. Reprinted, with permission, from [26].

return to their original unexcited state. Bottom line: Shielding against cosmic rays does you no good at all, and mitigation must be done at the part or system level.

As we have seen, SEE may be divided into nondestructive (soft error) effects, destructive (hard error) effects [27], and transient effects. There are a number of effects in each of these categories [28], with that number growing as complexity increases and further research is carried out. We will discuss each key effect briefly; note this is by no means a complete listing.

Single-event upset (SEU). This effect is a soft error consisting of a bit flip that can be readily corrected by a rewrite or reboot. SEU applies to bistable digital elements only, but is often (and quite incorrectly) used as a catchall for *all* SEE. An SEU in a continuous time analog function or in a digital function without bistable elements is a nonsensical concept. SEU may be mitigated by redundancy, memory scrubbing and error correction, by specialized circuit design approaches or by the use of passively isolated thin-film fabrication processes such as silicon on insulator (SOI) [29] or silicon on sapphire (SOS) [30].

Multi-bit upset (MBU). This effect is a soft error similar to SEU; it is defined as an upset [31, 32] of two or more digital bits, which are necessarily adjacent to each other in the part layout (but not necessarily adjacent in the bitmap of for example a random-access memory). MBU is caused by the charge track diameter approaching the memory cell size or by an ion strike coming in at a very low angle, hitting two or more adjacent bits (flip-flops or latches) and upsetting both. As in SEU, MBU is mitigated by error correction or by the use of a hardened process such as SOI or SOS; however, error correction is much more difficult for the MBU case.

Single-event transient (SET). This effect is a soft error manifesting as a pulse on the output of an analog or digital function caused by a heavy ion strike. This is mostly an analog phenomenon, but can cause formal upset in the digital sense in deep submicron processes if the analog transient is captured in a bistable element. Mitigation is through error correction, triple redundancy, or filtering techniques to eliminate the effects of the transient. SET has evolved into a major issue in power management functions, in which even a momentary output overvoltage can immediately destroy sensitive low-voltage devices.

Single-event functional interrupt (SEFI). This effect is a soft error in which an ion strike to the control circuitry of a digital or mixed-signal device causes a functional interrupt to the part. Mitigation again uses error correction or triple redundancy/voter techniques in critical control circuitry to eliminate the interrupt. However, recovery from SEFI can be a long process with time constants to the order of a reboot of the part.

Single-event latchup (SEL). This effect may be either a hard or a soft error and is similar to conventional electrical latchup but is caused by an ion strike instead of by electrical conditions. SEL can be destructive or nondestructive, depending on the part's latchup characteristics and application. Destructive failure usually occurs through interconnect destruction. There is a risk of latent interconnect damage, in which the part may fail later (or much later) in the system. SEL may be mitigated by specific layout design, by specialized dielectrically isolated (DI) processes or by external current sense/crowbar circuitry. High-temperature operation is the worst-case condition for SEL.

Single-event burnout (SEB). This is an invariably destructive error related to SEL and is caused by an ion strike turning on PNP-NPN parasitic BJT four-layer pairs that exist in the substrate of power metal-oxide semiconductor field effect transistor (MOSFET) devices. SEB is most often addressed by voltage derating of the power MOSFET, which is usually the only option, and high-temperature operation is the worst-case condition. Both SEB and SEL are emerging issues in integrated MOSFET point-of-load (POL) converters, in which the switching devices are integrated on the single POL chip and are lateral structures much different from discrete power MOSFET devices.

Single-event gate rupture (SEGR). This is also an invariably destructive error and is caused by heavy ions creating a momentary conductive path through the gate oxide of a power MOSFET or other MOS transistor. The charge stored on the drain-to-gate or body-to-gate capacitance (which can be quite considerable in large, high-voltage power MOSFET devices) then discharges through the conductive path, causing a permanent gate-to-body, gate-to-drain, or gate-to-source short. SEGR normally results in a degraded or nonfunctional MOSFET device, with no latency issues noted. A beam that is normal to the surface and room temperature operation are worst-case conditions for SEGR. The effect can be mitigated through the use of a thicker gate oxide but is usually addressed at the part level by derating the power MOSFET operating voltage. SEGR is an emerging issue in integrated MOSFET POL converters, as described above.

These SEE may be mitigated by a number of approaches. The basic semiconductor fabrication process may be changed to bound and minimize charge collection volume, using isolation techniques such as SOI [29], SOS [30] or various types of dielectric isolation [33]. Such specialized processes will be fabricated in much lower volume than commercial processes, which will decrease yield and increase cost. In junction-isolated processes, the use of thin or retrograde doped epi layers, heavily doped substrates, or selective or nonselective buried layers [34] has been shown to improve SEE performance. The detailed circuit design and device layout present opportunities as well: for digital functions redundancy, voters, error detection and correction (EDAC), and temporal latches [35] have been used. The chip layout may be improved [36] through the use of heavily doped guard rings and closed-geometry MOS devices. This hardened by design (HBD) approach [37, 38] has been increasingly attractive as it enables the use of unmodified high-volume commercial processes and offers cost reduction as a result. As somewhat of a last resort, and usually in the case of nonhardened commercial devices, latchup detection and circumvention may be applied externally to the part. This technique detects rapid increases in supply current and powers the part down for a preset time, enabling the use of otherwise nonhardened commercial parts. At the system level, there has historically been extensive use of part derating for "in beam" or "on orbit" operation, with multiple levels of derating leading to a very conservative approach.

In any case, extensive heavy ion testing is required to validate the circuit design, layout, and processing as well as to provide a periodic monitor of SEE vulnerability. The interpretation of SEE data in order to develop on-orbit error rates that can be used in mission planning is an important topic that has seen extensive research [39]. An SEE test and subsequent data analysis typically result in a curve plotting the SEE cross section as a function of ion LET, which may then be used as input data for simulation of the on-orbit error rates [40, 41] for the specific device and environment. Simulation methods such as scanning picosecond laser excitation have been proposed and are getting traction for diagnostic work, as discussed below.

5.2.4 Single-Event Effects (SEE) Sources and Facilities

Most current heavy ion testing for research and production work is performed using either cyclotron ion sources as found in facilities such as Texas A&M and Lawrence Berkeley Laboratories or tandem van de Graaff facilities such as Brookhaven. Most facilities can provide ions with LET values of up to 86 MeV cm^2/mg, with higher values attainable by irradiation at an angle of other than normal to the die surface. Many require in-vacuum testing, which complicates fixturing and electrical monitoring, with the Texas A&M facility at an advantage because of its in-air testing capability. Again, as in our ^{60}Co testing discussion, we are dealing with an accelerated test, with all the potential pitfalls inherent in such testing; Heavy ion irradiation also leads to incidental TID and DD, and high-flux testing (fluxes of 5×10^4 ions/cm^2 s are not at all unusual) is fraught with nonlinearities and data interpretation issues. The experimenter must be wary of flux sensitivity effects, as they can mask responses of interest.

The Texas A&M University (TAMU) Cyclotron Institute is a leading US heavy ion testing facility, with some 2000 hours of beam time a year used by government and industry. TAMU uses a K500 superconducting cyclotron and ion source to produce a variety of beams of the desired species at the experimental area. Ion energies for light ions such as protons can range from 8 to 70 MeV, while energies for heavier ions such as gold can range from 500 MeV (2 MeV per nucleon) to 3.5 GeV (15 MeV per nucleon). In the TAMU cyclotron, a strong magnetic field developed by a superconducting magnet constrains the ions in orbits between its poles. The 50 kilogauss magnetic field is generated by an 800 Ampere DC current through a coil of 5500 turns (25 mi!) of superconducting Nb-Ti wire. The ion acceleration is performed by an alternating electric field generated by a 240 kW RF source driving two hollow copper structures called dees (because of their half-circular shape), located between the poles of the magnet. The generation, acceleration, and beam extraction of the ions takes place in high vacuum, while the trajectory and composition of the beam outside the cyclotron is controlled by electromagnets. The SEE testing samples are held in a test board mounting fixture supported by a rotating stage to allow irradiation at angles other than normal to the die surface. The entire test station is located at the end of the beam line and is at atmospheric pressure. This "in-air" approach is generally preferred over in-vacuo testing due to greatly improved ease of access. Sample responses to ion irradiation are monitored on a continuous basis with appropriate thresholding functions to determine whether an "interesting" event has taken place. The monitoring is performed in a remote (to the order of 30 ft) instrumentation room, for safety reasons, and line drivers and coaxial cables are frequently required to assure the integrity of high-speed digital signals. Figure 5.8 shows the K500 cyclotron installation at Texas A&M, while Figure 5.9 shows an in-air rotating stage test station located at the end of a TAMU beam line.

A note on synchrotrons is appropriate. Cyclotrons were the leading high-energy particle accelerator technology until the 1950s, when synchrotrons became available for higher-power applications. A cyclotron uses a constant magnetic field and a constant-frequency electromagnetic field to constrain and accelerate the ions of

Figure 5.8 The K500 Cyclotron installation at Texas A&M. A 50 kilogauss superconducting magnet constrains the ions in a spiral trajectory between the magnet poles. The ions are accelerated by a radio frequency electric field generated by a 240 kW RF generator driving hollow copper structures (called "dees" because of their shape) located between the magnet poles. Operation takes place in high vacuum and the beam is controlled by electromagnets after exiting the cyclotron. Source: Image reproduced with permission of The Cyclotron Institute, Texas A&M University.

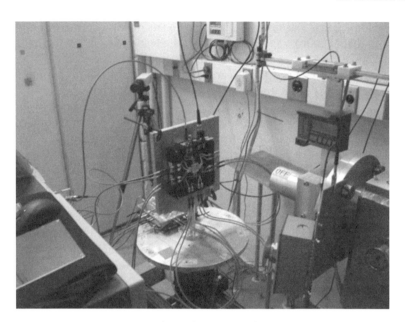

Figure 5.9 An in-air test station located at the end of a TAMU beam line. The station includes a rotating stage and test board mounting fixture to allow irradiation at angles other than normal to the die surface. The in-air approach is generally preferred over in-vacuo testing due to greatly improved ease of access. Source: Image reproduced with permission of The Cyclotron Institute, Texas A&M University.

interest; a synchrotron uses localized variations of the guiding magnetic field to perform the constrain and accelerate functions while compensating for relativistic particle mass increase during acceleration. The adaptation is thus done by temporal rather than spatial variations of the magnetic field strength. By varying these parameters as the particles are accelerated, the circulation path can be held at a constant radius rather than the spiral trajectory seen in cyclotrons. This permits a toroidal vacuum chamber rather than a disk, as seen in cyclotron designs. Lengthening the accelerator path will then become more cost-effective due to the thin toroidal configuration, which improves magnetic efficiency as well. Note also that the toroid does not necessarily have to be circular, permitting somewhat noncircular shapes such as round-cornered polygons to be used. As in most accelerators, the particle energy is limited by synchrotron radiation emitted while the particles are constrained to follow a curved path. The linear accelerator (LINAC) is, of course, an exception, and there are trends toward very large linear accelerators to increase ion energy; we will encounter this device in our transient gamma testing discussion. Synchrotrons are pervasive in high-energy particle accelerators and synchrotron light sources, but the ion beam in these devices is delivered in a cyclic manner that complicates SEE testing of electronic devices. The synchrotron duty cycle is typically 15%.

If we consider that SEE is caused by energy deposited in a sensitive device structure by a heavy ion, ionizing the material and generating mobile charge, it follows that other, possibly more economical means of introducing the energy may be of interest. This reasoning and the extreme cost of heavy ion facilities and beam time has led to the development of pulsed laser excitation [42, 43] as a more convenient and economical alternative to heavy ion testing. This test can be performed using less expensive equipment, which enables manufacturers and users to have the testing facility on-site. The actual physical mechanisms are different and the pulsed laser cannot be used to determine the cross section vs. LET curve, but it can determine the spatial and temporal aspects of SEE in a more or less qualitative manner. The laser does not damage the sample and may be focused to spot sizes to the order of a few microns; it may be raster scanned as well using a precision *x-y* stage, which enables efficient determination of specific layout features that are responsible for the responses observed during actual heavy ion testing. This makes the pulsed laser more of a troubleshooting tool than a basic research tool and results in the two methods being somewhat complementary, but it also results in the necessity for actual heavy ion beam validation testing at least somewhere in the product development cycle.

An actual heavy ion interacts directly with the semiconductor lattice for less than a picosecond, but the subsequent charge collection and device response processes take much longer. It follows that the pulse width of a laser used for heavy ion simulation must be significantly shorter than the device response time. If the pulse width approaches or exceeds the device response time the results may be inaccurate. This has led to the use of pulsed laser beams of very short duration. There are two approaches: linear single-photon absorption and nonlinear two-photon absorption.

Let us consider the pulsed laser requirements. The wavelength of the pulsed laser must be consistent with efficient ionization of the material under irradiation. In our case, this is likely silicon, which has a bandgap of some 1.12 eV. Photons with energies greater than the bandgap energy will be absorbed directly, creating e–h pairs in the process. We can now put a lower boundary on the laser wavelength using the familiar relationship

$$E = hc/\lambda \tag{5.3}$$

For an ionization energy of 1.12 eV for undoped silicon, we thus obtain a maximum laser wavelength of 1107 nm, which is well into the infrared part of the visible spectrum. At this wavelength, the incident photons will be absorbed directly in a linear process known as single-photon absorption (SPA). Laser simulation facilities use a number of different laser wavelengths, and the penetration depth into the silicon (which is defined as the depth at which the incident photons retain 1/e of their surface energy) will vary greatly as well. Two representative wavelengths are 590 and 1064 nm with corresponding 1/e penetration depths of 1.7 and 676 µm. The majority of the absorption will be near the surface, and the pulse is attenuated exponentially as it penetrates the silicon; this leads to limitations for picosecond laser testing of deep structures and for simulations of high-LET events. As an alternative substrate thinning [44] may be used, but this is a most difficult process and results in serious limitations on the sample operating power dissipated during laser testing.

To address this issue, the two-photon absorption (TPA) technique has been developed [45]. In TPA the laser pulse is shortened further and is focused to a very small spot size; absorption in this region of high-energy density will be by the simultaneous absorption of two photons to create a single e–h pair and will no longer follow a 1/e relationship. The spot can be focused at any depth in the silicon, which enables irradiation from the back of the wafer without wafer thinning; this is a major advantage, as frontside irradiation is sensitive to masking by metallic packaging and interconnect materials. TPA concentrates the carrier injection inside the confocal region where the laser intensity is

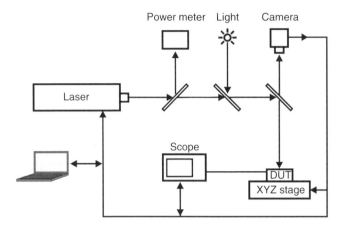

Figure 5.10 Block diagram showing the components in a typical TPA laser simulation system.

greatest, permitting control of the charge injection at any depth in the sample. TPA is thus a more versatile method but has a number of added variables and is more complex, requiring monitoring of pulse energy, pulse width, spot size, and focus depth. The high laser intensity required for efficient TPA carrier generation leads to a number of nonlinearities that must also be taken into account.

A typical laser simulation system consists of a laser source system with a wavelength range of 720–850 nm and about 2 ps pulse width, focused through a standard optical microscope system onto the device under test (DUT). A digital camera images the sample. Typical peak laser power is in the 500 mW to 1 W range, and the energy at the DUT typically ranges from 1 to 500 pJ. The sample position is computer-controlled using a precision x-y stage with a typical step size of 0.1 μm, as required for investigating samples fabricated in submicron processes. A computer-controlled data acquisition system is used to detect and record the transients generated by the laser. Figure 5.10 shows a representative laser simulation system block diagram.

A second alternative to actual heavy ion testing uses ^{252}Cf. This is a radioactive isotope of californium that undergoes spontaneous fission, producing mostly fission fragments and neutrons through an alpha decay process. The neutron energy range is 0–13 MeV, with a mean value of 2.3 MeV and a most probable value of 1 MeV. The isotope is useful as a compact neutron source for portable neutron spectroscopy and in medical radiotherapy as an alternative to radium. The isotope was also used in early efforts [46, 47] to develop a substitute for costly and difficult to access accelerator testing. At a distance of 1 cm a 5 μCi ^{252}Cf source will emit approximately 1×10^3 particles/cm^2 s with uniformity to the order of 20%. A typical irradiator uses a vacuum chamber, source holder, source shutter, and test fixturing.

^{252}Cf is highly toxic so stringent safety precautions must be used, including documented safety procedures and verified dosimetry. The flux value is changed by varying the sample to source distance or by filling the chamber with gas at a known pressure and composition. The LET of the neutrons and fission fragments must be reduced in order to be able to develop a usable cross section vs. LET curve, and this is accomplished through the introduction of thin degrader foils between the source and the sample.

^{252}Cf as a replacement for conventional heavy ion testing has not gotten a lot of traction as it is limited to LET values of up to $45 \, \text{MeV} \, \text{cm}^2/\text{mg}$ and is also severely limited by the low energy and short range of the emitted neutrons and ions. The low energy leads to limited penetration into the sample [48], and deeper sensitive volumes may not be reached at all. Further, the large number of metal and dielectric overlayers found in modern devices may not allow the particles to reach the sensitive volumes. These limitations have resulted in significant lack of correlation between ^{252}Cf and accelerator results, and this approach is not used frequently.

5.2.5 Neutron Testing

Neutron testing can be separated into DD testing, in which the samples are irradiated at high flux, and into neutron SEE testing in which the effects of single neutrons are evaluated. DD testing has historically been performed to simulate nuclear weapons environments but there has also been a great deal of interest in neutron testing as an adjunct to TID testing. Recall that protons are the dominant threat environment in space, and further recall our discussion of the use of TID testing as a means of bounding the proton response of devices. A proton causes both ionization *and* DD, and total dose testing addresses only one part of the device response. Neutron DD testing is used to complement TID testing for better simulation of proton environments by simulating the DD effects that are caused by protons. Neutron DD testing is quite considerably simpler than the complex procedures used for total dose and SEE testing, in which the samples have to be biased or actually operating while under irradiation. In DD testing, the samples are electrically characterized and are then irradiated to the specified neutron fluence. Neutron DD fluence specifications for electronic components depend on the system application and typically run from a low level of perhaps $1 \times 10^{10} \, \text{n/cm}^2$ to a high value of $1 \times 10^{14} \, \text{n/cm}^2$. For testing at multiple fluences, a different set of samples must be used for each level; this is in contrast to total dose testing, where the damage is cumulative and the same samples are used throughout the test. Samples are almost always irradiated in an unbiased configuration with all leads shorted together, see Test Method 1017 of MIL-STD-883 [5]; this electrically inactive test is colloquially known as a *bag test*. As neutron irradiation activates many of the heavier elements found in a packaged integrated circuit, samples exposed at higher neutron levels will nearly always require significant "cooldown" time before being shipped back for electrical testing. A typical generic set of neutron fluences for integrated circuit testing would be $5 \times 10^{11} \, \text{n/cm}^2$, $2 \times 10^{12} \, \text{n/cm}^2$, $1 \times 10^{13} \, \text{n/cm}^2$, and $1 \times 10^{14} \, \text{n/cm}^2$.

In a somewhat simplified model of DD [49], a high-energy neutron, proton, or heavy ion collides with a semiconductor lattice atom, displacing it and creating a vacancy defect and an interstitial defect. The two defect types are collectively known as a Frenkel pair [50]. The displacement interactions cascade at very high energies and fluences, creating a larger damage cluster (defect). These are stable defects that anneal only minimally and create deep and shallow traps in the material. The shallow traps compensate (trap) majority carriers, which removes these carriers and shifts the material resistivity upwards. The deep traps act as generation-recombination centers and decrease the minority carrier lifetime and mobility. The decrease in minority carrier lifetime results in different DD for bipolar and MOS technologies. DD effects on MOS technology are minimal as these are majority carrier devices, while DD effects on bipolar technology are severe; the bipolar transistor is a minority carrier device and depends on minority carrier lifetime, which is decreased by the lattice damage and recombination sites caused by DD. This results in increased carrier recombination in the base, which leads to increased excess base current and lower current gain. The common-emitter current gain (hFE) degradation can be estimated by the empirical Messenger-Spratt equation [51, 52]:

$$1/h_{FE} \text{ (post-neutron)} - 1/h_{FE} \text{ (pre-neutron)} = \Delta h_{FE} = \Phi/[K\,(2\pi f_T)] \quad (5.4)$$

in which Φ is the neutron fluence and $[K\,(2\pi f_T)]$ is a damage factor that recognizes the link between base width, transition frequency and DD hardness; high f_T devices will generally be more neutron hard due to shallower junctions and higher doping levels. The BJT current gain degradation results in effects such as increased input bias current and decreased open-loop gain in bipolar analog parts. This degradation is interesting as it is worse at low collector current, with the high current gain at the peak of the beta vs. Ic curve is only minimally affected. This results in a design issue: in a hardened design, the transistors have to be operated at much higher current than commercial designs.

A second and emerging area of interest is neutron SEE testing. A neutron of course has no charge so any SEE caused by these particles is due to indirect ionization, in which the incident neutron triggers nuclear reactions resulting in SEE by the (charged) reaction products. The key environment here is that of atmospheric neutrons, which are abundant all the way down to sea level. Atmospheric neutrons originate from nuclear reactions between very high energy cosmic rays and the upper layers of the atmosphere. These energetic reaction products then lead to further reactions, resulting in a "shower" propagating downward into the atmosphere. Terrestrial sea-level neutron SEE [53] has been a rapidly emerging concern driven by submicron and deep submicron feature sizes and especially by large-scale integrated circuit applications such as server farms.

5.2.6 Neutron Sources

FBRs were first developed for investigations into the behavior of reactor systems. The White Sands Missile Range (White Sands, New Mexico) FBR is routinely used for DD testing of semiconductor devices and provides a controlled 1 MeV equivalent neutron flux. It is an unmoderated cylindrical assembly of highly enriched (93.5% ^{235}U) uranium and molybdenum alloy and produces high-yield neutron pulses of microsecond width, as well as steady-state radiation, to effectively simulate the neutron radiation environment produced by a fission weapon. A key consideration in any neutron source is the "cleanliness" of the produced neutron flux, as there may be a substantial amount of for example gamma ray contamination, which may affect data fidelity through total dose damage.

An alternative neutron source is the TRIGA (Testing, Research, Isotopes, General Atomics) reactor. The Texas A&M TRIGA is a 1 MW water-cooled "swimming pool" research reactor. At 1 MW operating power the thermal neutron flux available at the sample locations will be in the 1×10^{12} n/cm^2 s to 1×10^{13} n/cm^2 s range. The gamma dose rate at the reactor face at 1 MW is about 2×10^7 rad/h (!), and this added gamma radiation must be accounted for in analyzing results. Other alternatives include high-energy particle beam approaches such as spallation neutron sources, quasi-monoenergetic neutron sources, and monoenergetic proton and neutron sources.

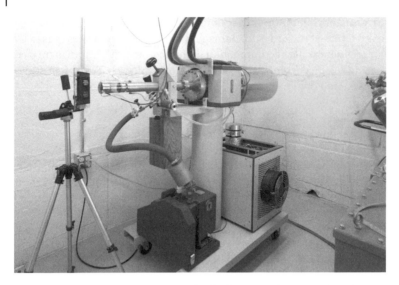

Figure 5.11 14 MeV neutron generator for displacement damage and single event upset testing. Source: © 2018 Cobham. Reprinted with permission from Cobham RAD Solutions, Colorado Springs, Co.

Neutron generators are commercially available compact neutron sources suitable for use in a laboratory environment. They contain a small linear accelerator and produce neutrons by nuclear reactions between hydrogen isotopes. This is accomplished by using the LINAC to accelerate deuterium, tritium, or a mixture of these two isotopes into a metal hydride target, which also contains these isotopes. The deuterium–tritium reaction is used more frequently than the deuterium–deuterium reaction, as the neutron yield of the former is 50–100 times higher. The nuclear reactions are

$$D + D \rightarrow \ ^3He + n, E(n) = 2.5 \ MeV \tag{5.5}$$

$$D + T \rightarrow \ ^4He + n, E(n) = 14.1 \ MeV \tag{5.6}$$

The deuterium–deuterium reaction results in the formation of a ^3He ion and a 2.5 MeV neutron, while the deuterium–tritium reaction results in the formation of a ^4He ion and a 14.1 MeV neutron. An example of a 14 MeV neutron generator and test fixture is shown in Figure 5.11. Neutrons produced by these reactions have moderate anisotropy along the accelerator beam axis. Neutron generators of the deuterium-tritium type have been used for DD testing [54], but as would be expected, there are correlation issues to overcome between the 14 MeV equivalent neutron generator results and the 1 MeV equivalent neutrons obtained from nuclear sources. 1 and 14 MeV neutron testing is covered in MIL-STD-883 [5] Test Method 1017.

Compact radioisotope neutron sources, including PuBe, AmBe, SbBe, and ^{252}Cf (see the earlier SEE discussion), deliver a broadly distributed "white" beam of fission neutrons, with a typical energy range up to 8 or 10 MeV and with a peak around 3 MeV. These sources provide a good approximation of the neutron energy spectrum encountered in many terrestrial environments but the fluences are low, which results in long irradiation times. There is also a significant gamma ray component that must be taken into account.

5.2.7 Proton Testing

As with heavy ions and neutrons, we are interested on two aspects of proton effects: the bulk effects of high proton fluxes [55], analogous to TID, and the effects of single protons or proton SEE [56]. Electrons are not penetrative enough for routine testing and their use in testing electronics is limited. Protons are the dominant environment in space applications and cause both ionizing and DD. Figure 5.12 shows proton fluence as a function of shielding thickness, and we note, as we did for cosmic rays, that these highly penetrative particles are difficult to shield against, at least at higher energies.

Proton testing is considerably more difficult than other radiation tests. Protons are very penetrative and will activate samples, requiring long cooldown periods before the samples can be safely handled. Proton effects on semiconductor devices are also highly dependent on proton energy and a range of energies must be explored [57, 58] in order to develop a complete assessment of proton damage. At this time there are few proton sources consistent with electronics testing, and this problem was exacerbated by the closing of the Indiana University (IU) proton facility in late 2014. The IU closure has driven an intensive program of development of alternative sources, driven by Aerospace Corporation and NASA [59]. These new sources are largely medical facilities, and it has been found that considerable facility development is required to enable proton testing of electronics; there are also serious beam-time access issues.

Historically proton environment validation had been approximated by the use of lower cost photon irradiation sources such as 10 keV X-rays and ^{60}Co gamma rays. This approach provides an approximation of proton response but has been extensively correlated to actual proton response, as we saw in our total dose discussion. The approach is based on the premise that the ionization damage from these photon

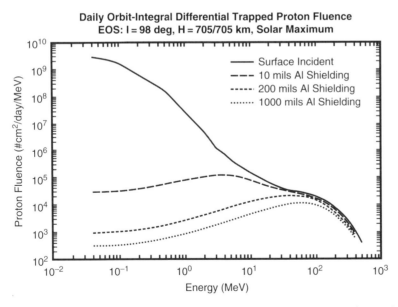

Figure 5.12 Trapped proton fluence as a function of proton energy in a 705 km Earth orbit for several thicknesses of aluminum shielding. Source: Barth 1997, Figure 5.1.3.7, p. I-36. © 1997 IEEE. Reprinted, with permission, from [26].

sources is comparable to the ionization damage caused by proton irradiation. For MOS circuitry, this is a valid assumption, as these majority carrier devices are not sensitive to the DD caused by massive particles such as protons and neutrons. For bipolar parts, this assumption breaks down as minority carrier devices depend on minority carrier lifetime for their operation, and the proton response of these devices cannot be accurately inferred from photon irradiation results. This discrepancy has been addressed by additional testing using neutrons to simulate the DD component of the proton response. As always, correlation is needed here as the response of a bipolar part is a nonlinear function of the displacement and TID damage. Still, the combination of DD and TID is routinely used to at least bound the proton response of bipolar parts.

In general, high-energy proton testing in the 200 MeV range is used for determining the SEE response of parts, with maybe 80% of high-energy proton testing used for this purpose. The proton is a low-LET particle (typically an LET of 2) but low-LET response is of great interest in space applications of commercial off-the-shelf (COTS) devices. Low-energy proton testing in the 50 MeV range is used for determining DD response, with maybe 70% of low-energy proton testing used for DD. These are very general bounds due to the lack of availability of 200 MeV protons for testing; you do what is consistent with the capabilities of the facility you have access, too, so DD testing may well end up at 200 MeV and SEE testing at 50 MeV. Low-energy protons are much more damaging in most cases [57, 58], leading to an *increase* in the damage constant with *decreasing* proton energy in DD testing. As a proton loses energy while moving through the lattice, it reaches a depth at which all its energy is lost; this is known as the Bragg peak, and if the peak falls in a sensitive volume of the device structure the damage will be much more severe than at higher energies. This results in a subtle and difficult challenge: the need for knowing a priori (or determining through destructive part analysis) the cross section and junction depth of the samples being tested in order to be able to correctly interpret the results.

Proton SEE testing, of course, requires in-situ monitoring as in heavy ion work, while DD testing can be electrically inactive. Proton SEE test is always a set of tradeoffs between absorbed total dose damage, DD and the desired SEE response; it is used on parts that are known or suspected of having a substantial response at LET values between $5 \, \text{MeV/cm}^2$ mg and $15 \, \text{MeV cm}^2/\text{mg}$. Hardened devices with heavy ion SEE LET of greater than $40 \, \text{MeV cm}^2/\text{mg}$ are usually exempt from proton SEE testing. Samples with known low TID tolerance must be carefully tracked and replaced if necessary.

5.2.8 Proton Sources

The primary proton source is the cyclotron, traditionally used for SEE testing and providing a uniform, constant beam through the use of single or double scattering foils and using defocusing magnets and collimators to achieve uniform beam sizes of typically 10 cm × 10 cm. Some sources use a synchrotron, which provides particles in "spills" extracted from the beam, with an approximate 10% duty cycle (e.g. 300 ms on, 2.7 s off), resulting in an instantaneous particle flux approximately 10 times higher than the average flux. As in heavy ion testing, the pulsed nature of the synchrotron significantly complicates dynamic testing. Proton therapy centers utilize two types of scanning beams for cancer treatment: a uniform, rapid rastered scan using a large "pencil" beam, and actual pencil beam scanning using a narrow beam and a scanning path tailored to the shape of the target to be irradiated. Neither scan method is particularly optimal for parts testing due to their small beam diameter and the dynamic characteristic of the moving beam with respect to the part under test.

The proton testing community is currently dealing with a diminishing number of sources that are adaptable to electronics testing. The dependence of device response on proton energy as well as fluence significantly complicates these tests. There are currently five facilities worldwide providing protons at energies of greater than 200 MeV. Three of these are in the United States (Loma Linda University, Francis H. Burr Proton Therapy Center, and the NASA Space Radiation Laboratory), one in Canada (TRIUMF), and one in Switzerland (Paul-Scherrer-Institut). Two of the three US sources are medical facilities that can typically be used only at night or on weekends and involve adaptation of medical treatment rooms for parts testing.

The Radiation Effects Research Program at Indiana University had provided 40–200 MeV proton beams at fluxes ranging from 10^2 to $>10^{11}$ p/s cm^2 in a beam of 30 cm maximum diameter, but this facility was closed in 2012. This large loss in available beam time has led to the development [59] of electronics testing capability at many other facilities, nearly all of which are largely used for medical applications. Investigated medical facilities include Northwestern Medicine Chicago Proton Center, Scripps Proton Therapy Center, and The Massachusetts General Hospital Francis H. Burr Proton Therapy Center.

The Proton Irradiation Facility (PIF) at the Tri-University Meson Facility (TRIUMF) at the University of British Columbia, Canada, has historically supported proton SEE testing. TRIUMF is cyclotron-based, as are nearly all proton sources, and was designed primarily for SEE characterization of electronic devices for space applications. The variable-energy capability of up to 500 MeV is well suited for SEE studies, as the energy spectrum of trapped protons peaks in the 10–100 MeV region and then drops by an order of magnitude at 500 MeV. PIF makes use of two separate beam lines for low-intensity radiation-damage studies. The lower energies, variable from as low as 5 MeV up to 120 MeV, are available from one beam line, which is also used on a regular basis for proton treatments of ocular melanoma. Energies from 120 to 500 MeV are available from a second beam line, which terminates in the same experimental area.

5.2.9 Transient Gamma Testing

In this section, we concern ourselves with the intense prompt X-ray and gamma radiation phase of nuclear detonations, which causes serious radiation effects in semiconductor devices. Transient gamma radiation is exclusively applicable to weapons environments, and is a part of the burst scenarios of both fission and fusion weapons. The detonation of a nuclear device initiates a characteristic sequence of effects, including an immediate ("prompt") X-ray and gamma radiation pulse, as well as neutron and delayed gamma pulses. The quantitative details of these scenarios are almost invariably classified, and our discussion will be somewhat limited as a result, as will our referencing. Transient gamma dose rates are not found in nature and can range up to as high as 1×10^{12} rad(Si)/s with very short pulse widths, resulting in nondestructive effects in semiconductor devices such as digital dose-rate upset and analog dose-rate transients and in destructive effects such as dose-rate latchup and burnout. Dose-rate latchup and burnout are similar to electrically induced latchup and its effects and junction isolated (JI) devices are particularly vulnerable. Parasitic NPN/PNP structures in the JI substrate are triggered resulting in very high local currents that may cause either immediately destructive damage or latent damage. The traditional and highly effective solution to dose-rate latchup is the use of DI [33] processes, in which devices are isolated by oxide or other dielectric layers rather than by reverse-biased junctions, eliminating the parasitic four-layer structures. DI processes do have a wafer fabrication cost penalty, which in these applications is not usually of great importance, and have also been vulnerable to diminishing sources of supply problems.

A second concern is primary photocurrent, in which reverse-biased device junctions act as photodetectors; these currents can run into tens of amperes and will burn out transistors, interconnects, and bond wires. Again, DI processing simplifies the problem (and makes it easier to predict performance as well) as the clearly defined and bounded generation volume of a DI transistor allows accurate photocurrent compensation using a diode of similar geometry in antiparallel with the collector-base junction. The resulting circuit configuration can be treated (and simulated) as a system of discrete devices, with minimal interaction between devices. There are several approaches to transient gamma hardening. The most ambitious is the "operate through" scenario, in which the system (really this will almost always be a subsystem) will be fully operational through the event. A second and much more manageable approach is transient gamma circumvention, in which one or more nuclear event detectors sense the event and trigger power-down sequences of the critical system functions and then reboots after a suitable delay. A third approach just allows the system to crash but then depends on fast recovery from that state.

The third concern is the ionization of the bulk silicon material. Due to the large amount of energy deposited by the prompt gamma pulse very large amounts of electron–hole pairs will be generated, resulting in downward conductivity modulation of the bulk silicon, which will persist until the carriers recombine. In JI processes this effect may effectively short the supply rails together temporarily and destroy the device. Again DI provides an effective solution: the equivalent resistance of each transistor during the prompt pulse is tractable to simulation and depends on the transistor geometry and the actual magnitude of the conductivity modulation. Even under the worst-case assumption of conductivity modulation to essentially zero resistance, the excess currents can be controlled by including either thin film resistors or heavily doped diffused resistors in all vulnerable circuit branches. Thin film resistors are either metallic or ceramic and sintered metal (CERMET) and are not vulnerable to conductivity modulation. Heavily doped resistors may be implemented using for example the emitter diffusion; these resistors will modulate during prompt gamma, but minimally so.

5.2.10 Transient Gamma Sources

Transient gamma testing simulates the high intensity X-ray and gamma spike associated with a nuclear detonation. Simulators include flash X-ray (FXR) machines (NSWC Crane, Boeing, Cobham, and Honeywell) and linear accelerator (LINAC) facilities (Stanford Linear Accelerator Center and White Sands Missile Range), with laser simulation as another option. Given the military context of this environment, open literature references are very limited. The simulation of pulsed ionizing radiation effects on electronic

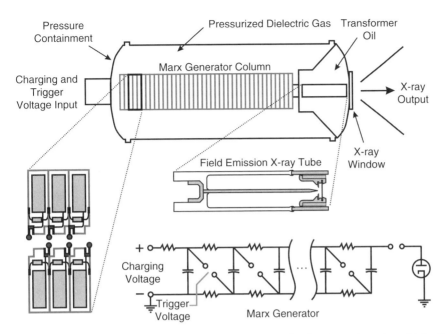

Figure 5.13 Exploded view of a typical flash X-ray system. The charging voltage is multiplied by the Marx generator and released into the X-ray tube when a triggering voltage is applied to one of the spark gaps. The Marx generator column is assembled from a number of identical disk-shaped units and its output is fed into a field emission X-ray tube. Everything operates at very high voltages, so the Marx generator column is inside a pressure container that is filled with a dielectric gas and the X-ray tube is surrounded by transformer oil to increase dielectric breakdown.

components merges those of total dose exposure leading to long-term ionizing radiation effects and the transient response to the prompt gamma pulse. Experimental facilities include a wide variety of FXR machines varying in energy spectrum and potential volume of high intensity exposure, linear accelerators in which high energy electron beams can be used to define response as a function of radiation intensity and pulse width, and pulsed laser sources, which bring laboratory convenience to the measurement of transient response.

A first method is the FXR generator [60], which consists of a large capacitor bank that is charged to a high voltage (to the order of several megavolts) by, for example, a van de Graaff electrostatic generator, and is then discharged in a high current electron beam, which is stopped in a high atomic weight target. The electrons interact with the target material in a bremsstrahlung process that produces a high-intensity, fixed-width X-ray pulse of high enough intensity to enable irradiation of very considerable size test articles. The FXR components are contained in a heavy steel enclosure. The dielectric medium inside the machine enclosure is a high-pressure gas. Figure 5.13 provides an exploded view of a typical FXR system.

FXR capability is available on a very limited basis in both industrial and government facilities and varies in the X-ray energy spectrum, the peak X-ray intensity, and the X-ray radiation pulse width available for testing. The exposure intensity is controlled by simply moving the test fixturing away from the target source. The radiation pulse width of the facilities is typically on the order of 20–100 ns. The Febetron 705/706, Fexitron 730, and

Figure 5.14 Pulserad 112A Flash X-ray system used for prompt dose testing and qualification. Source: © 2018 Cobham. Reprinted with permission from Cobham RAD Solutions, Colorado Springs, Colorado.

Ion Physics FX-25 are representative systems; they are now out of production but a number are still in operation. The FX-25 can supply gamma dose rates from 1×10^6 rad(Si)/s to 3×10^{11} rad(Si)/s with a pulse width of 20 ns, using a 2.5 MV maximum output Marx generator voltage. Cobham RAD Solutions in Colorado Springs, Colorado, operates an Ion Physics FX25 and a Pulserad 112A FXR system; the latter is shown in Figure 5.14.

An FXR alternative is provided by the linear accelerator (LINAC), which also generates high-energy, short-duration pulses of gamma ray radiation. In a LINAC, an electron beam is accelerated to high energy by subjecting the charged particles to a series of RF electric potentials along a linear beamline, producing a high-energy electron pulse that can be varied in pulse width from typically less than 20 ns to greater than a few μs. Note the use of a linear rather than a circular or other closed geometry beamline, which avoids synchrotron radiation losses incurred while bending the beam. As with a FXR machine, the resultant electron pulse is stopped in a high-Z target to produce bremsstrahlung X-ray output. Alternatively, the direct electron beam can be used for the ionizing radiation exposure. The beam current of a linear accelerator is substantially less than that of a FXR machine, which limits the available sample volume. The advantages of the linear accelerator facility are the ability to vary the gamma radiation pulse width and the ability to use high energy electrons (typically greater than 10 MeV) for direct exposure with the practical elimination of dose enhancement effects.

FXR and LINAC facilities provide extensive capability in the simulation of pulsed ionizing radiation effects, but at high capital cost, resulting in high beam-time charges and access difficulties. An alternative approach is the use of low-energy pulsed laser sources as discussed in the SEE section above. The use of pulsed laser irradiation enables more practical laboratory facilities to be used. Also, as in SEE testing, the laser will not penetrate the packaging materials and the sample must be delidded for testing. With surface exposure, the uniformity of carrier generation in the bulk semiconductor must be taken into account. In semiconductor devices, the surface will be covered with as many as nine metallization layers, which are opaque to the pulsed laser radiation. Alternatively, the sample can be irradiated from the back surface, which results in difficult sample preparation but which will get closer to uniform carrier generation throughout the bulk semiconductor. Gallium arsenide and neodymium-YAG lasers have been used for simulation of pulsed ionizing radiation effects. The GaAs laser allows output pulse control by varying the electrical input pulse but has limited intensity. The Nd-YAG laser pulse width is more difficult to adjust but provides higher radiation intensity. Dosimetry for laser irradiation must be measured directly from carrier generation in a semiconductor device, either as part of the DUT or by simultaneous exposure of a separate test device such as a P-i-N diode.

Dose enhancement effects may be an issue for pulsed X-ray exposure but can be minimized by careful test fixturing design. Dosimetry is usually performed using P-i-N diodes under high reverse bias to fully deplete the intrinsic region. The photoresponse of the diodes will be determined by carriers generated in the high-field depletion layer in the intrinsic region and the photocurrent will be proportional to the radiation intensity. Dose enhancement effects in both the sample under test and in the P-i-N diode may need to be accounted for, particularly when using X-ray energy below 1 MeV.

As discussed previously the objective of transient gamma testing is the determination of the part response at the specified dose rate. A transient gamma test must thus be electrically active, with the sample connected to power supplies in a static or dynamic configuration typical of the application. The sample needs to be driven by appropriate signals and the functionality and power supply current must be monitored. This is a difficult procedure because the samples and the test equipment are a relatively long distance apart for safety reasons. Additionally, the RF environment inside or near the irradiator is dominated by a pulsed electron beam of hundreds of Amperes at well over 1MV, leading to a difficult noise environment. In addition to electrically induced noise the beam exposure of the test fixturing and cabling produces large electron emission currents which appear as radiation-induced noise. For repeatable results, careful attention must be paid to the layout, shielding, and grounding of the test fixturing, cabling, and test equipment, and direct experience [61] is a major contributor to successful testing.

5.3 Summary and Closing Comments

This chapter has provided a very short introduction to a hugely complex field. Radiation testing is a key part of space mission assurance, and we have covered only the more common tests. Radiation testing is fraught with pitfalls, which seem to be only rarely documented due to security or competitive considerations. This chapter's references have a number of review papers in them, providing a good place to start further inquiry. The IEEE Transactions on Nuclear Science (TNS) is the publication of record in the field, and the December issue of this journal serves as the conference proceedings of the yearly Nuclear and Space Radiation Effects Conference (NSREC), which is the leading conference on radiation effects in semiconductors in the United States. The

EU equivalent is Radiations et ses Effets sur Composants et Systèmes (RADECS), sponsored by several organizations including the European Space Agency (ESA) and with some papers also published in TNS. Both conferences have one-day short courses on all aspects of radiation effects and publish a Short Course Notebook each year. The notebooks are an invaluable resource, but unfortunately, it is generally required to attend the conference to gain access to this information. Another very valuable resource is the large number of standards and guidelines, such as the military standards like MIL-STD-883 [4], MIL-STD-750 and the top-level MIL-PRF-38535 drawing. These are military in origin but are used internationally by civilian organizations and projects as well. Also of great value are standards issued by the Joint Electron Device Engineering Council Electron Devices (JEDEC), and, in particular, by that organization's Solid State Technology Association, an independent semiconductor engineering trade organization and standardization body. JEDEC standards and guidelines may be obtained through their website at www.jedec.org. Additionally, the American Society for Testing and Materials (ASTM) (www.astm.org) publishes a large number of guidelines, which in many cases provide more detailed test and analysis procedures. Table 5.1 provides a representative listing of representative radiation effects standards and guidelines.

Problems

5.1 You decide to go prospecting for uranium. Armed with an old Geiger counter that measures activity using the Röntgen scale, you identify an outcropping that deserves a little more investigation. Before continuing, you do a quick calculation to determine the absorbed dose based on the activity you are seeing and whether it is safe to proceed.

    a) Explain the quick calculation you performed from the Geiger counter reading. After verifying the Geiger counter reading, you feel safe to proceed, but you fiddle with the Geiger counter, as it has a separate wand containing the Geiger-Müller tube. You note that there is a metal sliding cover over the tube that, when opened, exposes a mica window, causing the Geiger counter reading to go much higher.

    b) What explains the difference in the two readings?

5.2 While reading up on radiation effects, you note that much of the literature describes absorbed dose in terms rad(Si) or rad(SiO_2). Your morbid curiosity causes you to convert these doses in terms of their effect on human tissue, rad(H_2O). Explain your conversion calculation between rad(SiO_2) and rad(H_2O).

5.3 Describe the radiation testing needed to validate circuit designs for the following radiation environments:

    a) Low Earth orbit, the integrated circuit contains CMOS and bipolar elements.

    b) Low Earth orbit, the integrated circuit uses CMOS digital logic only.

    c) High-density terrestrial server farm.

    d) Hydrogen bomb explosion.

Table 5.1 Representative US industry and government radiation effects standards and guidelines.

JEDEC JESD57	Test procedures for the measurement of SEE in semiconductor devices from heavy-ion irradiation
JEDEC JESD234	Test standard for the measurement of proton radiation SEE in electronic devices
ASTM F1192	Standard guide for the measurement of single event phenomena (SEP) induced by heavy ion irradiation of semiconductor devices
ASTM F1892	Standard guide for ionizing radiation (total dose) effects testing of semiconductor devices
ASTM F1190	Practice for the neutron irradiation of unbiased electronic components
MIL-STD-750-1	Environmental test methods for semiconductor devices Test method 1017: Neutron irradiation Test method 1019: Steady-state total dose irradiation procedure Test method 1080: SEB and SEGR
MIL-STD-883	Test method standard, microcircuits TM 1017: Neutron irradiation TM 1019: Ionizing radiation (total dose) test procedure
MIL-HDBK-814	Ionizing dose and neutron hardness assurance guidelines for microcircuits and semiconductor devices
MIL-PRF-38535	General specification for integrated circuits (microcircuits) manufacturing
MIL-HDBK-814	Ionizing dose and neutron hardness assurance guidelines for microcircuits and semiconductor devices
Sandia Nat'l Lab. SAND 2008-6983P	Radiation hardness assurance testing of microelectronic devices and integrated circuits: test guideline for proton and heavy ion SEE
Sandia Nat'l Lab. SAND 2008-6851P	Radiation hardness assurance testing of microelectronic devices and integrated circuits: radiation environments, physical mechanisms, and foundations for hardness assurance
NASA/DTRA	Field programmable gate array (FPGA) single-event effect (SEE) radiation testing
ESA-ESCC-25100	SEE test method and guidelines
ESA-ESCC-22900	Total dose steady-state irradiation test method

5.4 Simulation fidelity:
  a) Explain how gamma radiation from Cobalt-60 or 12 MeV electron irradiation each can produce the same fractional yield in silicon dioxide (Figure 5.1).
  b) Why can cobalt-60 irradiation be used to simulate many different TID environments?

5.5 Dose rate sensitivity:
  a) Calculate the time difference between high dose rate and low dose rate testing for a 1 Mrad(Si) total dose. Use dose rate values obtained from US industry and government radiation effects standards and guidelines.
  b) Why is ELDRS seen in bipolar transistors and circuits but not generally in MOS transistors and circuits?

5.6 Develop a test plan to provide post-radiation device models, including several total dose model corners. The device set includes both bipolar and MOS transistors.
  a) What precautions should be taken during testing of each type of device?
  b) Should devices tested during down points be put back up for further irradiation?
  c) When doing low-dose testing of bipolar transistors, what steps can be taken to provide simulation corner accuracy to circuit-level low-dose results?
  d) Should devices be under voltage bias during irradiation? Explain how each terminal would be biased, including polarity and voltage level for NMOS, PMOS, NPN, and PNP transistors.

References

1 Johnson, M.H. and Kierein, J. (1990). Combined release and radiation effects satellite (CRRES): spacecraft and mission. *Journal of Spacecraft and Rockets* 29 (4): 556–563.
2 Titus, J.L., Combs, W.E., Turflinger, T.L. et al. (1998). First observations of enhanced low dose rate sensitivity (ELDRS) in space: one part of the MPTB experiment. *IEEE Transactions on Nuclear Science* 45 (6): 2673–2680.
3 Holmes-Siedle, A.G. and Adams, L. (2002). *Handbook of Radiation Effects*. Oxford: Oxford University Press.
4 McLean, F. B. and Oldham, T. R. (1987). Basic Mechanisms of Radiation Effects in Electronic Materials and Devices. Report No. HDL-TR-2129. U.S. Army Laboratory Command, Harry Diamond Laboratories, Adelphi, MD.
5 US Military Standard MIL-STD-883 (n.d.). Test method standard for microcircuits. U.S. Defense Logistics Center, Land and Maritime, Columbus, OH.
6 Enlow, E.W., Pease, R.L., Combs, W. et al. (1991). Response of advanced bipolar processes to ionizing radiation. *IEEE Transactions on Nuclear Science* 38 (6): 1342–1351.
7 McClure, S., Pease, R.L., Will, W., and Perry, G. (1994). Dependence of total dose response of bipolar linear microcircuits on applied dose rate. *IEEE Transactions on Nuclear Science* 41 (6): 2544–2549.

8 Johnston, A.H., Rax, B.G., and Lee, C.I. (1995). Enhanced damage in linear bipolar integrated circuits at low dose rate. *IEEE Transactions on Nuclear Science* 42 (6): 1650–1659.

9 Johnston, A.H., Lee, C.I., and Rax, B.G. (1996). Enhanced damage in bipolar devices at low dose rates: effects at very low dose rates. *IEEE Transactions on Nuclear Science* 43 (6): 3049–3059.

10 Carriere, T., Ecoffet, R., and Poirot, P. (2000). Evaluation of accelerated total dose testing of linear bipolar circuits. *IEEE Transactions on Nuclear Science* 47 (6): 2350–2357.

11 Boch, J., Saigne, F., Schrimpf, R.D. et al. (2004). Effect of switching from high to low dose rate on linear bipolar technology radiation response. In: *Proceedings of the 7th European Conference on Radiation and its Effects on Components and Systems (RADECS).* 15–19 September 2003, 537–545. Noordwijk, NL: European Space Agency (ESA).

12 Winokur, P.S., Sexton, F.W., Schwank, J.R. et al. (1986). Total-dose radiation and annealing studies: implications for hardness assurance testing. *IEEE Transactions on Nuclear Science* 33 (6): 1343–1351.

13 Witczak, S.C., Lacoe, R.C., Mayer, D.C. et al. (1998). Space charge limited degradation of bipolar oxides at low electric field. *IEEE Transactions on Nuclear Science* 45 (6): 2339–2351.

14 Fleetwood, D.M., Riewe, L.C., Schwank, J.R. et al. (1996). Radiation effects at low electric fields in thermal, SIMOX, and bipolar-base oxides. *IEEE Transactions on Nuclear Science* 43 (6): 2537–2546.

15 Adell, P., McClure, S., Pease, R.L. et al. (2007). Impact of hydrogen contamination on the total dose response of linear bipolar microcircuits. In: *Proceedings of the 9th European Conference on Radiation and its Effects on Components and Systems (RADECS).* 10–14 Sept 2007, 537–545. Deauville, FR: European Space Agency (ESA).

16 Shaneyfelt, M.R., Pease, R.L., Schwank, J.R. et al. (2002). Impact of passivation layers on enhanced low-dose-rate sensitivity and thermal-stress effects in linear bipolar ICs. *IEEE Transactions on Nuclear Science* 49 (6): 3171–3179.

17 Pease, R.L., Platteter, D.G., Dunham, G.W. et al. (2007). The effects of hydrogen in hermetically sealed packages on the total dose and dose rate response of bipolar linear circuits. *IEEE Transactions on Nuclear Science* 54 (6): 2168–2173.

18 Pease, R.L., Adell, P.C., Rax, B.G. et al. (2008). The effects of hydrogen on the enhanced low dose rate sensitivity (ELDRS) of bipolar linear circuits. *IEEE Transactions on Nuclear Science* 55 (6): 3169–3173.

19 Wallmark, J.T. and Marcus, S.M. (1962). Minimum size and maximum packing density of nonredundant semiconductor devices. *Proceedings of the IRE* 50 (3): 286–298.

20 Binder, D., Smith, E.C., and Holman, A.B. (1975). Satellite anomalies from galactic cosmic rays. *IEEE Transactions on Nuclear Science* 22 (6): 2675–2680.

21 May, T.C. and Woods, M.H. (1979). Alpha-particle-induced soft errors in dynamic memories. *IEEE Transactions on Electron Devices* 26 (1): 2–9.

22 Pickel, J.C. and Blandford, J.T. (1978). Cosmic ray induced errors in MOS memory cells. *IEEE Transactions on Nuclear Science* 25 (6): 1166–1171.

23 Vargas, F. and Nicolaidis, M. (1994) SEU-tolerant SRAM design based on current monitoring. *Digest of Papers, IEEE Twenty-Fourth International Symposium on Fault-Tolerant Computing*, FTCS-24.

24 Normand, E. (1993). Single event upset in avionics. *IEEE Transactions on Nuclear Science* 40 (2): 120–126.

25 Barth, J.L., Dyer, C.S., and Stassinopoulos, E.G. (2003). Space, atmospheric, and terrestrial radiation environments. *IEEE Transactions on Nuclear Science* 50 (3): 466–482.

26 Barth, J. L. (1997). Applying computer simulation tools to radiation effects problems. 1997 IEEE Nuclear and Space Radiation Effects Conference Short Course. Snowmass Village, Colorado.

27 Sexton, F.W. (2003). Destructive single-event effects in semiconductor devices and ICs. *IEEE Transactions on Nuclear Science* 50 (3): 603–621.

28 Dodd, P.E. and Massengill, L.W. (2003). Basic mechanisms and modeling of single-event upset in digital microelectronics. *IEEE Transactions on Nuclear Science* 50 (3): 583–602.

29 Bruel, M. (1995). Silicon on insulator material technology. *Electronics Letters* 31 (14): 1201–1202.

30 Manasevit, H.M. and Simpson, W.I. (1964). Single-crystal silicon on a sapphire substrate. *Journal of Applied Physics* 35 (4): 1349–1351.

31 Kelly, A.T., Alles, M.L., Ball, D.R. et al. (2014). Mitigation of single-event charge sharing in a commercial FPGA architecture. *IEEE Transactions on Nuclear Science* 61 (4): 1635–1642.

32 Buchner, S.P., Campbell, A.B., Meehan, T. et al. (2000). Investigation of single-ion multiple-bit upsets in memories on board a space experiment. *IEEE Transactions on Nuclear Science* 47 (3): 705–711.

33 Krieg, J.F., Neerman, C.J., Savage, M.W. et al. (2000). Comparison of total dose effects on a voltage reference fabricated on bonded-wafer and polysilicon dielectric isolation. *IEEE Transactions on Nuclear Science* 47 (6): 2561–2567.

34 Fuller, R., Morris, W., Gifford, D. et al. (2010). Hardening of Texas Instruments' VC33 DSP. In: *IEEE 2010 Radiation Effects Data Workshop Record*, 153–157. Institute of Electrical and Electronics Engineers (IEEE).

35 Knudsen, J.E. and Clark, L.T. (2006). An area and power efficient radiation hardened by design flip-flop. *IEEE Transactions on Nuclear Science* 53 (6): 3392–3399.

36 McLain, M.L., Barnaby, H.J., Esqueda, I.S. et al. (2009). Reliability of high performance standard two-edge and radiation hardened by design enclosed geometry transistors. In: *IEEE Trans. Int. Rel. Phys. Symp*, 174–179. Institute of Electrical and Electronics Engineers (IEEE).

37 Mavis, D.G. and Alexander, D.R. (1997). Employing radiation hardness by design techniques with commercial integrated circuit processes. In: *Proc. 16th DASC AIAA/IEEE Digital Avionics Systems Conference*, vol. 1, 15–22. Institute of Electrical and Electronics Engineers (IEEE).

38 Rodbell, K.P., Heidel, D.F., Pellish, J.A. et al. (2011). 32 and 45 nm radiation-hardened-by-design (RHBD) SOI latches. *IEEE Transactions on Nuclear Science* 58 (6): 2702–2710.

39 Petersen, E. (2011). *Single Event Effects in Aerospace*. Hoboken, NJ: Wiley.

40 Petersen, E.L., Pickel, J.C., and Adams, J.H. (1992). Rate prediction for single event effects-a critique. *IEEE Transactions on Nuclear Science* 39 (6): 1577–1599.

41 Reed, R.A., Kinnison, J.D., Pickel, J.C. et al. (2003). Single-event effects ground testing and on-orbit rate prediction methods: the past, present, and future. *IEEE Transactions on Nuclear Science* 50 (3): 622–634.

42 Buchner, S.P., McMorrow, D.P., Melinger, J.S., and Campbell, A.B. (1996). Laboratory tests for single-event effects. *IEEE Transactions on Nuclear Science* 43 (2): 678–686.

43 Buchner, S.P., Miller, F., Pouget, V., and McMorrow, D.P. (2013). Pulsed-laser testing for single-event effects investigations. *IEEE Transactions on Nuclear Science* 60 (3): 1852–1875.

44 Kanyogoro, N., Buchner, S.P., McMorrow, D.P. et al. (2010). A new approach for single-event effects testing with heavy ion and pulsed-laser irradiation: CMOS/SOI SRAM substrate removal. *IEEE Transactions on Nuclear Science* 57 (6): 3414–3418.

45 Hales, J.M., Roche, N.J.-H., Khatchatrian, A. et al. (2015). Two-photon absorption induced single-event effects: correlation between experiment and simulation. *IEEE Transactions on Nuclear Science* 62 (6): 2867–2873.

46 Blandford, J.T. and Pickel, J.C. (1985). Use of Cf-252 to determine parameters for SEU rate calculations. *IEEE Transactions on Nuclear Science* 32 (6): 4282–4286.

47 Mapper, D., Sanderson, T.K., Stephen, J.H. et al. (1985). An experimental study of the effect of absorbers on the LET of the fission fragments emitted by Cf-252. *IEEE Transactions on Nuclear Science* 32 (6): 4276–4281.

48 Johnston, A.H. and Hughlock, B.W. (1990). Latchup in CMOS from single particles. *IEEE Transactions on Nuclear Science* 37 (6): 1886–1893.

49 Srour, J.R., Marshall, C.J., and Marshall, P.W. (2003). Review of displacement damage effects in silicon devices. *IEEE Transactions on Nuclear Science* 50 (3): 653–670.

50 Frenkel, Y. (1926). Über die Wärmebewegung in festen und flüssigen Körpern (*About the thermal motion in solids and liquids*). *Zeitschrift für Physik* 35 (8): 652–669.

51 Messenger, G.C. and Ash, M.S. (1992). *The Effects of Radiation on Electronic Systems*, 2e. New York: van Nostrand Reinhold.

52 Rax, B.G., Johnston, A.H., H., A., and Miyahira, T. (1999). Displacement damage in bipolar linear integrated circuits. *IEEE Transactions on Nuclear Science* 46 (6): 1660–1665.

53 JEDEC standard JESD89 (2006). *Measurement and reporting of alpha particle and terrestrial cosmic ray-induced soft errors in semiconductor devices*. Arlington, VA: JEDEC Solid State Technology Association.

54 Cobham (n.d.). Cobham rad solutions. http://ams.aeroflex.com/pagesfamily/fams-rad .cfm.

55 Barnaby, H.J., Schrimpf, R.D., Sternberg, A.L. et al. (2001). Proton radiation response mechanisms in bipolar analog circuits. *IEEE Transactions on Nuclear Science* 48 (6): 2074–2080.

56 Schwank, J.R., Shaneyfelt, M.R., and Dodd, P.E. (2013). Radiation hardness assurance testing of microelectronic devices and integrated circuits: test guideline for proton and heavy ion single-event effects. *IEEE Transactions on Nuclear Science* 60 (3): 2101–2118.

57 Schwank, J.R., Shaneyfelt, M.R., Dodd, P.E. et al. (2009). Hardness assurance test guideline for qualifying devices for use in proton environments. *IEEE Transactions on Nuclear Science* 56 (4): 2171–2178.

58 Schwank, J.R., Shaneyfelt, M.R., Baggio, J. et al. (2005). Effects of particle energy on proton-induced single-event latchup. *IEEE Transactions on Nuclear Science* 52 (6): 2622–2629.

59 Wie, B.S., LaBel, K.A., Turflinger, T.L. et al. (2015). Evaluation and application of U.S. medical proton facilities for single event effects test. *IEEE Transactions on Nuclear Science* 62 (6): 2490–2497.

60 Miller, R.B. (1982). *An Introduction to the Physics of Intense Charged Particle Beams.* New York: Springer US.

61 Passenheim, B.C. (1988). *How to Do Radiation Tests.* San Diego: Ingenuity Ink.

6

Device Modeling and Simulation Techniques

6.1 Introduction

While Chapter 5 provided guidance for radiation testing of devices and circuits, the focus of this chapter is to show how radiation affects devices, examine methods for modeling the effects of radiation on devices, and look at some ways to implement radiation models. The first section of this chapter provides an overview of device modeling. This includes discussion of analysis programs, circuit simulators, intrinsic models, and composite models. The second section provides examples of pre- and post-radiation characteristic curves for the more common devices that a circuit designer is likely to encounter, with discussion of the reasons for changes to device operation. This is not meant to be an exhaustive "effects" section, but only to provide some guidance on likely model parameter changes needed to match post-radiation characteristics. The final sections discuss implementing post radiation models and single-event transients (SET) simulation.

Providing an accurate prediction of circuit response to radiation environments can substantially reduce the nonrecurring engineering (NRE) costs and timeline for circuit qualification. This is easiest to accomplish by taking careful measurements during device testing and developing accurate models. If designing for space applications, cosmic rays must be taken into account. Predicting single-event effects (SEE) behavior can be much more difficult, but considering that a large portion of the NRE is for the design and tooling for IC fabrication, predicting functionality through SEE can become very important to the circuit development time to market.

Circuit designers depend on accurate device models, but often the device models in the foundry process design kit (PDK) will fall short in one way or another. Device models may not cover the temperature range of interest, for example providing only the commercial ($-25°C$ to $125°C$) rather than the military ($-55°C$ to $150°C$) temperature range. Electromigration rules may not cover the worst-case temperature of interest for the reliability required for some applications. Process technologies that are not specifically radiation tolerant or hardened will not have model corners for post-radiation conditions, so these must be developed in-house. On the other hand, if the foundry process has radiation corners, they may not be at the correct level or for the correct source of radiation. Process technologies are sometimes rated for a lower level of radiation due to export restrictions. These processes are often more radiation hard than advertised, and in-house provided corners can provide the assurance for meeting the correct mission life for the circuit being developed. Often, however,

Integrated Circuit Design for Radiation Environments, First Edition.
Stephen J. Gaul, Nicolaas van Vonno, Steven H. Voldman and Wesley H. Morris.
© 2020 John Wiley & Sons Ltd. Published 2020 by John Wiley & Sons Ltd.

the difference between a non-radhard and a radhard process technology is not only the provision for post-radiation corners but the design of the devices.

Throughout this chapter the role of device intrinsic models, parasitic devices, and process features will become much clearer, and in some cases will point to better device design. The pre- and post-radiation plots in this chapter are mostly of a generic nature and are not derived from any one specific process technology. The modeling techniques derive from standard approaches in the industry and the simulation methods are well known to anyone familiar with any of the commercially available design tools.

6.2 Device Modeling

Creating a device model or model corner requires some specialized tools and knowledge. Aside from the ability to develop test structures, take measurements, and organize data, it is very useful to be able to program in a statistical or analysis language that has good graphics capability. Another important skill is familiarity with the SPICE (Simulation Program with Integrated Circuit Emphasis) simulator being used for circuit simulations. For example, accessing the simulator from the command line provides better flexibility in netlisting and the collection of operating point data. This is especially important if a scripting language is used to help automate model development and very helpful when comparing measurement to simulation data. Most vendor simulation GUIs are designed for circuit simulation rather than model extraction, but their usage during model development can help ensure accuracy of the models, once installed in the PDK. The simulator will likely support many different intrinsic models, and familiarity with these models, including their schematics and parameter settings, is a must. Usually the engineering design automation (EDA) software vendor will provide this information in a user's guide or other document and this may be the only guidance for intrinsic models that are specific to the EDA vendor.

It is likely that intrinsic models for most submicron processes will be industry standard models developed and promoted through the Compact Modeling Coalition (CMC). The CMC was originally called the Compact Modeling Council and was formed in 1996 to promote collaboration and standards for the implementation of semiconductor device models. These models have largely replaced the proprietary models that were not portable across different circuit simulators. Still, there is the possibility that a specific intrinsic model will not provide a good fit to some devices. For this reason, some device models will actually be comprised of small schematics that use several intrinsic devices and models. Such composite or subcircuit models are very common, even a simple resistor can be implemented as a subcircuit model to properly distribute parasitic capacitances.

As there are many different device simulators and device intrinsic models provided by the various EDA software vendors, the simulation sections are written in the context of the frequently used Cadence Spectre® simulator and intrinsic models, including many that are supported by the CMC, including BSIM3, BSIM4, BSIMSOI, PSP, HICUM, and MEXTRAM. It is expected that such examples can be applied to simulators and intrinsic models that will be encountered in practice.

Table 6.1 provides an overview of the types of modeling efforts that can be done to provide post-radiation simulation capability in a design kit that is provided by a foundry

Table 6.1 Model effort for the various types of radiation and effects.

Radiation level	Radiation effect	Model effort
Cumulative dose	Total ionizing dose	Model corner
	Displacement	Model corner
Single event	Transient	Single-event device model
	Catastrophic	Overvoltage warning

or developed in-house. Exposures to radiation affecting all devices in a circuit that define a mission lifetime can be simulated using post-radiation corners, for example. This includes exposure to ionizing radiation as well as displacement damage from neutron irradiation.

The simulation of nondamaging SEE is more difficult, as only one or a few devices will be affected at any moment in time in some scenarios. The influence from a single event is transient in nature and requires a transient simulation for each device tested or suspected to be capable of an upset to the circuit operation. For large circuits, this can be a daunting simulation as well as bookkeeping task. Some EDA vendors provide simulators with built-in SEE simulation capability, but without some kind of automation, the simulation task can be difficult. For smaller process nodes, a single event can lead to proximity effects in nearby devices. This also complicates the simulation task and sometimes calls for detailed SEE simulation of the affected area. The difficulty of simulating single events is the main reason qualification of an integrated circuit for single-event immunity can end up being an iterative task through design, fabrication, and testing.

Catastrophic failures like single-event gate rupture (SEGR) fall into a simulation category where warning flags can be added to models. These provide feedback during the design process when gate voltage and other node voltages exceed the design limits, helping the designer to avoid voltage excursions during circuit operation that could lead to failure.

6.2.1 Circuit Simulators

Most commercially available simulation software for integrated circuit development will have a simulator based on SPICE [1, 2]. SPICE was developed by graduate students under the guidance of Dr. Donald Pederson and Dr. Ronald Rohrer at the University of California at Berkeley. The program known as SPICE today derived from a predecessor simulation program called CANCER (Computer Analysis of Nonlinear Circuits, Excluding Radiation) [3] that after extensive use by undergraduate and graduate students evolved into SPICE1 [4, 5]. CANCER was written with emphasis on linear IC design, and circuit size was limited to 400 components, including only 100 bipolar transistors and diodes and 100 nodes [6]. SPICE was originally written in FORTRAN and was widely distributed as public domain software in the 1970s and 1980s. In 1985 SPICE was rewritten in the C programming language but many early academic and industrial users had already developed proprietary versions. Eventually, commercial versions of SPICE surpassed most open-source and proprietary versions so that only the largest IC design companies could still support internal development of simulators.

Many IC design companies use one or more commercially available SPICE-based simulators, including, for example HSPICE® (Synopsys, Inc.), Spectre (Cadence Design Systems), Eldo® (Mentor, a Siemens Business – formerly Mentor Graphics), and SmartSpice™ (Silvaco, Inc.).

HSPICE is a commercial version of SPICE available from Synopsys, Inc. It was one of the first commercially available circuit simulators and was originally provided by Meta-Software in the 1980s. Meta-Software was acquired by Avant! in 1996 and Synopsys acquired HSPICE through its acquisition of Avant! in 2002.

The original Spectre simulator, like SPICE, was developed at the University of California at Berkeley but was targeted for microwave circuits. It used harmonic balance to obtain the steady-state solution of nonlinear circuits in the frequency domain. Cadence Design Systems developed a version of Spectre that utilized the learning from many years of SPICE optimization. Cadence replaced the harmonic balance algorithms with transient analysis algorithms, making it useful for other types of circuit simulation. The Cadence Spectre was three to five times faster than traditional SPICE but more accurate and reliable [7].

6.2.2 Intrinsic Models

Early versions of SPICE included built-in models, but as commercial versions of SPICE became available these built-in device models or intrinsic models became more proprietary and less portable between EDA software vendors. Industry leaders recognized the need for standards but also for model development leadership from academia and formed the Compact Model Council (CMC) in 1996. The CMC is now called the Compact Model Coalition and is a working group in the EDA industry focused on the standardization of SPICE device models (http://www.si2.org/cmc). The CMC includes members from many of the IC design and manufacturing companies as well as the major EDA software companies. Table 6.2 provides a timeline of device model development, including the delivery of many industry standard models. Most of the commercially available SPICE simulators will include most if not all the CMC-provided models provided in Table 6.2. The CMC is also involved in setting standards for the SPICE simulator language, enhancements to existing models, and the development of an application programming interface (API) to support extensions of the standard compact models. The TSMC (Taiwan Semiconductor Manufacturing Corp.) Modeling Interface (TMI) API is a good example, as it allows foundries to provide reliability aging parameters or functions into existing compact models.

6.2.3 Composite Models and Inline Subcircuits

While it is convenient to use a single intrinsic model to describe a device, it is often necessary to combine intrinsic models to model more complex behavior. A good example is the resistor model. It is possible to use a single netlisting line to define a resistor with a resistance value calculated from a function of the resistor width, length, sheet resistance, and temperature coefficient. In fact, many simulators come with library elements (e.g. Cadence analogLib) for the various schematic devices like resistors, capacitors, inductors, etc. The standard two-terminal resistor provided by all simulators requires only a

Table 6.2 Device model development timeline.

Device type	Model name	Year	CMC	Notes
MOSFET	Shichman/Hodges	1968		Quadratic model [8]
	BSIM1	1985		
	BSIM2	1990		
	BSIM3	1996	*	BSIM3 3.3.02005 (BSIM3v3)
	BSIM4	2000	*	BSIM4 4.8.1 (2017)
	PSP	2006	*	
	HiSIM	2011	*	
	BSIM-CMG	2012	*	Multigate MOSFET
SOI MOSFET	BSIMSOI	2002	*	
	HiSIM_SOI	2012	*	
LDMOS	HiSIM_HV	2007	*	
BJT	Ebers-Moll	1954		[9]
	Gummel-Poon	1970		[10]
	VBIC	1996		Open-source GP extension [11]
	MEXTRAM	2004	*	
	HiCUM	2004	*	
Resistor/JFET	R2_CMC	2005	*	2-terminal resistor
	R3_CMC	2007	*	3-terminal resistor/JFET
Capacitor	JUNCAP2	2005		Junction capacitor (NXP Semiconductor)
	MOSVAR	2006	*	PSP Varactor [12]
Diode	DIODE_CMC	2009	*	

Most new models are now reviewed and released by the Compact Modeling Coalition (CMC) (http://www .si2.org/cmc).

resistance value. However, if the capacitance of the resistive element to a well or body implant area is a concern an inline subcircuit can be used. This type of element builds up a composite model of the resistor by combining components in a mini-netlist.

Figure 6.1 shows the comparison between netlisting for a simple polysilicon resistor model and the inline subcircuit representation for the Cadence Spectre simulator. The rpoly function depends on l, w, and ρ_{poly} (length, width, and sheet resistance of the resistive element), and perhaps additional parameters specific to resistors. The cpoly functions will depend on the area and perimeter of the resistor. Of course, the better accuracy of the composite model comes at a cost – the capacitive elements as well as the resistive element need to be characterized and an additional node connection is added to the model.

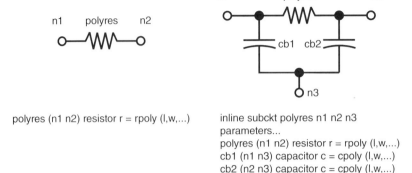

polyres (n1 n2) resistor r = rpoly (l,w,...)

inline subckt polyres n1 n2 n3
parameters...
polyres (n1 n2) resistor r = rpoly (l,w,...)
cb1 (n1 n3) capacitor c = cpoly (l,w,...)
cb2 (n2 n3) capacitor c = cpoly (l,w,...)
ends polyres

Figure 6.1 Netlist comparison between single component (left) and composite model or subcircuit (right) netlisting (Cadence Spectre).

6.2.4 Analysis and Statistics Programs

The essence of device modeling is matching measured results to simulated results, and that requires some mathematical and graphical help. Fortunately, there is software available that can ease the more difficult tasks so the practitioner can concentrate on getting the best fit to the data – S, S+, and open-source R. The S programming language was developed at Bell Laboratories in the 1970s [13]. It was first made available outside Bell Laboratories in the early 1980s and source code was licensed through AT&T Software sales in 1984. Updates were made to the language in 1988 [14] that align with modern versions of S-PLUS (first produced by a Seattle-based start-up company called Statistical Sciences, Inc.). The R programming language is an implementation of the S programming language under the GNU open-source license agreement (The R Project for Statistical Computing, https://www.r-project.org). Both the S and R languages are scripting languages for data manipulation, statistics, and graphics.

ICCap/MBP/MQA – The well-known IC-CAP (Integrated Circuit Characterization and Analysis Program) provided by Keysight Technologies (formerly Agilent Technologies) is a front-to-back device characterization and modeling product. It includes automated instrument control, data acquisition, parameter extraction, graphical analysis, simulation, optimization, and statistical analysis. There are turnkey extraction solutions for industry standard CMOS models as well as bipolar junction transistor (BJT), heterojunction bipolar transistor (HBT), and metal-semiconductor field-effect transistors (MESFETs). The Model Builder Program (MBP) fully automates model extraction but provides an open interface so that model extraction strategies can be customized. MBP supports all CMC model parameter extraction.

Utmost IV – Silvaco's Utmost device characterization and SPICE modeling solution is another well-known platform for device modeling. Utmost IV consists of four modules, including data acquisition, optimization, scripting interface, and model checking. The acquisition module includes all that is needed for device testing, collection, and manipulation of data. The optimization module is used to extract and optimize SPICE model parameters to the measured data. The script module allows the flexibility of controlling the data measurement, extraction and optimization from a scripting interface. The model-check module provides an easy-to-use tool to explore and test existing MOSFET device models. There is support for HSPICE, Eldo, Spectre, as well as Silvaco's Smart-Spice simulators.

6.3 Radiation Effects on Semiconductor Devices

This section examines the effects of radiation on many semiconductor devices used in IC fabrication. The section starts with simple devices like metal-oxide semiconductor (MOS) capacitors and proceeds to more complex devices. The radiation effects presented in previous chapters are applied to each device so as to provide a bridge between physics and practical application.

6.3.1 MOS Capacitors and Transistors

Perhaps the simplest of devices, the MOS capacitor can best illustrate some of the effects from radiation exposure and naturally lead to an understanding of such effects in MOS transistors. It would be a mistake, however, to assume that such simple devices behave simply in real life before and after irradiation. There are many types of MOS capacitors in integrated circuits including intentionally placed devices and parasitic devices. Figure 6.2 provides some examples of MOS capacitors that are often found in semiconductor processes. There are two that derive from their respective MOS transistors in that they have the same cross-sectional structure and there are two that are more likely candidates for use as a capacitor. These have the opposite body polarity from the expected MOS transistor so that there is a better connection to the substrate side of the capacitor.

MOS transistors are the primary element found in most integrated circuits. They make up the bulk of the devices in modern CPUs. It is clearly important to understand the effects that radiation may have on these devices and to minimize effects by using better device design and/or modeling post radiation corners that are not found in commercially available processes.

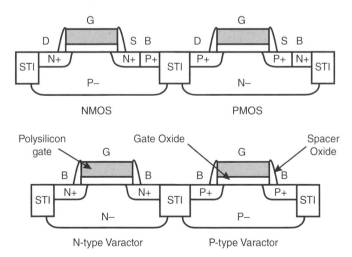

Figure 6.2 Cross section of four integrated MOS capacitors (not to scale).

6.3.1.1 MOS Capacitors

MOS capacitors are frequently found in analog integrated circuits. They are often the only available capacitor in some processes that offer a high capacitance per area if there is no provision for a PIP (poly-insulator-poly) or a MIM (metal-insulator-metal) capacitor. They are available for free in the process in that they usually do not need additional process masking steps, using instead the already-available masking steps needed to make MOS transistors. There are often capacitors available in bipolar transistor technologies, but these are usually at the cost of a mask step and utilize one of the bipolar device implants, forming a much higher doped substrate. These capacitors do not usually operate as MOS capacitors in the range of the supply voltages for bipolar processes and so the following sections will not strictly apply to such capacitors in their normal voltage operating range.

The MOS capacitors in Figure 6.2 include the NMOS-type and PMOS-type capacitors at the top of the figure, as well as the n-type and p-type varactors shown in the lower half of the figure. It is not likely that designers would use NMOS or PMOS capacitors, however, as their characteristics are less desirable than their varactor cousins. The discussion of the NMOS and PMOS capacitors is more for introduction to MOS transistors and for radiation characterization and measurement. They can provide, for example, scaled-up values for gate to body and gate to drain/source capacitances.

6.3.1.1.1 C-V Plot Basics

The most likely source of data for MOS capacitors will come from device measurements using some kind of capacitance-voltage (C-V) measurement technique. It is assumed that quasistatic conditions correspond closely to equilibrium during measurements so that when a voltage is applied to the MOS capacitor gate connection and swept from one value to another, the displacement current can be measured as a function of time. Integration of the charging current over time results in the measurement of the charge stored in the capacitor, yielding the capacitance

$$C = Q/V. \tag{6.1}$$

An example C-V plot for the n-type varactor is shown in Figure 6.3 and shows the influence of accumulation, depletion, and inversion on the capacitance of the device as

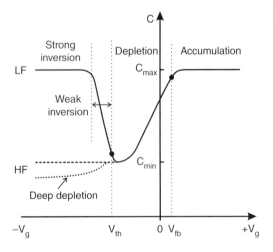

Figure 6.3 C-V plot for the n-type varactor.

the voltage is swept. During either inversion or accumulation, the maximum capacitance C_{max} is defined by the gate oxide thickness. In the subthreshold region of operation the capacitance is reduced to a minimum value C_{min} due to the series combination of the gate oxide capacitance and the capacitance of the depletion region. There is still a charge in the depletion region due to impurity doping atoms, but this charge is fixed in place and so it doesn't contribute to the displacement current during the capacitance measurement. When the body contact of the n-type varactor is tied to the highest supply voltage the capacitor will operate in its inversion mode. For a constant value of capacitance, the designer should avoid the subthreshold region of operation. The device can also be used in its accumulation mode if the body contact is at a lower voltage than the poly gate contact. This may be the better region of operation for the device especially if there is no source of minority carriers such as a nearby p-type region. Varactors are commonly used as the tunable element in a VCO, and in this case the voltage dependent region of operation is important. A p-type varactor can also be integrated in CMOS processes but operation in its accumulation mode would require the body to be tied to the highest voltage rail.

The capacitance of the MOS capacitor can also be measured directly by an impedance meter if a small sinusoidal signal is superimposed on the voltage sweep. The same behavior is expected as conditions remain at or near equilibrium, but depending on the frequency chosen for the sinusoidal signal the inversion region capacitance measurement will not be the same as the quasi-static measurements. The reason for this is the nonequilibrium nature of the inversion layer, which is formed by minority carriers swept into the region by an electric field. Modern semiconductor processes will have a high quality of starting material and also employ internal gettering early in the processing, resulting in a high-quality crystal. Carrier generation-recombination processes occur very slowly in such material with minority carrier lifetime in the millisecond range. Carriers in the inverted region simply cannot respond to or follow the superimposed sinusoidal signal if it is in the megahertz range, for example. The carriers in the inversion layer thus appear as fixed charges and, like the impurities in the crystal lattice, do not contribute to the displacement current. Figure 6.3 shows the resulting capacitance for this type of C-V measurement.

The time-dependence for the formation of the inversion layer can also affect C-V measurements that use a superimposed AC signal, making the measurements sensitive to the direction and speed of the voltage sweep. When the voltage is swept from accumulation to inversion, the voltage sweep can outpace the minority carrier generation rate, affecting the formation of the inversion layer. In fact, for high-quality substrates, the inversion layer may not form at all and the voltage sweep will just cause more depletion. This is called *deep depletion* and is indicative of a low defect substrate but it is usually something to be avoided in C-V measurements as it is a nonequilibrium condition.

To avoid deep depletion, high-frequency C-V measurements are usually made by sweeping voltage from inversion to accumulation, but this a questionable approach for automated testing. The model value for C_{min} may be influenced by deep depletion during such testing, especially if there is no source of minority carriers nearby (a p+ diffused region for the n-type varactor, for example). Another look at the n-type varactors in Figure 6.2 indicates that the mask layout of the devices may not provide a source of minority carriers near the gate region. This would cause operation in the inversion mode to be unreliable as populating the inversion layer would depend on generation/recombination in the n-well region. This could limit frequency response of the varactor and its actual capacitance could be much lower than the design capacitance due to deep depletion.

6.3.1.1.2 MOS Capacitor Response to Ionizing Radiation

Now as an example, the n-type varactor of Figure 6.2 is exposed to ionizing radiation, but it is not under electrical bias of any sort following its initial C-V measurement until it is again tested after irradiation. The resulting C-V plot will be shifted as shown in the post-radiation curve in Figure 6.4 due to the trapped charge in the oxide. In this case, the trapped holes behave like fixed charges so they provide a potential shift in the C-V plot but do not contribute to the capacitance measurement. Figure 6.4 shows only the high frequency C-V plot. The ionizing radiation results in hole and H^+ transport, breaking hydrogen-passivated bonds at the Si/SiO_2 interface and causing an increase in the interface trap density and border trap density. These trap locations can be confused because both can change their charge state rapidly and will show up on C-V plots using a measurement frequency in the 1 kHz or lower range. The slowest traps will not contribute to high frequency (>1 MHz) capacitance measurements, however. Changes at or near the Si/SiO_2 interface can affect the subthreshold slope. For very poor starting material or poor gate oxide growth conditions, this effect can be quite obvious in a C-V plot.

As discussed in Chapter 3, electric fields, temperature, and time contribute to the migration of trapped charge in the oxide to either the polysilicon gate or to the silicon–silicon oxide interface where it will recombine or occupy traps. This is obviously a simplified view of the annealing behavior shown in Figure 6.4, where the flatband voltage, V_{fb}, recovers toward its pre-rad value.

A likely intrinsic model for varactors is the PSP model mosvar [12]. This model is consistent with the PSP MOSFET model [15] but adds frequency dependency to the inversion charge. This is helpful when using the device in its subthreshold region, but the adjustment is a simple time constant approximation to keep the model compact. The flatband voltage parameter (VFBO) of the mosvar model can be adjusted for a post-radiation total ionizing dose (TID) corner, and perhaps the well doping level

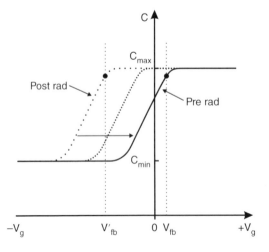

Figure 6.4 Post-radiation high-frequency C-V plot for the n-type varactor.

(NSUBO) could be adjusted to account for carrier removal as well. There is not a way to adjust for other changes that may occur, such as changes in subthreshold slope.

6.3.1.2 MOS Transistors

Following the ascendance of CMOS processes in the 1980s, the MOS transistor is almost universally associated with one of two complementary enhancement mode devices. The cross-sectional views of typical devices (not to scale) are shown at the top of Figure 6.2. To aid clarity, contacting and interconnect metal layers and other features of modern CMOS processes are not shown. Like the MOS varactor, CMOS devices are gate (G) controlled so that the appropriate applied voltage produces an inverted, conducting channel between the source (S) and drain (D) contacts. Figure 6.2 shows CMOS devices that have a body (B) connection integrated into the source contact. The body contact can also be separated so that a bias different from the source can be applied.

   In operation, CMOS devices conduct from the drain to source when the inversion layer is present. For an NMOS device, connections to drain and source are n-type and the inversion layer or channel is n-type. The same is true for the PMOS device, except all conduction regions are p-type. The point of this rather obvious observation is that CMOS devices use majority carriers for conduction, so they are less sensitive to radiation that affects minority carrier lifetime. The inversion layer is formed in the gate-controlled part of the body region and is readily populated by carriers supplied from either the source or the drain, so the formation of the inversion layer is not dependent on generation/recombination as with the MOS varactor.

6.3.1.2.1 MOS Transistor Response to Ionizing Radiation   A plot of drain current I_{ds} (logarithmic scale) versus gate voltage for a typical NMOS device is shown in Figure 6.5. In this plot, the drain to source voltage V_{ds} is positive but much lower than the threshold voltage V_{th} during measurement, typically around 50 mV. As the gate to source voltage V_{gs} is swept positive from 0 V, conduction from drain to source passes through three regions. The region below V_{th} is the subthreshold or exponential region. This region is due to channel to source p-n junction current, as some electrons can diffuse into the channel from the source if the source to body diode is forward biased. Once V_{gs} reaches the threshold voltage, the device enters a quadratic region, which then transitions into a linear region at higher gate voltages. These three regions are the result of

Figure 6.5 Pre- and post-radiation I_{ds} versus V_{gs} log-linear plots for NMOS devices (V_{ds} < 100 mV).

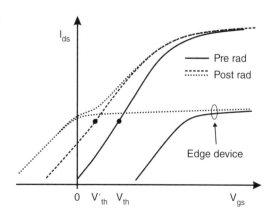

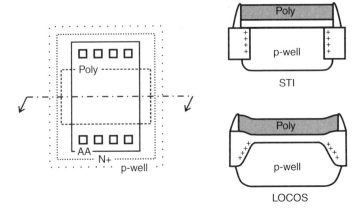

Figure 6.6 NMOS layout view and cutline cross sectional views for STI and LOCOS isolation. Trapped holes in the isolation oxide along the edge of the NMOS device reduce the field threshold leading to subthreshold leakage after irradiation.

two overlapping carrier transport processes. In subthreshold the diffusion of carriers dominates I_{ds} whereas in the linear region carrier drift accounts for most of the drain current.

MOS devices will experience a threshold voltage shift after exposure to ionizing radiation, and this is shown as the post-radiation curve in Figure 6.5. The reasons for this threshold shift are identical to those stated for the MOS varactor; namely, an accumulation of holes in the gate oxide adds a fixed positive charge that shifts the curve in the negative direction. This means that for CMOS elements, the PMOS and NMOS thresholds will become skewed. The same ionizing radiation exposure that reduces the positive threshold voltage of the NMOS shifts the PMOS threshold more negative.

Figure 6.5 show two representative post-radiation curves for NMOS devices. One looks almost identical to the pre-radiation plot except for the shift in V_{th} and a change in subthreshold slope; the other is very similar but also has a higher subthreshold leakage. For NMOS transistors, the p-type well region comprising the body of the NMOS can be prone to inversion in places where there is thicker oxide, for example at the edges along the oxide that isolates the thin oxide gate region of the device. This is shown in the planar view of the NMOS layout in Figure 6.6 as the leakage paths at the edges of the field oxide. In most CMOS processes, this oxide is either a thick local oxide (LOCOS – local oxidation of silicon) or shallow trench isolation (STI). Both types of isolation have a thermal oxidation of silicon that will segregate boron from the p-type well of the NMOS into the oxide so that regions under the oxide are depleted of boron and more prone to inversion. Additionally, the isolating oxide is thicker than the gate oxide and so will accumulate more holes relative to the thin gate oxide for the same ionizing radiation dose. This makes the regions at the sides of the NMOS much more likely to become inverted causing post-radiation subthreshold leakage. The influence of this edge device before and after ionizing radiation is shown in Figure 6.5.

Providing a good post-rad MOSFET model requires some forethought. MOSFET models have built-in geometry dependence, but the assumption for most is a scaling by channel width. It can be difficult to build both channel and sidewall conduction into

Figure 6.7 Separation of channel and sidewall devices in the MOS model.

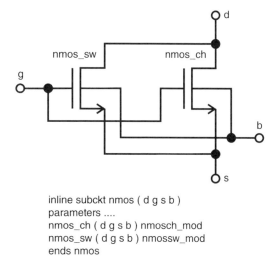

```
inline subckt nmos ( d g s b )
parameters ....
nmos_ch ( d g s b ) nmosch_mod
nmos_sw ( d g s b ) nmossw_mod
ends nmos
```

one device because these components have different device geometry dependencies. It is best to model these components as separate, parallel devices, as shown in Figure 6.7. Planning test chips with a suitable number of width/length device combinations will allow parameters for the channel and sidewall devices to be separated. In some ways, the more modern STI behaves worse than local oxidation (LOCOS) and layout or processing methods are the best approaches to minimize the effect from the sidewall device [16]. Even in hardened technologies, sidewall conduction should be modeled, however. A model extraction methodology such as outlined in [17] could be used, for example.

Both PMOS and NMOS transistors can have a change in subthreshold slope after radiation, and this is due to an increase in interface states. This occurs because most CMOS processes utilize forming gas (typically 5% H_2 in N_2) to passivate many of the Si dangling bonds at the Si – SiO_2 interface, which would otherwise contribute to the interface state density. Ionizing radiation and the corresponding transport of holes in the gate oxide will release hydrogen ions, which transport to the interface and break Si—H bonds producing H_2 and a dangling Si-bond according to McLean's two-stage model [18]. The increase in interface states after ionizing radiation increases the surface recombination during the operation of both PMOS and NMOS devices. This can reduce mobility, affecting the MOS transconductance and output conductance.

As gate oxide thickness is reduced, the cross section presented to radiation is correspondingly reduced, so that thinner oxides accumulate fewer trapped holes for the same radiation levels used on thicker oxides. However, thinner oxides are more prone to leakage due to tunneling, and the probability of tunneling increases exponentially with decreasing gate oxide thickness. While such gate oxides can be very robust to ionizing radiation, this exposure can increase gate leakage. A radiation-induced leakage current (RILC) was found by Scarpa et al. in 1997 [19] after a 5.3 Mrad(Si) dose on 4.4 nm thick oxides. It had many of the same features when compared to stress-induced leakage current (SILC) measurements in the same study and also as reported previously by other researchers [20–23].

6.3.1.2.2 MOS Transistor Models All MOS models used for simulation must ultimately relate an input voltage (Vgs) to an output current (Ids). For modern MOSFETs, this is not the simple task that it would seem. It involves relating an applied voltage in one dimension to a resulting current in another dimension, essentially a 2D problem. This requires certain simplifying assumptions to reduce the scale of the calculations so that the model is compact enough for circuit simulation. An early simplification dealt with the electrostatic solution of the input voltage and the resulting electrical conditions at the semiconductor surface by assuming the potential along the channel varies gradually enough for the input voltage to be valid. The output current equation is complicated by the need to account for the inversion layer thickness. To avoid the numerical overhead to integrate the inversion layer charge, another simplification is introduced, the so-called charge-sheet approximation. This approximation assumes the inversion layer is so thin that the potential doesn't vary through its thickness.

Models for MOS transistors can be classified into three modeling methods spanning five generations of model development [24]. The modeling method is based on the physical quantity used to derive the current equation of the device. In the first three generations of models I_{ds} is based on the threshold voltage. This formulation has the benefit of being very simple and compact but is difficult to reduce to a single current expression for both weak and strong inversion, so a smoothing function needs to be used to connect the two regions. The first-generation models were developed at UC Berkeley in the 1970s and were based in part on the quadratic model of Shichman and Hodges [8]. MOS model development over the ensuing years kept pace with process improvements and the quest for smaller device geometries leading to the introduction of the second-generation models, BSIM1 (Berkeley Short-channel Insulated-gate field-effect transistor Model) in 1985 and BSIM2 in 1990. A third generation of models sought unification of the two current functions into a single analytical expression over all regions of operation. BSIM3 [25, 26] was the first such model and was released in 1996 and was very successful. The BSIM3v3 version was the de-facto standard for process characterization by all foundries over a period of more than 10 years, covering technologies from 0.6 μm to 65 nm [24]. BSIM4 was introduced in 2000 and supported physical effects such as gate current, encountered in smaller process nodes. The last update to BSIM3 was in 2005, whereas the BSIM4 model was more recently updated in 2017. A MOS model suitable for use with silicon-on-insulator (SOI) substrates, BSIM-SOI, was selected by the CMC as the standard SOI MOSFET model in 2001. The latest release, BSIM-SOI 4.6.1 was released in late 2017 (the BSIM Group at Berkeley maintains a BSIM page at http://bsim.berkeley.edu/models/).

The next generations of MOS transistor models can be split into two groups [27], both of which are physics-based and result in a formulation for I_{ds} that is continuous over the entire range of operation. Both of these approaches to modeling would not be possible without the improvement in computing power and speed, as they provide implicit rather than explicit expressions for the device current. Each approach starts with the boundary conditions at the source and drain and use an iterative method to calculate the terminal potentials and currents. The first of the newer types of model solves for the surface potential at the two ends of the channel and then calculates the terminal charges. These are called *surface potential models*. The second type of model solves for the density of the inversion charge at either end of the channel so that both conductance and capacitance calculations are based on charge. Such models are called charge-based models.

Surface potential models represent the fourth generation of MOS models and provide accuracy for geometries $0.1\,\mu m$ and lower. The MOS current is a function of surface potential and provides a continuous I-V characteristic. The PSP MOS model [15] is a good example. It was jointly developed by Pennsylvania State University and Philips Research Laboratories and made available via the CMC in 2006. Both the intrinsic and extrinsic MOS features are based on surface-potential formulations in the PSP model. It provides an accurate description of the accumulation region and includes all relevant small-geometry effects, making it capable down to the 65 nm node and beyond. The HiSIM (Hiroshima University Starc Igfet Model) model [28] is a more recent surface potential–based CMC addition, standardized in 2011 after about a decade of release activity. HiSIM is suitable for modeling sub-100 nm scale MOS devices. The PSP and HiSIM models differ in their approach to the calculation of the surface potential. HiSIM uses an iterative solution using the exact surface-potential equation whereas PSP uses a noniterative algorithm.

Charge-based models are considered generation 5 MOS models and include BSIM5/6 (2004/2012) [29, 30] and the EKV MOSFET model [31]. They address several issues with surface potential (SP) models. For example, SP models require extremely high accuracy for the surface potential solution, and this accuracy isn't required when solving for the inversion charge. Solving for the inversion charge takes advantage of the direct link between inversion charge and transconductance as well as the fact that most of the important device characteristics can be expressed in terms of source and drain inversion charges. The efforts of the BSIM and EKV groups were combined in 2011 to focus on charge-based compact models. These models are aimed at the smallest device structures including multi-gate MOS devices such as FinFets. At very small dimensions, the gate can no longer control the leakage paths that are far from the gate. The gate control problem is addressed by reducing the size of the bulk/body region. FinFets use a fin-shaped body region that is made thin using an oxidation process. The gate surrounds the bulk region on three sides to increase the influence of the gate. The CMC-approved BSIM-CMG is a model based on BSIM6 that addresses the shape and characteristics of FinFets. Another approach to the gate influence issue is to start with a very thin SOI layer so that the bulk region can be controlled by front and back gates. There is also a BSIM6-based model for this structure, BSIM-IMG. Both of these models address the leakage currents caused by the structure of small MOS devices but other problems arise from scaling to such small geometries. There are process capability issues such as the resolution and accuracy of lithographic equipment. For the very smallest geometries lithographic accuracy affects the distribution of gate lengths. Another scaling issue is the distribution of dopants in the body region no longer appears uniform and can vary device to device. Both of these issues can cause a larger-than-expected mismatch between devices.

We now return to Figure 6.5 to explore the parameters available from the various MOS models that can be used to model the post-radiation behavior. Table 6.3 provides a cross reference for the various MOSFET models and the parameters each has available to model radiation effects in simulation corners.

An obvious omission from Table 6.3 is a column for subthreshold leakage. If the post-radiation characterization indicates substantial subthreshold leakage current the

Table 6.3 MOSFET compact model parameters for post-radiation corners.

MOS Model	V_{th} Shift	Subthreshold slope	Mobility	Gate leakage
BSIM3	v_{th0}	nfactor, cit	u0, ua, ub	AIGBACC, BIGBACC
BSIM4	v_{th0}	nfactor, cit	u0, ua, ub	AIGBACC, BIGBACC
BSIMSOI	v_{th0}	nfactor, cit	u0, ua, ub	AIGS, BIGS
PSP	V_{FBO}		P0MUE, U0, THEMU0	GCOO, CHIBO
HiSIM HiSIM-HV HiSIM SOI	V_{FBC}, N_{SUBC}		MUECBO, MUECB1, MUEPH0, MUEPH1, MUESR0, MUESR1	GLEAK1, GLEAK2
BSIM-CMG		CIT, PHIG	U0, UA, UC, UD	AIGBNV, BIGBINV, CIGBNV

parasitic element responsible for the leakage must either be eliminated or added to the model. For the latter, a composite model can be developed with the regular device in parallel with the edge device. There are some ways to reduce sub-threshold leakage in NMOS devices, including layout changes [32–34].

6.3.2 Diodes and Bipolar Transistors

It is worth noting that although bipolar devices were the first to be reduced to practice, the conceptual ideas underlying the field-effect transistor predated those of the bipolar transistor. Much of the experimental work done to investigate the physics of semiconductor surfaces benefited the development of the point-contact transistor in 1947, which, of course, lead to the first BJT quickly thereafter [35]. Most modern integrated circuit processes can be rightly identified as CMOS processes, but bipolar processes were mainstays of IC manufacture into the early 1980s. Since that time, the ability to scale size and power consumption has boosted the development of CMOS processes. Bipolar transistors are commonly integrated into CMOS processes using existing masking and processing steps, but their performance will be somewhat lacking unless there are additional features such as buried layers, collector sinker diffusions, or dedicated processing to form the emitter, base, and collector regions.

6.3.2.1 Diodes

Diodes are ubiquitous in semiconductor processes. They form the isolation in junction isolated (JI) processes, which include any process that starts with a bulk wafer. They are part of every active device in an integrated circuit. They are used to set reference voltages, provide programming paths, protect circuits from electrostatic discharge (ESD), and protect digital logic from process-induced charging (antenna effects on polysilicon and metal layers). Such small rectifying diodes appear to be relatively unaffected by ionizing radiation. Some early researchers [36] found that discrete diodes exhibit "in-herent radiation hardness" when exposed to doses of about 300 Krad(Si) with the main effect being an increase in leakage current.

For the most part, few changes should be needed in post-radiation diode models, but this will depend on the usage of the diode. Antenna diodes could be characterized for any changes in leakage current, but the same junctions are generally used in the MOS devices comprising the digital circuitry where such diodes are used. It is more likely that such changes in leakage current would be collected at the circuit level and result in a post-rad quiescent current specification for the production circuit. As most diodes are used to stand off voltage, perhaps the breakdown voltage should be characterized? Typically, most breakdown voltages do not change much after ionizing radiation unless there is a weakness in the construction of the device. The "washed emitter" bipolar transistor (Section 6.3.2.2) is a good example of a construction weakness that would translate to a poor diode design. Furthermore, hot electrons or holes generated during measurements of breakdown voltage tend to change the oxide charge near the junction so that a true post-radiation characterization isn't possible.

Neutron irradiation reduces the minority carrier lifetime according to Eq. (6.3). This can affect the frequency dependence of the diode junction capacitance. Post-neutron irradiation low frequency capacitance ($f < 1/\tau$) measurements will track with pre-rad, but capacitance at higher frequencies will be lower than pre-rad. The trapped charge lifetime is long compared to the measurement frequency, and the charge cannot fol-low the variation [37]. Neutron irradiation can also affect the forward voltage of diodes depending on the pre-rad series resistance. The lightly doped part of the junction would be most sensitive (Figure 3.37), becoming more resistive and adding to the forward drop of the diode.

There are some special-purpose diodes that provide exceptions to the above discus-sion. Zener diodes that are used in voltage references may have higher noise after irradi-ation especially if the junction breakdown occurs at or near the surface of the device. A buried Zener construction is a better choice for such applications. Some diodes are pur-posefully constructed to be sensitive to radiation. Diodes that have a very low doping or intrinsic region are sensitive to carrier removal and other displacement damage caused by neutron or other particle bombardment (Figure 3.39). Laser diodes are not partic-ularly affected by ionizing radiation but are sensitive to neutron irradiation. The main effect is an increase in threshold current density caused by the creation of nonradiative recombination centers that compete with radiative recombination sites [38].

6.3.2.2 Bipolar Transistors

Once the mainstay of the electronics industry, the bipolar transistor and bipolar semi-conductor processes saw declines in use as MOS processes and devices were developed and put into production. The largest impact was the transition of transistor logic due to the easy scalability of MOS devices and the low power consumption in CMOS processes. Still, bipolar transistors have their place and advantages over MOS devices. Their low output impedance means they are better at driving low impedance loads. They are bet-ter choices for low noise amplification because they can provide a higher gain per stage. In CMOS processes used for analog applications, bipolar transistors are still used in bandgap circuitry to provide stable, temperature compensated voltage references. While bipolar-only processes and circuits have their place in the analog world, bipolar transistors are more likely to be found as incidental or add-on devices to CMOS processes. They are also found as parasitic elements that can be serious problems in radiation environments, and this is covered in later chapters of this book. ESD protection is an application where parasitic elements are put to better use. ESD clamps based on silicon-controlled rectifier (SCR) or MOS snapback are good examples. Again, later chapters of this book are aimed at providing better understanding on how ESD circuitry behaves in radiation environments.

Due to the long history of bipolar transistor development and their integration into more modern processes, the practitioner will encounter many different geometries and a wide variety of mask artwork that can be challenging to both characterization and modeling. Some examples of integrated NPN transistor cross sections are shown in Figure 6.8. The topmost cross section is representative of planar junction isolated bipolar technology, where successive regions of the transistor are created by either solid source diffusion or ion implantation through openings in relatively thick thermally grown oxides. The middle cross section is typical of later generations of bipolar processing where a recessed local oxidation is used in addition to junction isolation to provide isolation. Typically, the sidewalls of the base diffusions are against the recessed oxide with only the floor area of the base being junction isolated. Another possible feature of this technology is the so-called washed emitter, which uses a self-alignment against the recessed oxide to align emitter to the base region. A super self-aligned polysilicon emitter NPN structure is shown at the bottom of Figure 6.8. This technology uses dopant out-diffusion from two polysilicon layers to create both base and emitter contacts. These three examples provide a timeline of bipolar technology scaling.

DI (dielectric isolation) was once the only option to provide electrical isolation from the substrate and was used extensively from the 1960s through the 1990s when radiation hardened bipolar circuits were in development and production for space and other applications. It is representative of a planar bipolar process except the isolation from the substrate is a dielectric (SiO_2) instead of a diode as in junction isolation. DI technology has been largely replaced by more modern approaches for SOI substrates but is still in production at Intersil Corporation (acquired by Renesas Electronics in 2017). During the early 1980s, bipolar processing took advantage of the scaling techniques used in CMOS processing resulting in the washed-emitter technique where base regions as well as emitter regions are self-aligned to an isolation oxide (typically a LOCOS oxide). The so-called oxide-isolated bipolar transistor allowed for smaller feature sizes and replaced many of the junction surfaces with oxide. This processing method was the swansong of the digital bipolar circuit, however, with most digital applications switching to CMOS processes during the latter part of the 1980s. As bipolar processes aimed at digital applications waned, they were still relevant for analog applications. Development of HBTs (heterojunction bipolar transistors) in the 1990s provided greater scaling and higher-frequency operation suitable for wireless applications. HBTs come in a variety of materials and structures, but most make use of polysilicon emitters and novel methods for forming or contacting the base region. HBTs are often integrated into CMOS processes, as many of the fabrication steps are common to both types of devices.

Lateral PNP devices are often available as a device selection in many foundry and proprietary processes. They are simple devices with a base width that is laterally scalable,

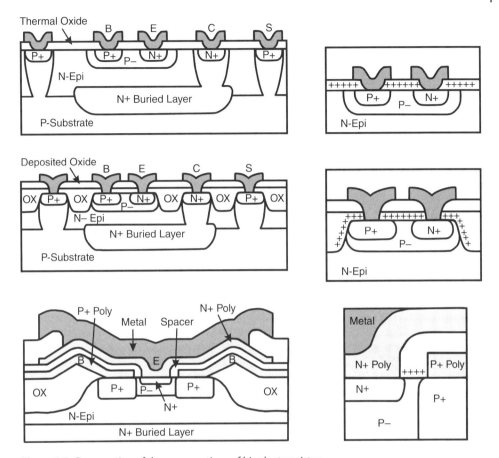

Figure 6.8 Cross section of three generations of bipolar transistor.

but must often be implemented as circular devices in order to increase the collector efficiency and this increases their size substantially. Lateral PNP devices are not generally good choices for use in radiation environments. The intrinsic base region has a large interface with an overlying oxide, and like a MOS device it is highly influenced by interface state density because current flow is near the interface. The effect is so pronounced that lateral gated PNP devices (Figure 6.9) are used to characterize the increase in interface states caused by radiation [39–43], especially for low-dose environments where Enhanced Low Dose Rate Sensitivity (ELDRS) is a concern.

6.3.2.2.1 Bipolar Transistor Response to Ionizing Radiation The primary effect from ionizing radiation on bipolar transistors is an increase in surface states and the associated increase in surface recombination velocity in the base region. The same mechanisms described in Chapter 3, namely, the transport of holes and hydrogen leading to the Si/SiO_2 interface, lead to the increase in interface states. The oxides in this case may be thicker and the result of multiple diffusions/oxidations during the device processing rather than carefully grown gate oxides. The increase in non-ideal base current shows up nicely on a Gummel plot as well as a beta vs. Ic plot as shown in Figure 6.10. For

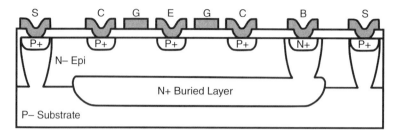

Figure 6.9 The gated lateral PNP (GLPNP) device.

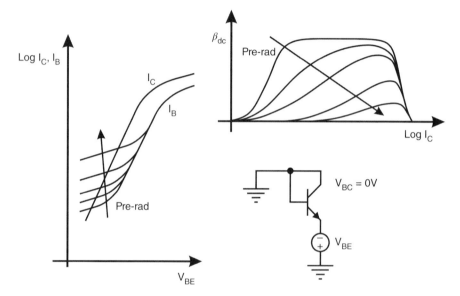

Figure 6.10 Gummel and beta versus Ic plots for an NPN transistor showing pre- and post-radiation response.

modeling purposes, these plots are generated with $V_{BC} = 0V$ and tested as shown in the inset schematic of Figure 6.10. These plots are representative of the response to gamma irradiation and show that increasing dose (direction of arrows) reduces the usable gain at low base injection. The gain degradation can be substantial for high total doses but because the amount of oxide interface over the base is defined by the geometry of the transistor, the degradation will saturate at some point.

A good example of how transistor geometry can affect total dose response is shown in Figure 6.11 [44] where NPN transistors from both a 15V planar process and a dual polysilicon emitter process are compared after a total dose of 1 Mrad(Si) (Aracor 170 Krad(Si)/min.). An inspection of the top and bottom right cross sections in Figure 6.8 shows the much reduced Si/SiO_2 interface over the intrinsic base region for the polysilicon-emitter NPN when compared to the planar device.

The use of local or STI oxides to define active regions for bipolar and other devices may provide greater scaling of the technology due to the reduction in junction capacitance, but charge collection in the thick isolating oxides can be a problem for MOS [16]

Figure 6.11 Comparison of 15V planar and polysilicon emitter NPN transistor response to 1 Mrad(Si) TID. Source: Adapted from Gaul et al. 1990, Figures 3 and 5 [44].

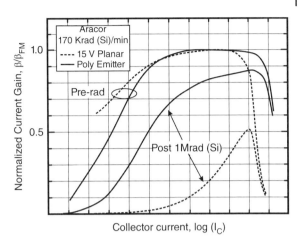

as well as bipolar devices. The middle right cross section in Figure 6.8 illustrates how this can come about. The base region of an NPN transistor is walled by the recessed oxide along its periphery. During device processing, the boron dopant in the base diffusion naturally segregates into the recessed oxide, depleting the surface of dopant. After ionizing radiation, charge in the recessed oxide depletes the sidewall of the base, changing its capacitance. For aggressively scaled NPN transistors the same active area opening is used to define both base and emitter diffusion edges. This can produce an NPN transistor that has a lateral weakness and a BVceo controlled by lateral instead of vertical breakdown. Such a transistor will have a higher gain at low base injection because the lateral base region is narrower than the base region in the floor area of the transistor.

Neutron fluence affects the minority carrier lifetime in the bases of bipolar transistors. Messenger and Spratt [45] found the change in current gain with neutron fluence is given by

$$\frac{1}{\beta} = \frac{1}{\beta_0} + \frac{\Phi}{\omega_T K} \tag{6.2}$$

where the 0 subscript denotes the pre-rad value of β, the transistor common emitter gain. The value of ω_T is taken as 2π times the common emitter gain bandwidth product, Φ is the neutron fluence, and K is the lifetime damage constant:

$$\frac{1}{\tau} = \frac{1}{\tau_0} + \frac{\Phi}{K}. \tag{6.3}$$

Equation (6.2) is accurate as long as $\beta > 3$. Equation (6.3) can be used for displacement damage caused by all kinds of radiation, including neutrons, protons, electrons, and gamma rays, but each will have its specific value for K [45, 46]. In practice, care must be taken when using Eq. (6.2) as ω_T includes dependencies on the values of R_E and $C_{BE,}$ which are geometry dependent.

6.3.2.2.2 *Bipolar Transistor Models* The first bipolar transistor models were very simple, for example the Ebers-Moll model with only 16 model parameters. It was introduced in 1954 by J. J. Ebers and J. L. Moll [9] and used in the first Spice simulators. An

Table 6.4 BJT compact model parameters for post radiation corners.

BJT model	Collector current	Forward beta	Reverse beta
SGP/VBIC	IS, NF	BF, ISE, NE	BR, ISC, NC
Mextram	IS	BF, IBF, MLF	BR, IBR, VLR
HICUM	C10 or IS	IBEIS, MBEI, IBEPS, MBEP, IREIS, MREI, IREPS, MREP	IBCIS, MBCI, IBCXS, MBCX

improvement over this model, the Spice Gummel-Poon (SGP) model, first described by Hermann Gummel and H. C. Poon at Bell Labs in 1970 [10], is still widely used. It captures most of the second-order effects, has current equations that are valid in all regions of operation, and all equations have continuous derivatives. Unlike MOSFET models, the SGP model does not include explicit geometry dependence other than a transistor area factor. Parasitic resistances such as Rc and Rb must be separately calculated based on the device geometry. The SPG model is well known for its limited accuracy at medium and especially high collector current densities which limits its application to high-speed circuitry [47]. Getreu [48] provides a good reference and guide for SGP model parameter extraction. The VBIC bipolar model is an open source model developed by IC and CAD industry representatives as a replacement for the SPG model [11]. It includes improved modeling of the Early effect, substrate current, quasi-saturation, and behavior over temperature.

The SiGe and BiCMOS processes used for RF applications benefit from the more modern Mextram [49–51] and HICUM [52–54] bipolar models. Mextram (Most Exquisite TRAnsistor Model) is a great choice for Si or SiGe processes. It includes high injection effects, has a dedicated epi-layer model, supports self-heating, and is formulated to minimize interactions between DC and AC characteristics to simplify model parameter extraction. It was originally developed at NXP Semiconductors by De Graaff and Kloosterman [49] for internal use but was released to the public in 1994. The HIgh-CUrrent Model (HICUM) has been a standard compact model for BJTs and HBTs for many years. Its development started around 1980 and was aimed at a more accurate and physics-based description for high-current effects when designing for fiber-optic applications. Table 6.4 lists parameters for each of SGP, Mextram, and HICUM compact models that may be affected or adjusted for post-radiation corners.

6.3.3 Power Devices

Many IC applications will require some kind of power control, for example motor control or buck/boost power conversion. This is usually accomplished by using either integrated or off-chip power devices. The most common device encountered is the double-diffused MOS (DMOS). These devices come in both n-type and p-type varieties and in vertical, quasi-vertical, and lateral geometries. DMOS devices are generally built to handle high currents and stand-off high voltages and so they present some special concern for space applications. As discussed in Chapter 4, such devices are prone to SEGR and so must generally be derated from their terrestrial operating voltages to guarantee safe operation in space, for example.

6.3.3.1 DMOS Composite Models

The normal power device modeling technique is to use a macromodel or schematic approach, using several elements to model the features of the DMOS device. One such approach is shown in Figure 6.12 for a quasi-vertical NDMOS device. The switching element of the DMOS device uses a conventional MOS device model whereas pinch-off of drain current near the source is controlled by a JFET element. Because DMOS elements often are of special geometries, parasitic resistances are calculated using geometry dependent functions. For example, the quasi-vertical DMOS will have a drain resistance that includes both lateral (N+ buried layer) and vertical resistances including the epitaxy layer as well as contacting layers to the N+ buried layer. Gate structure may be different than the simple structures used for low voltage MOS so that some overlap capacitances, for example cgs in Figure 6.12, are calculated separately for better accuracy. For discrete devices, the resistance in the contacting metals may be included in the macromodel. This is not generally done for devices integrated into the semiconductor process. The resistance of the metal for integrated devices is usually calculated using one of several parasitic extraction tools so that connections can be optimized for low resistance.

Aside from the danger of SEGR, the post-irradiation model for the DMOS device is much the same as for conventional MOS with the likely corner parameters given in Table 6.3, depending on the device model. Most of the other elements in the quasi-vertical DMOS macromodel are buried, and not in contact with device oxide layers, leaving them mostly unaffected by total dose effects. This is not true for all DMOS devices, however. Semiconductor processes that are designed for power regulation/conversion or motor control often have integrated power devices. These can

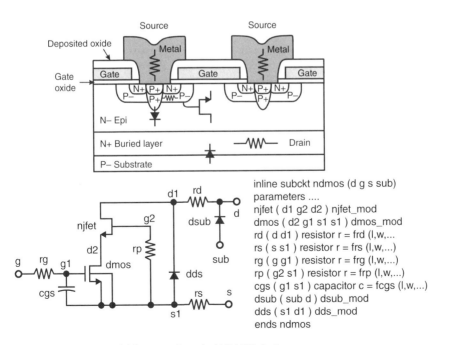

```
inline subckt ndmos (d g s sub)
parameters ....
njfet ( d1 g2 d2 ) njfet_mod
dmos ( d2 g1 s1 s1 ) dmos_mod
rd ( d d1 ) resistor r = frd (l,w,...
rs ( s s1 ) resistor r = frs (l,w,...)
rg ( g g1 ) resistor r = frg (l,w,...)
rp ( g2 s1 ) resistor r = frp (l,w,...)
cgs ( g1 s1 ) capacitor c = fcgs (l,w,...)
dsub ( sub d ) dsub_mod
dds ( s1 d1 ) dds_mod
ends ndmos
```

Figure 6.12 Macromodel for a quasi-vertical NDMOS device.

be quasi-vertical DMOS devices but are more often lateral devices. This typically puts the drain diffusion in contact with the isolation oxide. This will change the model for the drain resistance and capacitance, especially for PDMOS devices.

6.3.3.2 Operating Voltage
DMOS and other output devices may not operate reliably at their full-rated voltage in a radiation environment. Large MOS devices can store large amounts of charge, which can be released suddenly by a cosmic ray causing gate rupture (SEGR) or catastrophic burnout (SEB) of the device (see Chapter 4). The only reliable way to determine the operating voltage for space-rated devices is to subject them to single-event testing. It is likely that the devices will be derated by 50% or more from their maximum design operating voltage.

A simple step can be taken to identify issues with SEGR and SEB during circuit design. It involves examining the device model to find the location of the safe operating area (SOA) parameters. All of the common MOS models have these parameters, which cause a warning line to be added to the simulation log file when device terminal voltages exceed the parameter setting. There may not be any in the model, as they are optional parameters. The default value for the SOA parameter is infinity unless it is one of the reverse voltage parameters. The reverse voltage SOA value is set to the value of the forward voltage SOA parameter, for example vgdr_max = vgd_max by default. If there are SOA parameters in the device model, they are probably set to the commercial operating voltage of the device. Some SOA parameters will have to be set to lower voltages that are consistent with the SEGR and SEB limits of the power device. Table 6.5 provides a listing of the available SOA parameters.

6.3.4 Other Devices

There are several other devices one may encounter that are worth a quick look. These include JFETs and passive devices like resistors and capacitors. The models for these devices haven't seen the same level of development as has been done with the MOS and BJT devices. The CMC released two resistor models in the 2000s, R2_CMC, and R3_CMC. The R2_CMC model is a two-terminal resistor and the R3_CMC is a three-terminal resistor that could also be used to model a JFET device used in its linear region. Most simulators come with standard models for JFETs, resistors, and capacitors. The Cadence Spectre simulator supports a JFET model derived from the FET model of Shichman and Hodges. Also supported are several types of three-terminal resistors that implement the Figure 6.1 subcircuit and a version for junction-isolated resistors.

6.3.4.1 Junction Field Effect Transistors (JFETs)
The construction of the junction field effect transistor (JFET) is naturally hard to total ionizing dose. Figure 6.13 illustrates the construction details of the typical NJFET structure found in semiconductor processes. It may use an additional masking operation to allow selective implant of the channel and top gate, integrated late in the semiconductor process after other devices are formed. The gate p+ diffusion connects to the p− diffusion (bottom gate) and to the p top gate (not visible in this cross-sectional view). Like its cousin the MOS transistor, the JFET is a majority carrier device. The NJEFT in Figure 6.13 operates in depletion mode, however. A negative bias on the gate relative

Table 6.5 MOS safe operating area model parameters.

SOA parameter	BSIM3v3	PSP	HiSIM	Description
vds_max	*	*	*	Maximum allowed voltage across source and drain
vgd_max	*	*	*	Maximum allowed voltage across gate and drain
vgs_max	*	*	*	Maximum allowed voltage across gate and source/bulk
vbd_max	*	*	*	Maximum allowed voltage across source/drain and bulk
vbs_max	*	*	*	Maximum allowed voltage across source and bulk
vgb_max	*	*	*	Maximum allowed voltage across gate and bulk
vgdr_max	*	–	–	Maximum allowed reverse voltage across gate and drain
vgsr_max	*	–	–	Maximum allowed reverse voltage across gate and source
vgbr_max	*	–	–	Maximum allowed reverse voltage across gate and bulk
vbsr_max	*	–	–	Maximum allowed reverse voltage across bulk and source
vbdr_max	*	–	–	Maximum allowed reverse voltage across bulk and drain

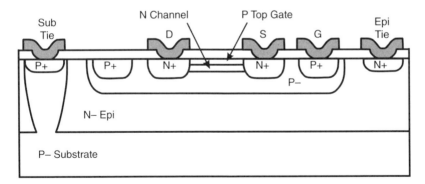

Figure 6.13 Cross section of the NJET device.

to the channel creates a depletion region that constricts the channel. The channel itself is formed in bulk silicon, so it is immune to charge trapping and transport in device oxides. This is not true for the top gate, however. The NJFET channel and gate implants are shallow and the p-type gate implant can be affected by charge trapping in the device oxides. The most sensitive JFET device parameter for total dose exposure is the gate to source leakage current, I_{GSS}.

JFETS are also relatively hard to neutron irradiation, depending on the channel and top-gate resistivities. The degradation mechanism is carrier removal, as discussed in Chapter 3. The important JFET parameters can be modeled simply for post-neutron fluence according to Zuleeg's analysis of GaAs FETs [55]

$$\frac{V_p}{V_{p0}} = 1 - a\Phi \tag{6.4}$$

$$\frac{I_{Dmax}}{I_{Dmax0}} = \frac{(1 - a\Phi)^2}{1 + b\Phi} \tag{6.5}$$

$$\frac{g_{max}}{g_{max0}} = \frac{1 - a\Phi}{1 + b\Phi} \tag{6.6}$$

$$\frac{f_c}{f_{c0}} = \frac{1 - a\Phi}{1 + b\Phi}. \tag{6.7}$$

Equations (6.4) through (6.7) use the subscript 0 to denote pre-rad parameter values and Φ for the neutron fluence. The parameters a and b are determined empirically. The relationships work well with silicon devices also, but start to lose accuracy at high neutron fluences.

The simulator support for JFET models is pretty slim. Most simulators support the R3_CMC resistor model, which can be used to model the JFET linear region. Many also support a JFET model based on the Shichman and Hodges FET model, for example.

6.3.4.2 Resistors

Most semiconductor processes support some kind of resistor. Resistors can use any of the device diffusion regions so even simple processes will likely have several "flavors" of resistor. MOS processes that utilize polysilicon gates will provide additional resistors that are constructed from the polysilicon layer along with device implants/diffusions. Some analog processes also support thin film resistors. These resistors are usually made from NiCr, SiCr, or other resistive metal alloy that are contacted directly by the first layer of metal in the semiconductor process. When looking at device models remember that there are resistive elements embedded in many of the inline subcircuits or macromodels, as well as bipolar and diode models.

Diffused resistors use the various device implant/diffusion layers to form resistive elements and the appropriate n+ or p+ regions (emitter diffusions or source/drain diffusions) as contacting regions at the resistor terminals. Resistors are generally hard to total dose, but there can be some issues depending on the sheet resistance of the resistive element and the resistor layout. As most diffused resistors have an interface with an oxide layer, they are affected by charge trapped in the oxide as well as the increase in interface trap density. Resistors constructed from lightly doped P-type diffusions (e.g. p-collector or p-well diffusions) can be affected by total dose causing a change in resistance and noise characteristics. N-type resistors defined in p-well regions can develop bridging between resistor legs if the surface becomes inverted. Neutron irradiation at high enough fluences will affect the resistivity, especially for lightly doped regions (Section 3.4). Using lower sheet resistance diffusions for resistors and taking care to eliminate leakage paths between resistor segments may be worth the increase in device area in order to avoid most of the effects from radiation. Still, diffused resistors are junction isolated, so bulk displacement damage will result in higher leakage currents at the isolating junction.

For CMOS processes, the polysilicon resistor is usually available as a design element. Often, there are various values of sheet resistance available, depending on the resistor construction. Polysilicon resistors have several advantages over diffused resistors. They can generally be made narrower then diffused resistors due to favorable ground rules for polysilicon linewidths, and they are isolated from the substrate so there is no junction capacitance or leakage.

Precision analog processes often have a thin film resistor. This is commonly made from NiCr or SiCr alloys using metal deposition and either lift-off or direct etch photolithography steps. NiCr and SiCr resistors can be very thin, on the order of 100 Å and with sheet resistances in the range of 50–200 $\Omega/\square$ for NiCr and 1000 to 2000 $\Omega/\square$ for SiCr. It is possible to get very low TCR resistor elements from such thin films by changing the composition of the alloy or film thickness. Thin film resistors are very hard to radiation, but care must be taken to avoid layouts that are sensitive to transient currents that can cause fusing of the resistor element. This is also true for polysilicon resistors.

6.3.4.3 Capacitors

There are many types of capacitors that can be integrated in modern CMOS and BiC-MOS processes. These include MIM, PIP, MOS, and finger capacitors. MIM capacitors use a deposited oxide between one of the interconnect layers and a special capacitor metal, so they require an extra process masking step. PIP capacitors utilize an insulator between two polysilicon layers. Often this insulator is a combination of thermal oxide and deposited oxide. MOS capacitors like the varactor were covered previously (Section 6.3.1). Finger capacitors use the process interconnect layers to form capacitors that rely on metal to metal capacitance as well as the fringing capacitance between metal layers. Instead of using large sheets of interconnect metal, the metal is laid out in narrow "fingers." All of these capacitors will see little change post-radiation, but some may need to have their voltage ratings reduced, depending on the size of capacitor that is used.

6.3.5 Some Modeling Challenges

There are many challenges to developing good post-rad model corners, but most fall into one of three categories:

1. *How should data be collected?* Measurements at the device level should align with the expected maximum voltage and current compliances. These may be set much lower than would be expected, but the aim is to avoid injecting charge into oxides. For example, when characterizing MOS devices, use a maximum V_{ds} much lower than the absmax for the device to avoid high fields during device operation. The order of device tests may be changed to put less stressful testing first. Avoid doing breakdown measurements, but if needed, do this testing last and do not reuse the device for other testing. Leakage measurements of gate oxides should be tailored to avoid confusion between SILC and RILC. *Some measurements are not very easy to do.* These include collecting temperature coefficients information and statistical data for Monte Carlo simulation. Taking measurements above room temperature will certainly show changes from lower temperature measurements, but is the behavior real or the result of annealing the effects of irradiation? The collection of enough data for meaningful statistical variation is unlikely during post radiation testing. Usually, the sample size

is too small or may have been comprised of screened units from the normal population of parts. Noise measurements are often left out, but devices get noisier after irradiation [44, 56].

2. *What is the worst case?* There are many options during radiation testing. Will parts be biased during irradiation, and if so, what polarity/voltage/current for each terminal? Will there be post-irradiation biasing or annealing, and for what length of time? What dose rate should be used? There may need to be more than one set of conditions during irradiation and afterward to identify the worst case.

3. *What if devices fail or do not meet post-irradiation specifications?* Some devices just are not built for radiation environments, and their usage may need to be avoided in circuit design. The usual weakness is with the NMOS device in a CMOS process. There are several ways to improve the device through layout, but this needs to be anticipated for its impact to existing IP including any digital cell libraries.

6.4 Circuit Simulation

The ability to modify foundry-provided device models may be limited in some cases, either by encryption or by lack of write permission. Either way, it is not advisable to go far from the foundry-provided files without a concerted effort to develop new models as replacements. Parts of the following sections describe how to implement post radiation models, but it is assumed that new models are being developed or the foundry originals are available for modification.

6.4.1 Corner Simulation

Adding a post-rad corner to existing model files can take several paths. The most direct approach is to simply add the corner the way other simulation corners are implemented. Unfortunately, there are many ways to implement model files and corners. The examples in this section show two ways to implement corners using the Cadence Spectre simulator. This will hopefully illustrates the process so as to be applicable to other simulators and other ways that corners may be organized.

Model files are usually found in the PDK installation location. Ideally, the files will be organized by device type (e.g. mos.scs, bjt.scs, etc.) as shown in Figure 6.14. There will also likely be one or more corner files and several other files. When a designer runs a simulation, the path to one of the corner files (e.g. corner_tt.scs) is provided to the simulator. The simulator uses the values defined by both the model file as well as the corner parameter settings during the simulation. Figure 6.15 expands on one of the corner files and one of the model files. The corner file takes advantage of the fact that the real corner information is held in "sections" contained in the model files. While the corner file specifies the section settings for each of the device types, the sections in each device model file define the section naming convention as well as the scope of the possible model variation. For the example in Figure 6.15, a MOS model has sections for tt (typical nmos, typical pmos), ff (fast nmos, fast pmos), etc., each of which defines a set of parameters that are later used in the model itself.

When a section is properly defined, it can be specified during simulation. Corner files can be crafted to use any of the sections in the various device model files so that

Figure 6.14 Process design kit (PDK) file organization.

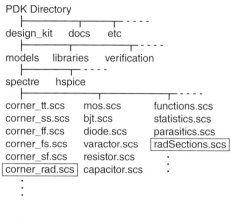

PDK Directory

design_kit   docs   etc   - - - -

models   libraries   verification   - - - -

spectre   hspice   - - - -

corner_tt.scs	mos.scs	functions.scs
corner_ss.scs	bjt.scs	statistics.scs
corner_ff.scs	diode.scs	parasitics.scs
corner_fs.scs	varactor.scs	radSections.scs
corner_sf.scs	resistor.scs	.
corner_rad.scs	capacitor.scs	.
.		.

```
// Example Corner File
// Generic PDK
// Set Corners
//
include "mos.scs" section = tt
include "bjt.scs" section = tt
include "diode.scs" section = tt
include "varactor.scs" section = tt
include "resistor.scs" section = tt
include "capacitor.scs" section = tt
include "functions.scs"
include "statistics.scs"
include "parasitics.scs"
```

corner_tt.scs

```
// Example MOS Model File
// Generic PDK
//
section tt
<parameter settings>
endsection tt
section ss
<parameter settings>
endsection ss
section ff
<parameter settings>
endsection ff
//
nmos (d g d b) nch_mod
  .
  .
  .
```

mos.scs

Figure 6.15 Corner and model file examples.

each device type can be controlled separately. Note that the corner file specifies the file location for each corner section. This allows a new file containing only the additional corner sections to be brought in via a new corner file. These additions to the PDK models directory are indicated by the boxed files in Figure 6.14. In order for the new sections to operate properly, all of the parameters required by the model files must be provided. This may mean that some of the parameter settings from the pre-rad corners need to be added to the parameter settings in the post-rad sections. Inspection of the model files and/or trial simulations of single devices using the post-rad sections will reveal any post-rad parameter settings that are being overwritten during the netlisting. Figure 6.16 shows an example organization of the corner_rad.scs and the radSections.scs files from Figure 6.14.

Adding post-rad simulation capability using corner files and sections has the advantage that existing device sections can be reused in the corner files. Many passive components will see little if any change post total dose, so the pose-rad corner file can just specify the 'tt' corner for these elements, for example. The post-rad testing sample size may be small or comprised of selected units, as discussed in Section 6.3.5.

```
// Post-Rad Corner File
// Generic PDK
// Set Corners
//
include "radSections.scs" section = mos
include "radSections.scs" section = bjt
include "radSections.scs" section = dio
include "radSections.scs" section = var
include "radSections.scs" section = res
include "radSections.scs" section = cap
include "functions.scs"
include "statistics.scs"
include "parasitics.scs"
```

```
// Post-Rad Sections File
// Generic PDK
//
section mos
<parameter settings>
endsection mos
section bjt
<parameter settings>
endsection bjt
section dio
<parameter settings>
endsection dio
        .
        .
        .
```

corner_rad.scs   radSections.scs

Figure 6.16 Implementation of post-rad corner and section files.

One drawback of the corner file method is that post-rad simulation will represent only that small sample size. It may be safe to assume that corner variation and scaling can also apply to the post-rad corner, but there is no way to do this using a corner implementation for the post-rad data.

There is another way to implement the post-rad corners while preserving the capability to select other simulation corners, but it is much more intrusive to device model files. It involves adding a table function to the equations used to calculate device parameters that change after irradiation. For example, an NPN transistor using the SGP model may calculate the forward beta (bf) using a function comprised of a parameter value and a function that provides the geometry dependence

$$\text{bf} = \text{kbfnpn} * \text{geomFunc(le, we, ne)}. \tag{6.8}$$

The parameter kbfnpn is the forward beta value for the npn supplied by the model section and le, we, and ne are the emitter length, emitter width, and number of emitters, respectively. The function geomFunc is developed based on measurements of a variety of npn emitter configurations. If the post-radiation model is provided by a separate corner/section file, the value of kbfnpn would be modified when the post-rad corner section is selected. This eliminates the possibility of using the other corners in tandem with the post-rad model. Instead, Eq. (6.8) can be modified to include a separate scaling function:

$$\text{bf} = \text{kbfnpn} * \text{bfnpnTable(rad)} * \text{geomFunc(le, we, ne)}. \tag{6.9}$$

The function bfnpnTable(rad) is a table function that provides a return value depending on the value of the variable rad:

$$\text{bfnpnTable(rad)} = \text{table}(0, 1.0, 1, 0.5, 2, 0.25). \tag{6.10}$$

Specifying rad $= 0$ results in a return value of 1.0 so that the pre-radiation condition of the model parameter is preserved. Specifying rad $= 1$ or rad $= 2$ results in return values of 0.5 or 0.25, respectively. These settings for the rad variable would represent post-rad parameter values that have been normalized to the pre-rad value, for example.

Table functions would have to be developed for each device parameter that changes in the pose-rad model. These functions could be specified using a separate file so that all of the post-rad device variation is contained in a single place. This provides an added level of security for access control as the table function file can be located in IC project areas instead of the PDK. The PDK would have "dummy" table functions defined that return a scaling factor of 1.0 no matter the value of the rad variable. The export-controlled developments would have access to the "real" table functions, which would overwrite the "dummy" ones from the PDK.

During post-rad simulation, the corner section selection can be made independently from the pre- or post-rad simulation setting of the rad variable. This would allow the post-rad scaling to be applied to any of the device corners. It would be advisable to verify that such scaling still applies to the post-rad model, and this would require a larger sample size during radiation testing. Another advantage of using table functions to provide post-rad scaling of parameters is that all of the post-rad conditions can be simulated by sweeping the rad variable. This is useful when there are many post-rad corners represented by the values of rad = 1, 2, 3, etc.

6.4.2   SEE Simulation

One of the more costly steps in developing an IC for use in space will often be the single-event testing during the IC development. It may take several passes through testing before a reliable part is made. The expense of the NRE and the delay to market can be substantial. Unfortunately, this is one problem that the design engineer will have to face without some way to predict circuit behavior during single events. The problem of SEE simulation is threefold:

1. *Device coverage.* Depending on the complexity of the IC design, it may not be feasible to test the reliability of every single transistor in a design. Most designs are built up from smaller schematic blocks, many of which may be repeated or reused throughout the design. It may not be possible to modify schematics for one block without affecting the schematics of repeat/reused blocks. This can make it difficult to single out individual devices or blocks in a design. The designer will be forced to simulate SEE at the block level without a strategy and methodology for a top-level simulation.
2. *Device topology.* Each device comprises separate regions that react differently to a single-event strike. Figure 6.17 shows a cross-sectional view of a NPN transistor. Vertical arrows indicate potential single-event strike locations, assuming a normal incidence. Charge collection will occur at any junction pierced by a single event, so the SET response will depend on the strike location.
3. *Layout topology.* There could be layout topologies that complicate the SEE response. The triggering of a parasitic SCR or circuit latchup would be good examples. For small process nodes, more than one device can be affected by a single event.

The layout topology aspect of SEE can be driven by developing and enforcing ground rules used in the completed circuit layout. Covering all of the devices for SET sensitivity in a top-level simulation seems like a big job, but it can be accomplished with some programming and automation of the simulation tasks. The device topology lends itself nicely to a simulation model based on what is known about the process and device cross sections, and it is possible to develop a circuit that will inject the simulated currents representative of SET. Take, for example, an 80 LET particle strike to the emitter of the transistor in Figure 6.17 and make the assumption that it is normally incident at the

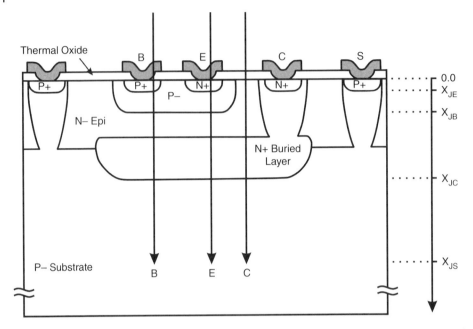

Figure 6.17 Cross section of an NPN transistor showing possible single-event particle paths through the device.

emitter. Using the ionization energy for silicon from Table 3.1 (3.6 eV/pair), we find the ionized charge density for the 80 LET particle in silicon:

$$Q_{80} = \frac{q\rho d}{E_{eh}} \cdot LET = 0.818 \cdot d \text{ pC}/\mu\text{m} \tag{6.11}$$

where q is the electron charge, ρ is the density of silicon, d is the path length of the particle, and E_{eh} is the ionization energy for silicon. The junction depths indicated to the right of the NPN cross section in Figure 6.17 can be used to estimate the charge generated in each region of the transistor. Using the generic values $X_{JE} = 0.25\,\mu\text{m}$, $X_{JB} = 1.00\,\mu\text{m}$, and $X_{JC} = 10\,\mu\text{m}$ the charges generated in each region of the transistor are

Emitter $Q_e = (0.818 \text{ pC}/\mu\text{m})(0.25\,\mu\text{m}) = 0.205 \text{ pC}$ \hfill (6.12)

Base $Q_b = (0.818 \text{ pC}/\mu\text{m})(0.75\,\mu\text{m}) = 0.614 \text{ pC}$ \hfill (6.13)

Collector $Q_c = (0.818 \text{ pC}/\mu\text{m})(9.00\,\mu\text{m}) = 7.362 \text{ pC}.$ \hfill (6.14)

The collector region will also collect charge from the p− substrate region, but the depth from which charge is collected along the path of the particle has to be estimated. Wagner [57] provides some guidance in this regard, based on the resistivity of the substrate. For the sake of illustration a depth of 10 μm is used yielding an additional charge $Q_s = 8.18 \text{ pC}$ injected into the collector. Now it might be argued that any currents due to funneling of the charge to device junctions would follow an equivalent circuit as in [58], but this is a worst-case model, so some liberties can be taken. For a strike in a base region (arrow B in Figure 6.17) the worst-case estimate has all of the 0.614 pC of base charge collected by the n− epi (collector) and all of the 7.362 pC of collector charge collected by the base region (p−). This yields a net charge transfer of about 8 pC from collector to

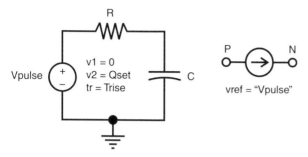

Figure 6.18 Single-event current pulse circuit.

base. At the same time, there will be 8.18 pC of substrate charge that is collected by the n− epi for a net transfer of 8.18 pC from collector to substrate. Using similar reasoning, a strike in the collector region (arrow C in Figure 6.17) will result in a net charge transfer of about 15.5 pC from collector to substrate. For the emitter strike (arrow E in Figure 6.17), charge collection in the emitter is small compared to other regions. The charge collection in the emitter can be ignored so the emitter strike is treated like another base strike.

Equations (6.12) through (6.14) are the starting point for crafting a circuit block that can be used to simulate the charge injection into a single device. One of the hurdles is the development of the current waveform. A simplistic approach can be taken where only the decay of the transient is important and based on the expectation from the highest resistivity region, the substrate. Despite the limitation in waveform storage time to about 500 ps in [57], it would appear that for lightly doped substrates in the 10 Ω-cm range and the most energetic particles (about 30 LET for 100 MeV iron), the decay transient could be 1 ns or longer. A simple RC circuit as shown in Figure 6.18 could be used to generate a reference transient for the current injection. This is an implementation of Roche's current model [59] but with the time constant set by the generator resistor and capacitor:

$$I(t) = I_0 e^{-t/RC}. \tag{6.15}$$

There will be a need to scale the reference (V_{pulse}) to model the correct charge collection (Q_{set}) for the various devices and their constituent junctions. For PDK device libraries that have many different devices, this will entail developing charge injection circuits for each device. The traditional modeling approach [60] entails assembling the PDK devices and the charge injection circuits into device "wrappers." A wrapped device is substituted for a suspect device in the schematic and a transient simulation is run. In practice, the V_{pulse} reference is set to a period long enough to encompass the transient simulation so that only one "SET" pulse is triggered. There may be several V_{pulse} blocks used in a device wrapper, depending on the device junctions being tested.

Messenger [61] developed an approximate analytic solution for the injected current of the form

$$I_{\text{SET}}(t) = \frac{Q}{\tau_f - \tau_r} \left[e^{-\frac{t}{\tau_f}} - e^{-\frac{t}{\tau_r}} \right] \tag{6.16}$$

where Q is the collected charge, τ_f is the time constant for charge collection at the junction (pulse fall time), and τ_r is the time constant for the initialization of the ion track (pulse rise time). The pulse rise time of the V_{pulse} source in Figure 6.18 can be adjusted to approximate the shape of Eq. (6.16) or a modeling language such as Verilog-A can be used to provide a more accurate current waveform [62, 63]. A good overview of modeling and simulation of SET at the circuit level is provided by Andjelkovic et al. [64].

If the aim is to calibrate the device response to single events, then a mixed-mode simulation can substitute for expensive beam time. Several of the commercially available device simulators will pair with circuit simulators to provide a mixed SPICE/TCAD simulation of single events to devices [65, 66], as well as circuits such as SRAM cells [59]. TCAD simulations are too slow to use during IC development, but can provide a better view of the single-event pulse and can aid the development of a better charge injection circuit. This is especially true for non-SOI substrates, where there can be significant collection from both funneling and diffusion.

6.5 Summary and Closing Comments

This chapter provided a practical review of some of the radiation effects seen in semiconductor devices building on knowledge gained in Chapters 2 and 3. The approach taken was from a device modeling perspective aimed primarily at silicon-based semiconductors. This included some background information on simulators, device intrinsic models, and model parameter extraction software. EDA tool vendors no longer provide only vendor-specific intrinsic models. Standardization in the industry has been driven by the CMC, and most simulators have CMC models available. This has allowed device model development to closely follow technology trends. This is most prevalent in the MOS models, where device scaling has driven the need toward more physics-based representations. The trend in BJT intrinsic models is generally to expand on the SGP model and to provide better geometry modeling.

MOS, BJT, and other devices were discussed in terms of construction, response to radiation, and modeling effort aimed at developing corner models for post-radiation simulation. There are clearly some limitations to what can be provided in a post-rad corner. Some post-rad measurements are just not practical or advisable. For example, sample sizes will be limited by the expense of the tests, including manpower and test chamber time. Considering the need for low-dose-rate testing to identify ELDRS sensitivity, test chamber time can become expensive. Some types of tests should be avoided. This includes post-rad measurements at elevated temperatures and testing that causes injection of electrons or holes into device oxides. This means post-rad corner models will likely depend on the pre-rad device parameter statistics and temperature dependence. Integration of model corners can affect the access to other corners, however. Approaches that provide a post-rad corner as a section, similar to other device corners, are easy to implement but will limit the ability to utilize the process variability represented in the other pre-rad corners. A more intrusive approach that integrates post-rad corners at the device model level provides this flexibility. Some post-rad corners will require adding parasitic devices, adding still more complexity to the modeling effort.

Finally, a traditional approach to SET simulation was outlined. It takes advantage of existing circuit primitives and some worst-case charge collection calculations to simulate a single event strike to a circuit device using the traditional "device wrapper" approach. Being able to predict SET performance at the circuit level is important in order to reduce NRE and improve product to market. The accurate simulation of charge collection and SET influence on circuit function can pay dividends during product development by reducing or eliminating circuit sensitivities to single events before validating function during single-event testing. Single-event testing is often an iterative process involving the expense of new tooling and repeat testing. Designing SET, immune circuits will require better models and simulation techniques, though. Using mixed-mode (SPICE/TCAD) simulation can help calibrate charge injection wrappers, but the smallest process nodes may benefit from better integration of SET simulation capability into compact models [60, 67].

Problems

6.1 Varactors (Figure 6.2) are not normally biased for depletion mode operation. Why is this?

6.2 A circuit design intended for operation in a total dose environment might use a varactor element. Explain your choice of either p-type or n-type varactor for such a design with the understanding that either device would be biased in its accumulation region during circuit operation.

6.3 Discuss how ionizing radiation affects the device parameters and operation of MOS devices. Are there differences in the post-ionizing radiation response between NMOS and PMOS devices?

6.4 Two types of silicon Zener diode undergo total dose exposure and post-irradiation corner model development. The first Zener diode type has a conduction/breakdown path at the silicon surface, whereas the other type has its conduction/breakdown path through the bulk region. Both diodes are characterized for 1/f noise. The surface conduction/breakdown Zener diode post-radiation 1/f noise is seen to increase substantially whereas the bulk conduction/breakdown Zener diode 1/f noise remains mostly unchanged. Explain this result.

6.5 A low-power bipolar amplifier circuit previously designed for commercial purposes is being redesigned for a total ionizing dose (TID) environment of 1 Mrad(Si). The circuit is fabricated using a planar bipolar junction isolated process and currently doesn't support a 1 Mrad(Si) TID model corner. What circuit and/or process issues might impact the ability to complete the redesign?

6.6 A bipolar circuit is built using a semiconductor process that supports vertical NPN and lateral PNP transistors. The design uses lateral PNP transistors only for current mirrors, so all PNP devices are close to the positive voltage rail. The design was originally for commercial applications but now needs to be redesigned to meet a TID requirement. You determine that the PNP layout can be redesigned to include a gate over the active base region, as in Figure 6.9. How do you connect the gate terminal? Explain your reasoning.

6.7 What layout changes are likely to result in a change to the SEGR voltage of the quasi-vertical DMOS device shown in Figure 6.12?

6.8 A circuit designer just completed the package testing of a bipolar circuit that uses thin-film NiCr for all resistive elements. The circuit works perfectly and meets all design criteria. He sends several functional units in unlidded hermetic packages (cerdips) to the marketing department for die photographs. When the units are returned, they no longer function. The designer finds several NiCr resistors that are blown and open circuited during failure analysis. What happened?

6.9 Section 6.4.2 provides a simple way to simulate a single-event strike to a device junction.
   a) Set up a simulation of a bipolar device with single-event current injection at the collector-substrate, base-collector, and emitter-base junctions. Are there any undesirable side effects from using the circuit in Figure 6.18?
   b) Does the circuit in Figure 6.18 provide a good simulation result for lightly doped junctions such as the collector-substrate junction in junction-isolated processes?

References

1 Nagel, L. W and Pederson, D. O. (1973). SPICE (Simulation Program with Integrated Circuit Emphasis). Memorandum No. ERL-M382. University of California, Berkeley.
2 Nagel, L. W. (1975). SPICE2: A computer program to simulate semiconductor circuits. Memorandum No. ERL-M520. University of California, Berkeley.
3 Nagel, L.W. and Rohrer, R.A. (1971). Computer analysis of nonlinear circuits, excluding radiation (CANCER). *IEEE Journal of Solid-State Circuits* 6 (4): 166–182.
4 Nagel, L.W. and Pederson, D.O. (1973). *Simulation Program with Integrated Circuit Emphasis*. Waterloo, Canada: Proc. 15th Midwest Symp. Circ. Theory.
5 Pederson, D.O. (1984). A historical review of circuit simulation. *IEEE Transactions on Circuits and Systems* 31 (1): 103–111.
6 Vladimirescu, A. (1994). *The SPICE Book*. Wiley.
7 Kundert, K.S. (1995). *The Designer's Guide to SPICE & SPECTRE*. New York: Kluwer Academic Publishers.
8 Shichman, H. and Hodges, D.A. (1968). Modeling and simulation of insulated-gate field-effect transistor switching circuits. *IEEE Journal of Solid-State Circuits* 3 (3): 285–289.
9 Ebers, J.J. and Moll, J.L. (1954). Large-signal behaviour of junction transistors. *Proceedings of the Institute of Radio Engineers* 42 (12): 1761–1772.
10 Gummel, H.K. and Poon, H.C. (1970). An integral charge control model of bipolar transistors. *Bell System Technical Journal* 49 (5): 827–852.
11 McAndrew, C., Seitchik, J., Bowers, D. et al. (1995). VBIC95: an improved vertical, IC bipolar transistor model. *IEEE Journal of Solid-State Circuits* 31 (10): 1476–1483.
12 Victory, J., Zhu, Z., Zhou, Q. et al. (2007). PSP-based scalable MOS Varactor model. In: *2007 IEEE Custom Integrated Circuits Conference*, 495–502. San Jose, CA.

13 Becker, R. A. (2018). *A Brief History of S*, AT&T Bell Laboratories, Murray Hill, NJ 07974. http://citeseerx.ist.psu.edu/viewdoc/download?doi=10.1.1.131.1428& rep=rep1&type=pdf (accessed 30 September 2018).

14 Becker, R.A., Chambers, J.M., and Wilks, A.R. (1988). *The New S Language.* Chapman and Hall/CRC.

15 Gildenblat, G., Li, X., Wu, W. et al. (2006). PSP: an advanced surface-potential-based MOSFET model for circuit simulation. *IEEE Transactions on Electron Devices* 53 (9): 1979–1993.

16 Shaneyfelt, M.R., Dodd, P.E., Draper, B.L., and Flores, R.S. (1998). Challenges in hardening technologies using shallow-trench isolation. *IEEE Transactions on Nuclear Science* 45 (6): 2584–2592.

17 Gennady, Z.I. and Gorbunov, M.S. (2009). Modeling of radiation-induced leakage and low dose-rate effects in thick edge isolation of modern MOSFETs. *IEEE Transactions on Nuclear Science* 56 (4): 2230–2236.

18 McLean, F.B. (1980). A framework for understanding radiation-induced interface states in SiO_2 MOS structures. *IEEE Transactions on Nuclear Science* 27 (6): 1651–1657.

19 Scarpa, A., Paccagnella, A., Montera, F. et al. (1997). Ionizing radiation induced leakage current on ultra-thin gate oxides. *IEEE Transactions on Nuclear Science* 44 (6): 1818–1825.

20 Olivo, P., Nguyen, T.N., and Ricco, B. (1988). High-field-induced degradation in ultra-thin SiO_2 films. *IEEE Transactions on Electron Devices* 35 (12): 2259–2257.

21 Dumin, D.J. and Maddux, J.R. (1993). Correlation of stress-induced leakage current in thin oxides with trap generation inside the oxides. *IEEE Transactions Electron Devices* 40 (5): 986–993.

22 Takagi, S., Yasuda, N., and Toriumi, A. (1996). Experimental evidence of inelastic tunnelling and new I-V model for stress-induced leakage current. In: *Proc. of the 1996 IEDM*, 323–326. San Francisco.

23 Schuegraf, K.F. and Hu, C. (1994). Metal-oxide-semiconductor field-effect-transistor substrate current during fowler-nordheim tunneling stress and silicon dioxide reliability. *Journal of Applied Physics* 76 (6): 3695–3700.

24 Maloberti, F. and Davies, A.C. (eds.) (2016). *A Short History of Circuits and Systems*, 195–202. Delft: River Publishers.

25 Cheng, Y., Jeng, M.C., Liu, Z. et al. (1997). A physical and scalable I-V in BSIMv3 for analog/digital simulation. *IEEE Transactionson Electron Devices* 44 (2): 277–287.

26 Cheng, Y. and Hu, C. (2002). *MOSFET Modeling & BSIM3 User's Guide*. Springer Science+Business Media, LLC.

27 Watts, J., McAndrew, C., Enz, C., et al. (2005). Advanced compact models for MOSFETs. Workshop on Compact Modeling at the 8th International Conference on Modeling and Simulation of Microsystems, Anaheim, California.

28 Mattausch, H.J., Miura-Mattausch, M., Sadachika, N. et al. (2008). The HiSIM compact model family for integrated devices containing a surface-potential MOSFET core. In: *2008 15th Int. Conf. on Mixed Design of Integrated Circuits and Systems*, 39–50. Poznan, Poland.

29 Xi, X., He, J., Dunga, M. et al. (2004). BSIM5 MOSFET Model. In: *Proc. 7th International Conference on Solid-State and Integrated Circuits Technology*, vol. 2, 920–923.

30 Chauhan, Y., Venugopalan, S., Karim, M. et al. (2012). BSIM – Industry standard compact MOSFET models. In: *Proc. ESSCIRC*, 30–33.

31 Enz, C.C., Krummenacher, F., and Vittoz, E.A. (1995). An analytical MOS transistor model valid in all regions of operation and dedicated to low-voltage and low-current applications. *Analog Integrated Circuits and Signal Processing Journal on Low-Voltage and Low-Power Design* 8: 83–114.

32 Gaul, S. J., Church, M. D., and Doyle, B. R. (2010). Radiation hardened device. U.S. Patent 7,804,143, filed Feb. 18, 2009 and issued Sep. 28, 2010.

33 Cherne, R. D., Clark, J. E., Dejong, G. A. et al. (1995). SOI CMOS device having body extension for providing sidewall channel stop and bodytie. U.S. Patent H1,435, filed Oct. 21, 1991 and issued May 2, 1995.

34 Morris, W. H. (2013). Method for radiation hardening a semiconductor device. U.S. Patent 8,497,195, filed Jan. 9, 2012 and issued July 30, 2013.

35 Warner, R.M. Jr. and Grung, B.L. (1983). *Transistors, Fundamentals for the Integrated-Circuit Engineer*. New York: Wiley.

36 Price, W. E., Martin, K. E., Nichols, D. K., Gauthier, M. K., and Brown, S. F. (1981–1982). Total dose radiation effects data for semiconductor devices. Report No. 81-66, Vols I to III, Jet Propulsion Laboratory, Pasadena, California.

37 Wilson, D.K. (1968). Capacitance recovery in neutron-irradiated silicon diodes by majority and minority carrier trapping. *IEEE Transactions on Nuclear Science* 15 (6): 77–83.

38 Phifer, C. C. (2004). Effects of radiatioin on laser diodes. Sandia Report No. SAND2004–4725, Sandia National Laboratories, Albuquerque, New Mexico.

39 Pease, R.L., Platteter, D.G., Dunham, G.W. et al. (2004). Characterization of enhanced low dose rate sensitivity (ELDRS) effects using gated lateral PNP transistor structures. *IEEE Transactions on Nuclear Science* 51 (6): 3773–3780.

40 Chen, X.J., Barnaby, H.J., Pease, R.L. et al. (2004). Radiation-induced base current broadening mechanisms in gated bipolar devices. *IEEE Transactions on Nuclear Science* 51 (6): 3178–3185.

41 Chen, X.J., Barnaby, H.J., Pease, R.L. et al. (2005). Estimation and verification of radiation induced not and nit energy distribution using combined bipolar and MOS characterization methods in gated bipolar devices. *IEEE Transactions on Nuclear Science* 52 (6): 2245–2251.

42 Chen, X.J., Barnaby, H.J., Schrimpf, R.D. et al. (2006). Nature of Interface defect buildup in gated bipolar devices under low dose rate irradiation. *IEEE Transactions on Nuclear Science* 53 (6): 3649–3654.

43 Chen, X.J., Barnaby, H.J., Vermeire, B. et al. (2007). Mechanisms of enhanced radiation-induced degradation due to excess molecular hydrogen in bipolar oxides. *IEEE Transactions on Nuclear Science* 54 (6): 1913–1919.

44 Gaul, S. J., Crandell, T. L., Davis, C. K., and Werner, J. W. (1990). Radiation Response of an Advanced Polysilicon Emitter Complementary Bipolar Process, 3rd Radiation-Hardened Linear Integrated Circuit Workshop, University of Arizona, Tucson, Arizona.

45 Messenger, G.C. (1958). The effects of neutron irradiation on germainium and silicon. *Proceedings of the IRE* 46 (6): 1038–1044.

46 Messenger, G.C. (1973). A general proof of the β degradation equation for bulk displacement damage. *IEEE Transactions on Nuclear Science* 20 (1): 809–810.

47 Rein, H. (1995). Very-high-speed Si and SiGe bipolar ICs. In: *ESSDERC '95: Proceedings of the 25th European Solid State Device Research Conference*, 45–56. The Netherlands: The Hague.

48 Getreu, I. (1976). *Modeling the Bipolar Transistor*. Beaverton, Oregon: Tektronix, Inc.

49 De Graaff, H.C. and Kloosterman, W.J. (1985). New formulation of the current and charge relations in bipolar transistor modeling for CACD purposes. *IEEE Transactions on Electron Devices* 32 (11): 2415–2419.

50 De Graaff, H.C., Kloosterman, W.J., Geelen, J.A.M., and Koolen, M.C.A.M. (1989). Experience with the new compact MEXTRAM model for bipolar transistors. In: *Proceedings of the Bipolar Circuits and Technology Meeting*, 246–249. Minneapolis, MN, USA.

51 Van Rijs, F., Bertonnaud, S., Vanoppen, R., and Dekker, R. (1996). RF power large signal modeling with MEXTRAM. In: *Proceedings of the 1996 BIPOLAR/BiCMOS Circuits and Technology Meeting*, 57–60. Mineapolis, MN, USA.

52 Stübing, H. and Rein, H. (1987). A compact physical large-signal model for high-speed bipolar transistors at high current densities – part I: one-dimensional model. *IEEE Transactionson Electron Devices* 34 (8): 1741–1751.

53 Rein, H. and Schröter, M. (1987). A compact physical large-signal model for high-speed bipolar transistors at high current densities – part II: two-dimensional model and experimental results. *IEEE Transactions on Electron Devices* 34 (8): 1752–1761.

54 Schröter, M. and Walkey, D.J. (1996). Physical modeling of lateral scaling in bipolar transistors. *IEEE Journal of Solid-State Circuits* 31 (10): 1484–1492.

55 Zuleeg, R. (1989). Radiation effects in GaAs FET devices. *Proceedings of the IEEE* 77 (3): 389–407.

56 Fleetwood, D.M., Meisenheimer, T.L., and Scofield, J.H. (1994). 1/f noise and radiation effects in MOS devices. *IEEE Transactions on Electron Devices* 41 (11): 1953–1964.

57 Wagner, R.S., Bordes, N., Bradley, J.M., and Magiore, C.J. (1988). Alpha-, boron-, silicon- and iron- ion-induced current transients. *IEEE Transactions on Nuclear Science* 35 (6): 1578–1584.

58 Peczalski, A., Bergman, J., Berndt, D., and Lai, J.C. (1988). Physical SEU model for circuit simulations. *IEEE Transactions on Nuclear Science* 35 (6): 1591–1595.

59 Roche, P., Palau, J.M. et al. (1999). Determination of key parameters for SEU occurrence using 3-D full cell SRAM simulations. *IEEE Transactions on Nuclear Science* 46 (6): 1354–1362.

60 Francis, A.M., Turowski, M., Holmes, J.A., and Mantooth, H.A. (2007). Efficient modeling of single event transients directly in compact device models. In: *2007 International Behavioral Modeling and Simulation Workshop*, 73–77. San Jose, CA.

61 Messenger, G.C. (1982). Collection of charge on junction nodes from ion tracks. *IEEE Transactions on Nuclear Science* 29 (6): 2024–2031.

62 Liu, J., Wang, Y., Chen, G. et al. (2016). Modeling of single event pulse with VerilogA: implementation and application. In: *IEEE International Nanoelectronics Conference (INEC)*, 1–2. Chengdu.

63 Zhao, X., Zhao, Z., Zhang, M., and Li, S. (2009). Verilog: A based implementation for coupled model of single event transients in look-up table technique. In: *2009 IEEE 8th International Conference on ASIC*, 666–669. Changsha, Hunan.

64 Andjelkovic, M., Ilic, A., Stamenkovic, Z. et al. (2017). An overview of the modeling and simulation of the single-event transients at the circuit level. In: *Proc. 30th International Conference on Microelectronics (MIEL 2017)*, 35–44. Nis, Serbia.

65 Wang, Y., Wang, W., Du, Y., and Cao, B. (2013). Modeling and analysis of analog single event transients in an amplifier circuit. In: *2013 International Conference on Optoelectronics and Microelectronics (ICOM)*, 94–97. Harbin.

66 Turowski, M., Raman, A., and Fedoseyev, A. (2009). Mixed-mode simulation of single event upsets in modern SiGe BiCMOS mixed-signal circuits. In: *2009 MIXDES – 16th International Conference Mixed Design of Integrated Circuits & Systems*, 462–467. Lodz.

67 Kauppila, J.S., Massengill, L.W., Ball, D.R. et al. (2015). Geometry-aware single-event enabled compact models for Sub-50 nm partially depleted silicon-on-insulator technologies. *IEEE Transactions on Nuclear Science* 62 (4): 1589–1598.

7

Radiation Semiconductor Process and Layout Solutions

7.1 Introduction

Solutions to improve radiation tolerance are achievable by providing semiconductor processes that lessen the sensitivity to ionizing particles. This chapter will focus on semiconductor process development and layout solutions that minimize both soft error rate (SER) and latchup. Technologies that provide immunity from ionizing particles are known as "radiation hardened" technologies. Radiation tolerance can be natural to the structural features of the devices in the technology. Radiation sensitivity can also be addressed using layout solutions as well. In this chapter, both radiation hardened technology and layout solutions are discussed.

The chapter will discuss silicon on insulator (SOI), silicon on sapphire (SOS), silicon on diamond (SOD), and other radiation hardening techniques in bulk complementary metal oxide semiconductor (CMOS). Bulk CMOS radiation hardening can include solutions in the substrate, epitaxial region, wells, isolation, and buried layers. The chapter will examine radiation-hardened semiconductor technologies. The chapter will also discuss SOI, SOS, SOD, and other radiation hardening techniques in bulk CMOS [1–58].

7.2 Substrate Hardened Technologies

Radiation hardening can be addressed by the choice of the substrate used in the technology. In radiation environment, the substrate plays a key role in interaction between the incoming ionized particle and the wafer that supports the active devices. Additionally, the substrate wafer has an influence on the carrier generation and the electron–hole pair (EHP) diffusion to the active devices.

Substrate hardened technologies can include process solutions of silicon on insulator (SOI), silicon on diamond (SOD), silicon on sapphire (SOS), to silicon on nothing (SON). In these technologies, the substrate region is isolated from the active devices, providing a barrier, which reduces the amount of carrier diffusion to the active elements. This provides an improvement for both minimizing charge generation and collection. Figure 7.1 illustrates the different types of substrate-hardened technologies.

Integrated Circuit Design for Radiation Environments, First Edition.
Stephen J. Gaul, Nicolaas van Vonno, Steven H. Voldman and Wesley H. Morris.
© 2020 John Wiley & Sons Ltd. Published 2020 by John Wiley & Sons Ltd.

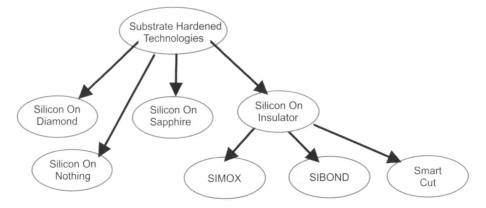

Figure 7.1 Substrate hardened technologies.

7.2.1 Silicon-on-Insulator (SOI) Technologies

A common form of substrate-hardened technology is SOI [1]. SOI technology has been used in space application due to its natural solution to reduce single-event upsets (SEUs). SOI technology reduces SEUs, single-event transients (SETs), and single-event latchup (SEL). Historically, it has been used in space applications to reduce both SEU and SEL. As a result, due to its usage in space applications, it was a low-volume niche technology for many years, and had no impact on base CMOS technology applications.

To remain on the Moore's law curve, it was found that SOI can improve the performance scaling needed for future mainstream CMOS technologies. Partially-depleted silicon on insulator (PD-SOI) was introduced as a mainstream technology by IBM in 2000 for advanced microprocessor applications. Today, it is a competitor to bulk CMOS in planar MOSFETs and FinFET technologies.

7.2.1.1 Separation by Implanted Oxygen (SIMOX)
A common process solution for SOI technology was the semiconductor process of separation by implanted oxygen (SIMOX) [1, 2]. The SIMOX semiconductor process forms a buried oxide (BOX) layer in the substrate under the active devices illustrated by an SOI MOSFET cross section (Figure 7.2). This is achieved through oxygen implantation into the substrate. The SIMOX wafer is a silicon substrate, a BOX layer, and a thin silicon film on top of the oxide layer (Figure 7.2). The substrate region is isolated from the active devices by the BOX layer. This provides a barrier to EHP diffusion, which reduces the amount of minority carrier diffusion to the active elements.

7.2.1.2 Silicon-Bonded (SIBOND) Technology
A second process of forming an SOI wafer was a process that used a bonding between two wafers, known as silicon-bonded (SIBOND) technology [1, 3]. A SIBOND wafer is formed by growing an oxide layer on a silicon wafer. A second wafer is then placed on top of the first wafer making a structure which is silicon, silicon dioxide, silicon structure (Figure 7.2b). The second wafer is then etched to a thin silicon layer on top of the silicon dioxide layer. This process method is not as widely used as the SIMOX process.

A new process method known as "Smart-Cut" achieves the same goal without the etching of the second wafer [4]. In "Smart-Cut" technology, a wafer is implanted in such a

Figure 7.2 (a) SIMOX SOI cross section illustrating a SOI MOSFET. (b) SIBOND structure.

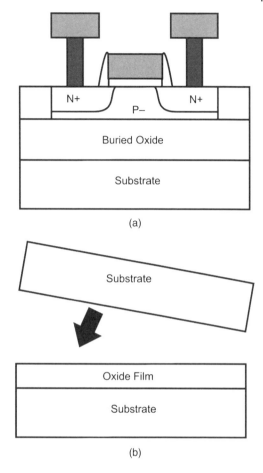

(a)

(b)

fashion, where the wafer is "snapped" off leaving behind a thin silicon layer and allowing reuse of the residual wafer. This technology has been successful, and being used by SOI wafer semiconductor wafer suppliers.

7.2.2 Silicon on Sapphire (SOS)

Substrate hardened technologies can include process solutions which is not a silicon wafer. In space and military applications, SOS is used [5]. SOS is a hetero-epitaxial process for integrated circuit manufacturing that consists of a thin layer (typically thinner than 0.6 µm) of silicon grown on a sapphire (Al_2O_3) wafer (Figure 7.3). SOS is part of the SOI family of CMOS technologies.

Typically, high-purity artificially grown sapphire crystals are used. The silicon is usually deposited by the decomposition of silane gas (SiH_4) on heated sapphire substrates. The advantage of sapphire is that it is an excellent electrical insulator, preventing stray currents caused by radiation from spreading to nearby circuit elements.

SOS faced early challenges in commercial manufacturing because of difficulties in fabricating the very small transistors used in modern high-density applications. This is because the SOS process results in the formation of dislocations, twinning, and stacking

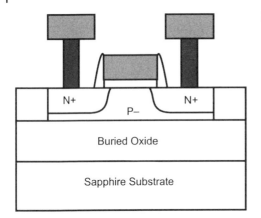

Figure 7.3 Silicon on sapphire (SOS) structure.

faults from crystal lattice disparities between the sapphire and silicon. Additionally, there is some aluminum, a p-type dopant, contamination from the substrate in the silicon closest to the interface.

7.2.3 Silicon on Diamond (SOD)

SOD technology is also proposed as an advanced alternative to conventional SOI technology [6]. An advantage of diamond is that it has a high thermal conductivity. In an SOD concept, the diamond film can be 100 μm thick and serves as an electrical insulator, heat distributor, and substrate. SOD can sustain up to 10-times higher power loads than SOI [6]. As a result of the thermal properties, an improvement in the SEL can be achieved due to the reduction of the active device temperatures. SOD has an advantage for SEU and SEL.

7.2.4 Silicon on Nothing (SON)

A new concept is to form a silicon device layer with no substrate wafer, known as SON technology [7]. SON technology allows for extremely thin (in the order of a few nanometers) buried dielectrics and silicon films to be fabricated with high resolution and uniformity guaranteed by epitaxial process. The SON process allows the buried dielectric (which may be an oxide but also an-air gap) to be fabricated locally in dedicated parts of the chip, which may present advantages in terms of cost and facility of system-on-chip (SOC) integration.

The SON stack itself is physically confined to the under-gate-plus-spacer area of a device, thus enabling extremely shallow and highly doped extensions, while leaving the HDD (highly doped drain) junctions comfortably deep (Figure 7.4). Therefore, SON embodies the ideal device architecture taking the best elements from both bulk and SOI and getting rid of their drawbacks.

SON enables excellent I_{on}/I_{off} tradeoff, suppressed self-heating, low S/D series resistance, close to ideal subthreshold slope, and high immunity to short channel effect (SCE) and drain-induced barrier lowering (DIBL) down to ultimate device dimensions of 30–50 nm [7].

The advantage of the SON concept is that there is no interaction between the single event particle, and the substrate wafer. No interaction occurs between the "wafer"

Figure 7.4 Silicon on nothing (SON) structure.

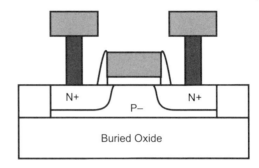

and the incoming particle. As a result, there is no nuclear spallation events, or carrier generation influencing the device. This semiconductor process methodology has not been widely used today.

7.3 Oxide Hardening Technologies

Hardening of insulators is a technique used for prevention of shifting during radiation. The following section discusses techniques to provide radiation hardening. These methods include fluorinating thin oxides and isolation.

7.3.1 Oxide Growth and Fluorination of Oxide

For radiation hardening of the MOSFET isolation, fluorine can be introduced to reduce the positive oxide charge density. Fluorine can introduced into the LOCOS field oxide (FOX) by high-energy (2-MeV) F implantation. Additionally, an anneal is performed after the implantation. The implant energy of the Fluorine can be a high energy implant (e.g. 2-MeV) and subsequent annealing (e.g. 950 °C for 60 minutes) leading to an improved radiation hardness of the FOX and its associated device parameters [11]. N-channel MOSFETs isolated by the fluorinated oxide will exhibit a lower radiation-induced source-drain leakage current due to the smaller density of radiation-induced positive oxide charge in the fluorinated FOX (e.g. compared to the nonfluorinated oxide). Threshold voltage shifts of the field-oxide FETs are also reduced. In addition, the radiation-induced leakage currents of reverse biased n/sup +/p-junction diodes fabricated with the F implantation process are suppressed, suggesting that the generation of interface traps at the gate SiO_2–Si and the field SiO_2–Si interfaces is also reduced in the fluorinated device [11].

7.3.2 MOSFET Gate Oxide Hardening

MOSFET gate oxides can shift due to radiation-induced threshold voltage shifts, ΔV_{th}, and interface-trap generation ΔV_{it}. In a thin oxide MOSFET, the radiation-induced threshold shift bears a dependence of the square of the oxide thickness (e.g. t^2_{ox}). As a result, the threshold shift of the isolation structures is significantly larger than the thin oxide of the MOSFET structure. As technologies are scaled, the oxide thickness is reduced, lowering the radiation sensitivity.

7.3.3 Recessed Oxide (ROX) Hardening

Recessed oxide (ROX) isolation can lead to radiation induced threshold shifts [11]. In a thin oxide MOSFET, the radiation induced threshold shift bears a dependence of the square of the oxide thickness (e.g. t^2_{ox}) leading to a larger shift due to the ROX isolation. ROX is made thick in order to reduce the electric field between the interconnects and the silicon substrate. But, as the ROX isolation is increased, the threshold shift is more significant.

7.3.4 LOCOS Isolation Hardening

Radiation-induced threshold shifts of the sidewall device occurs due to oxide- and interface trap charge generation under the LOCOS isolation [11]. Experimental data exhibits excess leakage current due to radiation-induced charge buildup.

Radiation-hardening solutions using structure and implantation also exist. These can include the following:

- Split FOX and p+ implant utilizing a notch in the isolation [12]
- Heavily doped buried layer (HDBL) under the LOCOS isolation [13–20]
- Buried guard ring (BGR) [14–20].
- Parasitic isolation device (PID) on the edges of the active channel [14–20].

Custode et al. introduced a "notch" in the isolation where a metal layer was placed to change the electrical potential under the isolation [12]. In addition, implantation of p+ dopants in the isolation notch was also introduced [12].

HDBL and BGR can be introduced under the devices for improved radiation hardness [13–21]. A PID can also be introduced on the edges of the MOSFET channel to prevent the generation of oxide traps near the edge of a MOSFET [13–20].

7.3.5 Shallow Trench Isolation (STI) Hardening

Radiation hardening for a technology with shallow trench isolation (STI) is more complex than a traditional hardened FOX. High electric fields at the edges of the STI limit the total dose radiation hardness.

Parasitic devices can occur on the trench sidewall that is influenced by radiation. Radiation-induced threshold shifts of the sidewall device occurs due to oxide- and interface trap charge generation. Corner devices on the trench sidewall can lead to "humps" in the MOSFET I-V characteristic.

Hardening of the STI can be achieved by increasing the threshold voltage of the parasitic sidewall device. This can achieved by STI "corner rounding" etch processes. Additionally, a sidewall implant can be placed in the edges of the STI.

Radiation hardening solutions include the following:

- STI edge implant
- STI corner rounding etch process [23]
- PID on the edges of the active channel [14–20]

7.4 CMOS Latchup Process Solutions

In mainstream CMOS technology, the technology goals are focused on circuit density, chip performance, and cost. Derivative technologies, such as radio frequency (RF) CMOS, bipolar, RF BiCMOS, and power technologies, have a significant larger set of semiconductor devices that contain many additional masks, implants, and structures. Many of these features can be used to improve SEL robustness of the CMOS circuitry within the derivative technology, or can be added to a base CMOS technology. In this chapter, the focus will be on utilization of these semiconductor features and structures for SEL prevention.

7.5 CMOS Substrates – High-Resistance Substrates

Today, high-resistance substrates are needed today to avoid noise coupling between circuits. In a mixed-signal chip, a concern exists between the digital CMOS circuits, analog and radio frequency (RF) circuitry. High-resistance substrates will be needed in mixed-signal CMOS (MS-CMOS), radio frequency CMOS (RF-CMOS), BiCMOS silicon germanium (SiGe), and gallium arsenide (GaAs) technologies. As technologies are scaled and as the application frequency increases, noise coupling, cross-talk, and noise rejection methods will be needed. Hence, a key issue in the CMOS latchup robustness is the role of the substrate resistance. Figure 7.5 shows a plot of latchup trigger voltage versus substrate resistance.

The trend in the industry is the migration from 1 to $50\,\Omega$-cm [23]. For example, CMOS migration to $50\,\Omega$-cm substrates will occur at 0.13-μm technology generation, whereas Boselli, Reddy, and Duvvury showed $50\,\Omega$-cm substrates occurs later at a 65-nm CMOS technology node [24]. The starting wafer substrate resistance has an effect on the substrate spreading resistance term. As a result, this influences SEL [23].

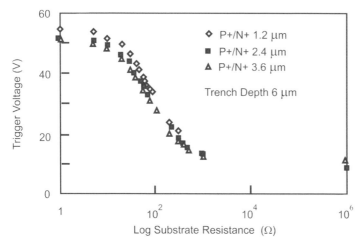

Figure 7.5 Latchup trigger voltage as a function of substrate resistance.

For latchup, the spreading resistance has a role in the cathode-to-substrate contact spacing. As the distance increases, the influence is more significant. Additionally, the spreading resistance does not saturate but continues to increase in latchup analysis. In order to initiate latchup, the product of the NPN and the PNP bipolar current gain, $\beta_n \beta_p$, needs to satisfy the following inequality:

$$\beta_n \beta_p \geq \frac{1 + \left(\frac{Isx}{I}\right) \beta_n}{\left[1 - \left(\frac{Iw}{I}\right)\left(\frac{\beta_n+1}{\beta_n}\right) - \left(\frac{Isx}{I}\right)\right]}$$

where

$$I_w = \frac{(V_{be})_{pnp}}{R_{nw}} = \frac{V_o}{R_{nw}} \ln\left[\frac{I - I_w}{(I_o)_p}\right],$$

and where the substrate current expression, Isx, can be modified as follows:

$$I_{sx} = \frac{(V_{be})_{npn}}{R_{pw}\|R_{sx}} = \frac{V_o}{R_{pw}\|R_{sx}} \ln\left[\frac{I - I_{sx}}{(I_o)_n}\right],$$

where the effective substrate resistance comprises of both the p-well resistance, R_{pw}, and the base wafer resistance R_{sx} spreading resistance terms:

$$\frac{1}{R_{pw}\|R_{sx}} = \frac{1}{R_{pw}} + \frac{1}{R_{sx}}.$$

In the generalized tetrode formulation, the latchup condition can be expressed as

$$\alpha * f_{ns} + \alpha * f_{ps} \geq 1$$

where

$$\alpha * f_{ns} = \frac{\alpha f_{ns}}{1 + \frac{r_{en}}{R_{sx}\|R_{pw}}}$$

and

$$\alpha * f_{ps} = \frac{\alpha f_{ps}}{1 + \frac{r_{ep}}{R_{nw}}}$$

where the substrate resistance is explicitly shown as the parallel resistance of the p-well resistance and the substrate base wafer resistance [23]. As the substrate wafer resistance increases, it influences the generalized NPN transport factor. As the substrate resistance becomes large compared to the p-well resistance, the effect on the differential generalized stability criteria is minimal. Note that the parallel resistance of the substrate and the well is approximately the p-well resistance:

$$\frac{1}{R_{pw}\|R_{sx}} \approx \frac{1}{R_{pw}} \qquad R_{sx} \gg R_{pw}.$$

From the $\beta_{PNP}\beta_{NPN}$ criteria, the solution of what is the substrate resistance value that initiates latchup can be derived. Given

$$\beta_n \beta_p \geq \frac{1 + \left(\frac{I_{sx}}{I}\right) \beta_n}{\left[1 - \left(\frac{I_w}{I}\right)\left(\frac{\beta_n+1}{\beta_n}\right) - \left(\frac{I_{sx}}{I}\right)\right]},$$

the expression for the substrate current can be factored out, and expressed as follows:

$$\frac{I_{sx}}{I} \leq \frac{\beta_n \beta_p \left[1 - \left(\frac{I_w}{I} \right) \left(\frac{\beta_n + 1}{\beta_n} \right) \right]}{\beta_n (1 + \beta_p)},$$

and

$$I_{sx} \leq I \left\{ \frac{\beta_n \beta_p \left[1 - \left(\frac{I_w}{I} \right) \left(\frac{\beta_n + 1}{\beta_n} \right) \right]}{\beta_n (1 + \beta_p)} \right\}.$$

From this expression, the relationship of the forward bias of the NPN and the resistance can be substituted into the equation. The effective substrate resistance can be expressed as the parallel product of the p-well and the substrate resistances. The effective substrate resistance needed in order to initiate latchup can be expressed as

$$R_{sx} \| R_{pw} \geq \frac{(V_{be})_{NPN}}{I} \left\{ \frac{\beta_n (1 + \beta_p)}{\beta_n \beta_p \left[1 - \frac{I_w}{I} \left(\frac{\beta_n + 1}{\beta_n} \right) \right]} \right\}.$$

Hence, the "effective substrate resistance condition" to initiate latchup can be expressed as

$$(R_{sx})_{eff} \geq \frac{V_o \ln \left(\frac{I - I_{sx}}{I_o} \right)}{I} \left\{ \frac{\beta_n (1 + \beta_p)}{\beta_n \beta_p \left[1 - \frac{I_w}{I} \left(\frac{\beta_n + 1}{\beta_n} \right) \right]} \right\}.$$

In future applications, high-resistance substrates will be sensitive to external events, such as single-event particles causing SEL. Hence, the condition for initiation of latchup from a SEU, in the "alpha formulation" is when [23]

$$(\alpha_p + \alpha_n) = 1 + \alpha_p \frac{(I_w)}{I} + \alpha_n \frac{(I_{sx})}{I} - \frac{I^*}{I},$$

or the current for latchup initiation for an external current source occurs when

$$I^* = I \left\{ 1 + \alpha_p \left[\frac{(I_w)}{I} - 1 \right] + \alpha_n \left[\frac{(I_{sx})}{I} - 1 \right] \right\},$$

where the well and substrate current expression can be shown to be (respectively)

$$I_w = \frac{(V_{be})_{PNP}}{R_w} = \frac{V_o}{R_w} \ln \left[\frac{I - I_w}{(I_o)_p} \right]$$

and

$$I_{sx} = \frac{(V_{be})_{NPN}}{R_{sx}} = \frac{V_o}{R_{sx}} \ln \left[\frac{I - I_{sx}}{(I_o)_n} \right].$$

Hence the condition for latchup from an external source can be expressed also as

$$I^* = I \left\{ 1 + \frac{\beta_p}{\beta_p + 1} \left[\frac{I_w}{I} - 1 \right] + \frac{\beta_n}{\beta_n + 1} \left[\frac{I_{sx}}{I} - 1 \right] \right\},$$

or

$$I_{inj} = I \left\{ 1 + \frac{\beta_p}{\beta_p + 1} \left[\frac{I_w}{I} - 1 \right] + \frac{\beta_n}{\beta_n + 1} \left[\frac{I_{sx}}{I} - 1 \right] \right\} - I_{cno} - I_{cpo}.$$

For the same circuit, this can be expressed as a function of the bipolar current gains. The "beta product" relation for undergoing latchup in the environment of an external injection can be written in the form

$$\beta_n \beta_p = \frac{1 + \beta_p \left[\frac{I_w - I^*}{I} \right] + \beta_n \left[\frac{I_{sx} - I^*}{I} \right] - \left[\frac{I^*}{I} \right]}{\left[1 - \frac{I_w + I_{sx} - I^*}{I} \right]},$$

where

$$I^* = I_{cpo} + I_{cno} + I_{inj}(x, t)$$

$$I_w = \frac{(V_{be})_{PNP}}{R_w} = \frac{V_o}{R_w} \ln \left[\frac{I - I_w}{(I_o)_p} \right]$$

$$I_{sx} = \frac{(V_{be})_{NPN}}{R_{sx}} = \frac{V_o}{R_{sx}} \ln \left[\frac{I - I_{sx}}{(I_o)_n} \right].$$

In an alpha representation,

$$(\alpha_p + \alpha_n) = 1 + \alpha_p \frac{(I_w)}{I} + \alpha_n \frac{(I_{sx})}{I} - \frac{I^*}{I}.$$

Solving for the substrate current needed to initiate latchup,

$$I_{sx} = I \left[1 + \left(\frac{\alpha_p}{\alpha_n} \right) \left[1 - \frac{I_w}{I} \right] + \left(\frac{1}{\alpha_n} \right) \left[\frac{I^*}{I} - 1 \right] \right]$$

and

$$I_{sx} = \frac{(V_{be})_{NPN}}{(R_{sx})_{eff}} = \frac{V_O}{(R_{sx})_{eff}} \ln \left[\frac{I - I_{sx}}{(I_o)_n} \right].$$

Hence the substrate resistance condition needed to initiate external latchup under a condition of an external injection source can be quantified.

$$(R_{sx})_{eff} \geq \frac{(V_{be})_{NPN}}{I \left[1 + \left(\frac{\alpha_p}{\alpha_n} \right) \left[1 - \frac{I_w}{I} \right] + \left(\frac{1}{\alpha_n} \right) \left[\frac{I^*}{I} - 1 \right] \right]}$$

or expressed as follows,

$$(R_{sx})_{eff} \geq \frac{V_o \ln \left[\frac{I - I_{sx}}{I_{on}} \right]}{I \left[1 + \left(\frac{\alpha_p}{\alpha_n} \right) \left[1 - \frac{I_w}{I} \right] + \left(\frac{1}{\alpha_n} \right) \left[\frac{I^*}{I} - 1 \right] \right]}.$$

From these formulations, the substrate resistance condition for the amount of resistance in the substrate in order to initiate SEL for the case of external injection. As the external injection from a ionized particle increases, the denominator of the expression increases in magnitude, where the effective resistance needed to undergo latchup is lower. Hence, as the external source increases, the substrate resistance requirement to undergo latchup decreases. In the case that the substrate resistance continues to increase with scaling, circuits will become more sensitive to the external sources. High-resistance substrates will influence the latchup design practice as follows:

- The impact of substrate resistance on the local PNPN latchup substrate shunt resistance will lessen as the substrate resistance is significantly less than the p-well resistance.
- A high-low p+/p− step junction transition will occur as the p− substrate doping concentration is lower than the p-well doping concentration, forming a n++/p+/p− diode structure; this will increase the vertical diode injection to the substrate.
- The substrate thermal impedance will increase, leading to a higher self-heating within the PNPN structure.
- Minority carrier recombination time will increase, leading to larger diffusion lengths and longer propagation distances of minority carriers.

First, since the shunt resistance is the parallel resistance of the p-well and the p− substrate, the p-well will play a greater role as the p− substrate resistance is significantly increased. Second, with the transition from a first substrate resistance to a higher second substrate resistance the change will be less than anticipated. For the second item, as the p- substrate decreases below the p-well, a p+/p− high-low junction is formed; this increases the vertical injection of carriers. As a result, the lateral NPN current gain should be reduced. For the third item, as the conductivity of the substrate decreases, the thermal conductivity also decreases; this leads to a higher thermal impedance. Hence, latchup events that introduce self-heating will have a higher junction temperature for the same amount of power dissipated. As a result, the rise in temperature and the change in parameters with the temperature will be more significant for high-resistance substrate wafers. For the fourth item, it is anticipated that transient and external latchup considerations will be worse. Minority carrier recombination time will increase, leading to longer diffusion lengths and longer distances that carriers traverse the substrate. Additionally, the ability to shunt the excess carriers has increased, making products more vulnerable to SEL to transient phenomena.

7.5.1   50 Ω-cm Substrate Resistance

Figure 7.6 is a plot of the β product (NPN and PNP product) versus the p+/n+ space as a function of the 10 and 50 Ω-cm substrate wafers. Measurements of the β_{NPN} show as the p+/n+ space increases, β_{NPN} decreases. At a small n+ diffusion to p− substrate contact space, and the presence of p-well implants, only a small change in the extracted NPN β_{NPN} is evident (e.g. β_{NPN} extraction is at a 1 mA current level) [23–25].

Figure 7.7 shows the latchup I-V characteristic vs the p+/n+ space [23]. The p+/n+ spacing comprises the space between the p+ diffusion and n-well edge, and the space between the n+ diffusion and the adjacent n-well. The plot shows the data for a p+/n+ spacing cases of 270, 380, 600, and 860 nm. The data provided is simulation for a 280 nm STI depth. The data illustrates that as the p+/n+ spacing increases, the trigger voltage and the holding voltage increases.

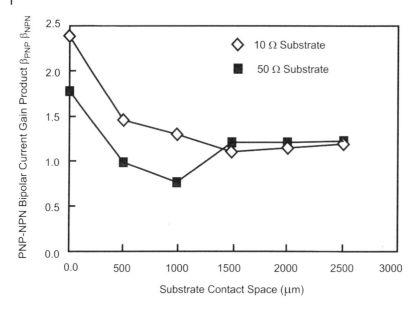

Figure 7.6 Parasitic NPN and PNP bipolar current gain product as a function of substrate resistivity.

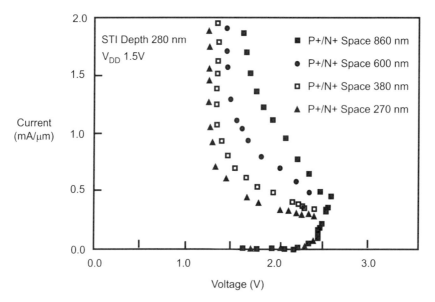

Figure 7.7 Latchup I-V simulation versus p+/n+ spacing.

7.6   Wells

Well structures influence the EHP generation, the EHP recombination, and minority carrier transport. Technologies support diffused wells, retrograde wells, and dual-well structures. These influence both the SER and latchup.

7.6.1 Single Well – Diffused N-Well

Diffused well structures were used prior to the retrograde well, but are continued to be used in power technologies. Diffused wells have the highest doping concentration at the surface of the substrate. Dopants are diffused into the wafer through a hot process. Diffused wells are typically deeper than retrograde wells (e.g. 6–12μm depth). This leads to a larger amount of minority carrier charge collection in the well structure. A diffused well technology has a built-in electric field away from the surface since the dopant is monotonically decreasing into the wafer. This leads to a built-in electric field driving the EHP toward the well-to-substrate junction.

7.6.2 Single Well – Retrograde N-Well

Retrograde well structures were introduced with the use of high energy implant tools. Retrograde wells have the highest doping concentration away from the surface of the substrate. Dopants are implanted deep into the well structure. Retrograde wells are typically 1.5–4.0 μm deep. This leads to a smaller amount of minority carrier charge generation in the well structure. A retrograde well technology has a built-in electric field that is nonmonotonically decreasing into the wafer. Figure 7.8 shows the PNP bipolar current gain as a function of retrograde wells for different doping concentrations [29–32].

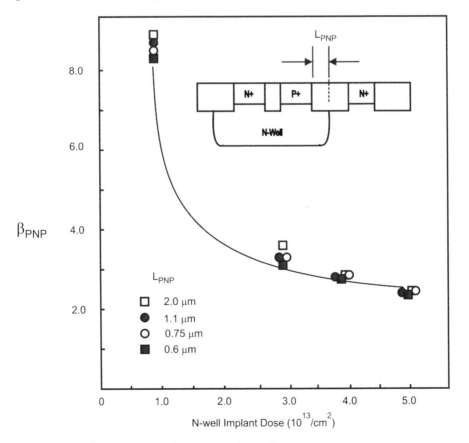

Figure 7.8 Bipolar PNP current gain vs retrograde n-well structure.

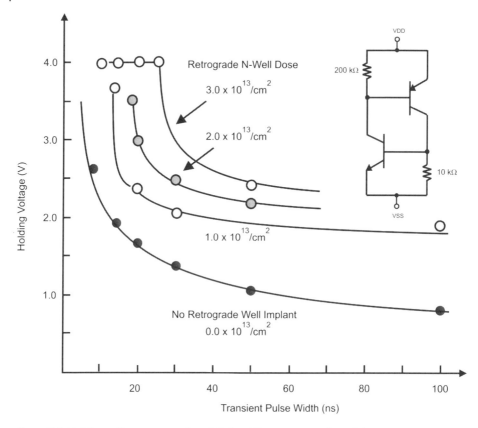

Figure 7.9 Holding voltage versus pulse width for different retrograde well dose.

In Figure 7.9, the plot illustrates the latchup holding voltage as a function of pulse width for different n-well structures. The lowest holding voltage occurs for no retrograde n-well structure. With the addition of the retrograde n-well implant, the holding voltage increases for all transient pulse widths. As the retrograde n-well implant dose increases, the holding voltage increases to a higher magnitude.

7.6.3 Dual-Well Technology

In more advanced CMOS technologies, dual-well technology was introduced [23]. Dual-well technology added a second well, of p-type dopant, to form a p-well (Figure 7.10). The addition of a p-well provides two improvements for radiation immunity. First, the higher doping concentration of a p-well versus p- epitaxial region leads to a higher minority carrier recombination. Second, it introduces a p+/p− step at the interface between the p-well to p− substrate interface providing a built-in field; this drives the minority carriers away from the sensitive junction areas.

7.6.3.1 P-well and P++ Substrate
In scaled dual-well technology, the p-well is scaled to a shallower depth. Figure 7.11 illustrates the cross section of a vertical bipolar in a scaled p-well on a p+ substrate.

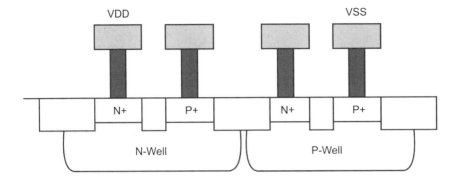

Figure 7.10 Dual well technology cross section.

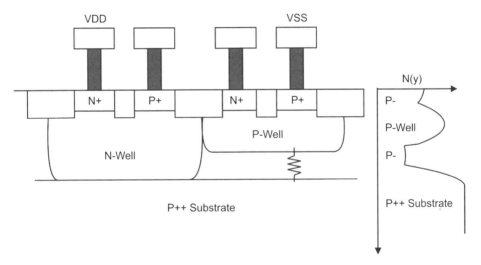

Figure 7.11 Cross section of the vertical bipolar in a scaled p-well and p++ substrate.

With this implementation, there exists a p-well/ p− epitaxy/p+ substrate structure that is detrimental to latchup and ionizing radiation. The resistance of the p− epitaxy introduces a series resistance that impacts latchup sensitivity. For ionizing radiation, the low recombination rate of the p− epitaxy increases the potential minority carrier collection.

7.6.3.2 P-Well and P+ Connecting Implant

In scaled dual-well technology, the SEU and latchup solution is to add a "connecting implant." Figure 7.12 illustrates the cross section of a vertical bipolar in a scaled p-well on a p+ substrate with the p-type connecting implant. With this implementation, there exists a higher dopant region in the p− epitaxial region, improving both the latchup and ionizing radiation immunity [23, 33–35].

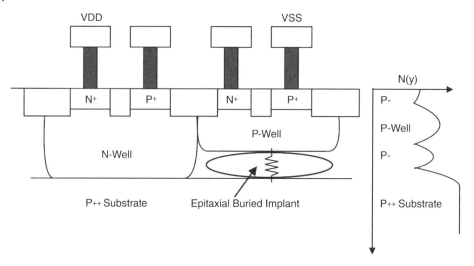

Figure 7.12 P-well and p+ connecting implant.

7.7 Triple-Well Technology

With the introduction of triple-well technology, another advantage to mitigate SEUs and SEL was introduced. There are two types of implementations of triple-well technology. The first triple-well implementation has full separate of the p-channel and n-channel transistors. The second type will be referred to as *merged triple well*.

7.7.1 Triple Well – Full Separation of Wells

With the introduction of triple-well technology came full separation of the p-channel and n-channel transistors [33–35]. In this case, there is no parasitic PNPN formed by the p-channel and n-channel transistors, mitigating the onset of SEL. Figure 7.13 shows the cross section of the triple-well technology with full well separation. It was this implementation that led semiconductor developers to believe that latchup would no longer be an issue. Unfortunately, design teams did not implement this layout but chose to use a "merged triple well" implementation.

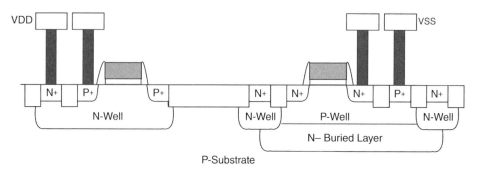

Figure 7.13 Fully separated triple-well technology.

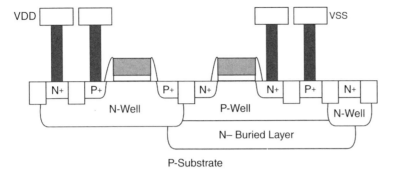

Figure 7.14 Merged triple-well technology.

7.7.2 Triple Well – Merged Triple Well

With merged triple well technology there is not a full separation of the p-channel and n-channel transistors [33–35]. In this case, a parasitic PNPN is still formed by the p-channel and n-channel transistors, with an additional vertical NPN transistor. This does not mitigate the onset of SEL. Figure 7.14 shows the cross section of the merged-triple-well technology without full well separation [33–35].

Figures 7.15 and 7.16 show a comparison of the merged-triple-well versus standard dual-well technology for the PNP transistor and NPN transistor current gain, respectively. The merged triple well PNP bipolar current gain is lower than the case of dual wells. For the triple well, the PNP bipolar current gain varies from $\beta_{PNP} = 2-1.5$ at $140\,°C$. For the merged-triple-well NPN current gain, it is higher than the dual-well case due to the vertical NPN, which is enhanced due to the buried layer. Although for the dual well, the bipolar NPN current gain is approximately unity, for the triple well, the bipolar current gain exceeds $\beta_{NPN} = 5.0$.

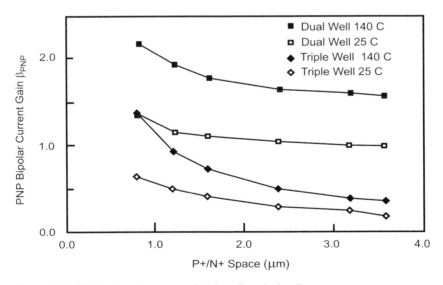

Figure 7.15 PNP bipolar gain – merged triple well vs. dual well.

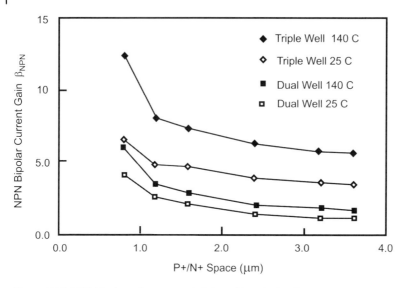

Figure 7.16 NPN bipolar gain – merged triple well vs. dual well.

7.7.3 Triple Well – Merged Triple Well with Blanket Implant

In another embodiment of the triple well is where the buried layer is a continuous blanket implant under both the p-well and the n-well [33–35].

7.8 Sub-Collectors

In CMOS processes, retrograde n-well design has shown to have a significant impact on latchup [29–35]. Bipolar and BiCMOS technologies have used sub-collectors in both homo-junction and hetero-junction bipolar transistors for transistor performance. Sub-collector have a number of advantages for bipolar transistors which can be utilized for latchup [33–36]. First, they provide a low-resistance collector. Second, at high currents they minimize the Kirk effect. Third, they establish a low sub-collector/substrate junction capacitance. Fourth, it minimizes minority carrier injection into the substrate. These fundamental characteristics of a sub-collector have natural advantages for latchup prevention.

7.8.1 Epitaxial Grown Sub-Collector

There are two methods of forming a sub-collector region – epitaxial and nonepitaxial. In a first process, sub-collectors are formed prior to an epitaxial growth step. A substrate wafer is implanted through a masked area, followed by an epitaxial growth process; in this process, the region can be placed deep into the substrate wafer with a very high doping concentration. The dose level of these sub-collectors are typically 10^{16} cm^{-2}. This doping concentration can be at a silicon saturation level (e.g. 10^{19}–10^{21} cm^{-2}) providing very low sheet resistance (e.g. 1–10 $\Omega/\square$ sheet resistance). These sub-collectors provide a deep sub-collector to p– substrate metallurgical junction on the order of 3–5 μm below the wafer surface. Additionally, the doping concentration is so significant, the minority carrier recombination time is Auger recombination dominant [33–36].

7.8.2 Implanted Sub-Collector

A second form of sub-collectors are a nonepitaxial process of using a high-energy MeV implant [36]. In this case, the implant depth is shallower, limiting the dose, concentration, and sheet resistance. From a CMOS latchup design practice, the addition of a sub-collector (added under CMOS devices) provides the following advantages:

- Wider PNP base width of vertical parasitic PNP transistor.
- Higher dopants in the base region of the vertical parasitic PNP transistor.
- Lower "n-well" shunt resistance between the n-well contact and the p-channel MOS-FET device.

Experimental work demonstrates CMOS latchup tradeoffs exist between the enhancement β_{NPN} and reduction of β_{PNP}.

The implanted sub-collector design point for the sub-collector was defined to be suitable for a low cost bipolar transistor technology for a 0.13-μm CMOS-base technology [36].

7.8.3 Sub-Collector – NPN and PNP Bipolar Current Gain

Sub-collectors can mitigate SEU and latchup events. The first important point result is that the addition of a sub-collector increases the bipolar NPN current gain, β_{NPN} (contrary to the desired result). At large p+/n+ space, the two cases converge toward a common result – but the case of the dual-well CMOS with the additional sub-collector is worse; the addition of the sub-collector increases the bipolar NPN current gain, β_{NPN} approximately 2× (e.g. 0.76–1.7). This latchup degradation effect can be associated with three possible issues: (i) increase in the effective collector area; (ii) sub-collector lateral out-diffusion leading to a smaller base width; and (iii) an improved electron transport with deep collector.

There are three factors that lead to an increased bipolar current gain:

1. The increased collector area (due to the sub-collector depth) increases the bipolar current gain.
2. The out-diffusion of the sub-collector decreases the base width.
3. The deeper structure leads to a change in the minority carrier trajectory.

In the case of a shallower well region the electron trajectory is of a more two-dimensional nature. Whereas, as the emitter or the collector extends below, the worst case trajectory path is decreased, leading to a shorter effective base width and higher bipolar gain.

The sub-collector implant provides a significant reduction in the bipolar PNP current gain, β_{PNP}. Note that the β_{PNP} for dual-well CMOS with and without sub-collector shows a weak sensitivity to the p+/n+ space.

7.8.4 Sub-Collector – Beta Product $\beta_{PNP}\beta_{NPN}$

First, the dual-well CMOS $\beta_{PNP}\beta_{NPN}$ is significantly higher. Second, as the p+/n+ space is reduced, the $\beta_{PNP}\beta_{NPN}$ increases significantly for the case of dual-well CMOS technology. An interesting result is that the sub-collector depth increases the total area leading to an increase in β_{NPN} by approximately 2×, but at the same time leads to a 10× decrease in β_{PNP}. But, the role of the β_{PNP} reduction is a greater effect, hence lowering the

"β product" term. Additionally, one can note that the dual-well CMOS case $\beta_{PNP}\beta_{NPN}$ is increasing above unity as the p+/n+ space decreases; yet, in the case of the n-well and sub-collector implant, the $\beta_{PNP}\beta_{NPN}$ remains well below a unity.

From a latchup design practice, the addition of a sub-collectors provide the following advantages:

- Significantly lower n-well shunt resistance is achieved (e.g. 10–100×).
- Lower β_{PNP} is achieved with sub-collector.
- Lower $\beta_{PNP}\beta_{NPN}$ is achieved providing improved latchup robustness.
- A higher undershoot and overshoot current is required to achieve latchup.

7.9 Heavily Doped Buried Layers (HDBL)

Similar to the concept of triple well technology, is the structure of a HDBL. HDBL provides robust structures for both minimizing and mitigation of SEU and SEL.

7.9.1 Buried Implanted Layer for Lateral Isolation (BILLI) Process

An early concept was the concept of implanting a discontinuous layer with a single implant, known as the BILLI process [13]. In this process, a high-energy implant was used to form a layer under the n-well region, and a second part in the p– epitaxial region [13] (Figure 7.17). This was suitable for single-well CMOS, or dual-well CMOS. In this method, the implant was performed using a mask. This lead to limitations of semiconductor processing.

7.9.2 Continuous HDBL Implant

A second method was implant a HDBL that was implanted without usage of a mask [14–21, 33–35]. A continuous implant layer was formed under both the p-well and n-well. Figures 7.18 and 7.19 show the vertical doping profile of the HDBL under the n-well, and p-well, respectively.

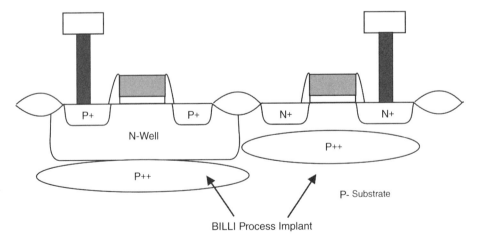

Figure 7.17 Cross section of the BILLI structure.

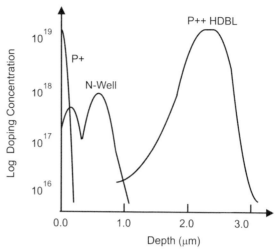

Figure 7.18 Heavily doped buried layer (HDBL) – vertical doping profile under the n-well.

Figure 7.18 illustrates that the p++ HDBL is placed deep under the retrograde well. This is to avoid lowering of the n-well breakdown voltage, and to avoid doping compensation of the n-well structure. Note that the doping concentration of the HDBL is significantly higher than the n-well leading to strong EHP recombination in the HDBL layer.

Figure 7.19 illustrates that the p++ HDBL is placed deep under the p-well. This is to avoid influence the n+ MOSFET device threshold voltage. Note that the doping concentration of the HDBL is significantly higher than the p-well leading to strong EHP recombination in the HDBL layer.

HDBL provides both a low-resistance shunt for reduction of the SEL sensitivity. Additionally, the doping concentration is high, leading to Auger and Shockley-Hall-Read (SHR) recombination (Figure 7.20). Experimental results show a lowering of the recombination time in the substrate region with the HDBL implant.

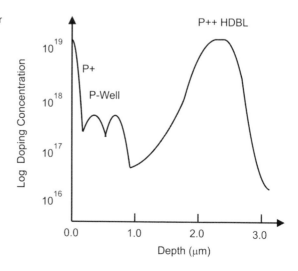

Figure 7.19 Heavily doped buried layer (HDBL) – vertical doping profile under the p-well.

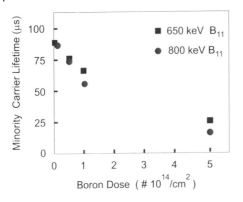

Figure 7.20 Minority carrier lifetime versus heavily doped buried layer (HDBL) dose.

Figure 7.21a–c illustrates latchup simulation with and without HDBL, as a function of HDBL dose, and with variation in STI depth.

7.9.3 Buried Guard Ring (BGR)

A further improvement in the prior process solutions is the addition of a reach-through implant that connects the source/drain implant to the HDBL forming a BGR structure [14–21, 33–35]. Figure 7.22 is a cross section of the BGR structure, implemented in a CMOS technology. The BGR structure consists of the p++ HDBL and the p++ implant.

Figure 7.23 is the latchup simulation of the BGR structure. The results show that the latchup I-V characteristic does not undergo a negative resistance state, avoiding a latchup condition.

7.10 Isolation Concepts

In this section, different isolation structures that influence SEU and SEL will be discussed.

7.10.1 LOCOS Isolation

In early technology development, an isolation structure known as LOCOS was prevalent in mainstream CMOS [29]. The LOCOS structure was typically shallower than the adjacent implants. This led to lateral transport of the EHPs. Additionally, lateral parasitic transistors were present between adjacent diffusions. From a SEU and latchup perspective, these technologies were not immune from SEU and latchup interaction.

7.10.2 Shallow Trench Isolation (STI)

In later technology generations, STI was introduced [9, 33–35]. In this technology, the isolation was deeper than the adjacent junctions. This lead to elimination of lateral carrier transport along the surface of the device. Figure 7.24a shows an example of the latchup holding voltage as a function of the STI depth.

Latchup simulation also shows that the latchup response is a function of the STI depth. This is illustrated in Figure 7.24b.

7.10.3 Dual Depth Isolation

With technology scaling, each technology generation is reducing physical dimensions; as a result, the p+/n+ spacing is being reduced. As the p+/n+ space is reduced, the STI width and depth is also being scaled. A solution to minimize the scaling impact to CMOS latchup is the usage of two isolation depths in the physical structure between p-channel and n-channel MOSFETs. This was first proposed by M. Bohr [39]. M. Bohr proposed a

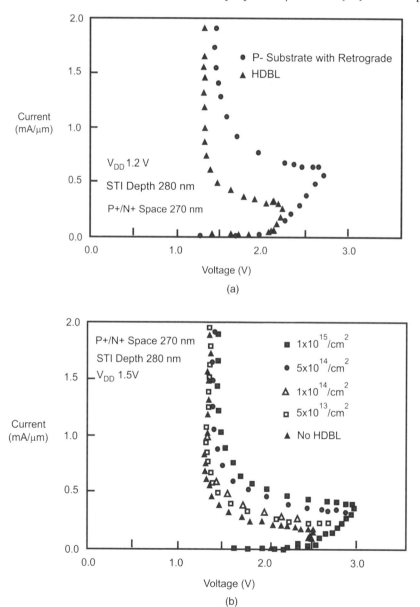

Figure 7.21 (a) Latchup with and without HDBL; (b) Latchup simulation versus heavily doped buried layer (HDBL) dose; (c) Latchup simulation versus STI depth with and without HDBL.

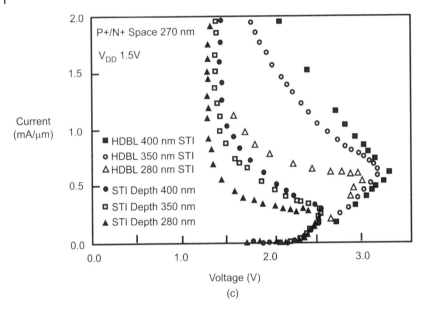

Figure 7.21 (*Continued*)

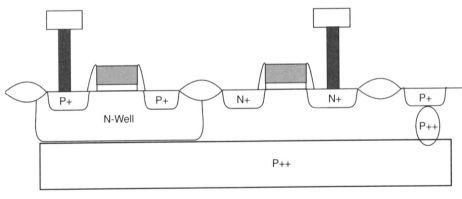

Figure 7.22 Buried guard ring (BGR) cross section.

"dual depth isolation structure," which has a scalable shallow depth, and a deeper second depth. By providing alternate process solutions and new isolation structures, the latchup robustness of CMOS circuits can be improved. There are second isolation structures that are normally intended for the bipolar transistor that can be utilized for this purpose. This concept was not implemented in mainstream technology.

7.10.4 Trench Isolation (TI)

Figure 7.25a shows a low-cost trench isolation (TI) structure [40, 41]. The TI structure is an isolation structure with no poly-silicon fill regions [40, 41]. The TI structure penetrates through the standard CMOS STI. The final structure formed is similar to a

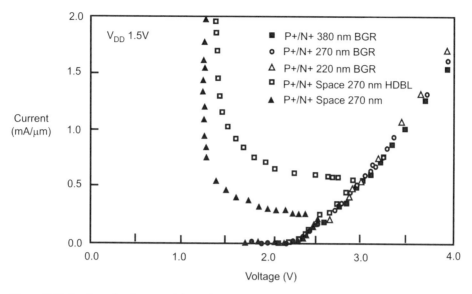

Figure 7.23 Latchup simulation with buried guard ring (BGR).

proposed "dual-depth" STI, but is dissimilar in that the integration is not completed as a dual-damascene process. As a result, the structure is not constrained to the STI process optimization (e.g. density, polishing requirements), nor is it constrained in the process flow placement. As an independent technology step, it can be independently optimized for width, depth, and density requirements. This isolation structure can mitigate SEU and SEL events.

For a CMOS latchup design practice, the trench isolation (TI) structure can be integrated into a mainstream base CMOS technology. Placement of the TI structure on the n-well edge, makes n-well breakdown voltage not a function of the p-well, or adjacent structures. From a latchup perspective, this improves the latchup trigger voltage, V_{TR}, for positive ramp test modes of the n-well. Additionally, the lateral parasitic PNP transistor gain will be significantly reduced because of the inability for the holes to flow from the p+ emitter to the p-well collector region. As a result, the lateral parasitic PNP bipolar device is eliminated, leaving only a vertical parasitic PNP bipolar device. An advantage of this feature is as the technology is scaled, there will be no p+/n+ space sensitivity. Additionally, the placement of the TI-bordered n-well, the lateral parasitic NPN bipolar is also impacted. The minority carrier electron flow from an NPN emitter structure to the trench-bound n-well collector structure is inhibited by the TI structure; this decreases the NPN bipolar current gain. Additionally, the dependence on spacing will also be weakened since it will be less dependent on lateral current flow and more dependent on the flow from the emitter to the collector under the isolation structure.

A latchup design practice is as follows:

- A TI perimeter to an n-well region lowers the lateral β_{NPN} scaling with p+/n+ space.
- A TI perimeter structure on a CMOS n-well region lowers the lateral β_{PNP} with p+/n+ scaling.
- A TI perimeter structure to a CMOS n-well region lowers the $\beta_{PNP}\beta_{NPN}$ scaling with p+/n+ space.

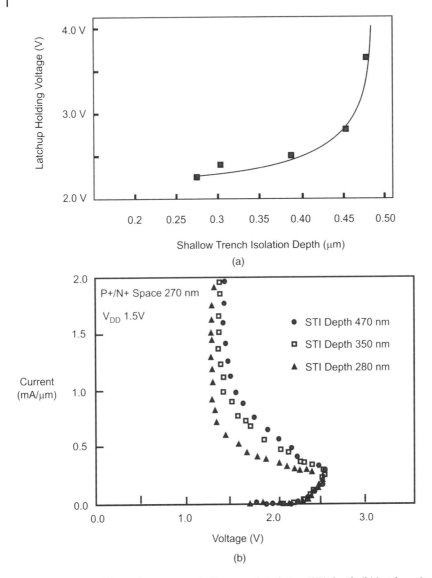

Figure 7.24 (a) Holding voltage versus shallow trench isolation (STI) depth; (b) Latchup simulation versus shallow trench isolation depth.

- A TI structure demonstrates an improvement in all latchup metrics (e.g. bipolar current gain, bipolar current gain product, overshoot, undershoot, turn-on voltage, and trigger voltage).

7.10.4.1 Trench Isolation (TI) and Sub-Collector
Latchup robustness can be further improved by integrating both the sub-collector implants and trench isolation (TI). The addition of implants to the well structure can improve latchup by decreasing the PNP bipolar current gain, and reducing the n-well

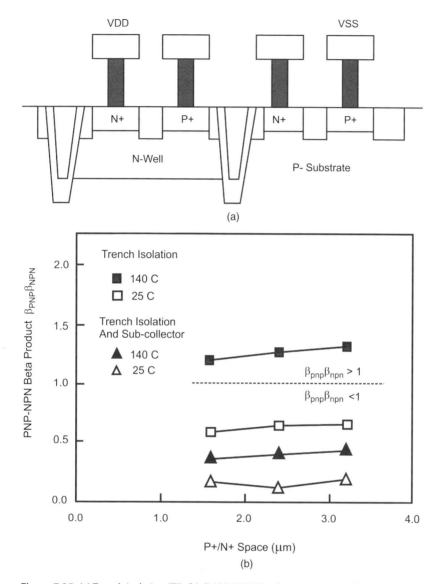

Figure 7.25 (a) Trench isolation (TI); (b) CMOS PNPN latchup structure with trench isolation and implanted sub-collector.

shunt resistance. In this study, an implanted sub-collector and the TI are integrated [23, 33–35]. Figure 7.25b shows the integration of the sub-collector implant and TI structure. The sub-collector implant is placed below the CMOS n-well implant. The sub-collector sidewall capacitance is also reduced using the TI structure. The TI depth and the sub-collector are chosen to optimize the sub-collector capacitance and the other device parameters.

With the addition of sub-collector and TI, the lateral $\beta_{NPN} = 1.02$ at large p+/n+ space and $\beta_{NPN} = 0.99$ at the smallest space (140 °C). Again, in contrast to the standard CMOS process, the lateral β_{NPN} decreases with p+/n+ space (e.g. instead of an increasing nature in the base CMOS technology). Additionally, the temperature sensitivity is weak in the case of the TI case, remaining at near unity values even at elevated temperatures [40, 41].

The lateral PNP gain is also influenced by the TI and sub-collector. Evaluation of the parasitic lateral PNP current gain for case (i), as the p+/n+ spacing decreases the bipolar PNP current gain increases. With TI, there is a decrease in the lateral β_{PNP} as the p+/n+ spacing decreases; as the p+ diffusion approaches the isolation sidewall, the vertical current decreases, providing a small decrease in the current gain. With the addition of an implanted sub-collector, the parasitic bipolar current gain significantly decreases well below unity (Table 7.1).

In the Table 7.2, the parameters are shown for three different processes: for the p+/n+ spacing of 1.6 µm and for the forward and reverse β_{PNP} . The results show that the "sub-collector only" case has a stronger influence on the PNP bipolar current gain (e.g. 3× lower) whereas the addition of TI, the β_{PNP} is reduced 30%.

The experimental results show that $\beta_{PNP}\beta_{NPN}$ with TI-only is below unity for ambient temperature, but above unity values for elevated temperatures, for all p+/n+ spacings. Measurements show that $\beta_{PNP}\beta_{NPN}$ decreases with the smaller p+/n+ spacings. For the case of the additional sub-collector, the latchup $\beta_{PNP}\beta_{NPN}$ results remain below unity for all structure sizes tested and maintaining a 3× improvement (over the TI case only).

Table 7.1 Bipolar parameters vs. p+/n+ space for CMOS PNPN structures with trench isolation (TI) and sub-collector.

p+/n+ Spacings	Forward PNP bipolar current gain, β_F	Reverse PNP bipolar current gain, β_R	Avalanche breakdown voltage, V_{AV} (V)
1.6	0.21	0.009	55
2.4	0.22	0.009	55
3.2	0.22	0.008	55
3.6	0.24	0.007	55

Table 7.2 Bipolar parameters for CMOS PNPN structures with combinations of trench isolation (TI) and implanted sub-collector.

Process	p+/n+ Spacing (µm)	Forward PNP bipolar current gain, β_F	Reverse PNP bipolar current gain, β_R
Trench isolation	1.6	0.95	0.055
Sub-collector	1.6	0.29	0.019
Trench isolation and sub-collector	1.6	0.21	0.009

For the undershoot analysis, it was shown that with TI, the latchup undershoot parametric does not decrease with spacing. With TI, but without the sub-collector implant, the undershoot current drops to $I_{under} = 70\,mA$ (e.g. for a 25 µm wide PNPN structure) at elevated temperatures.

This same effect is evident with the I_{OVER} analysis of the PNPN structures. With elevated temperature, the wider spacings shows evidence of I_{OVER} degradation in the TI case. With the addition of the implanted sub-collector, the maximum test level of 100 mA overshoot (e.g. 4 mA/µm) is preserved at 140 °C.

The introduction of TI has significant implications for the future of prevention of latchup in advanced semiconductors for CMOS, as well as power, automobile, and space applications. In essence, the integration of TI structure with the STI structure is structurally equivalent to a "dual depth" isolation structure that has been proposed for mainstream CMOS technology, but never implemented. More recently, in the 1990s, to address STI scaling, Bohr proposed the concept of a dual depth isolation structure [19]. From a scaling perspective, the TI dimensions can be scaled with the scaling of the sub-collector implant. The width scaling of the TI structure will allow smaller p+/n+ space, as this structure is utilized for latchup solutions. Hence, the scaling of the structure is natural to CMOS MOSFET scaling and BiCMOS SiGe HBT device scaling objectives.

A CMOS latchup design practice, combining both trench isolation and sub-collector, is as follows:

- A TI perimeter to an n-well region with a sub-collector lowers the lateral NPN current gain scaling and eliminates any sub-collector NPN enhancement concerns.
- A TI perimeter structure and sub-collector implant on a CMOS n-well region significantly lowers the lateral PNP current gain.
- A TI structure and sub-collector demonstrates an improvement in all latchup metrics (e.g. bipolar current gain, bipolar current gain product, overshoot, undershoot, turn-on voltage, and trigger voltage) and eliminates the dimensional scaling impacts observed in CMOS technology.

7.11 Deep Trench

Deep trench (DT) structures are utilized in trench DRAM capacitor elements, and embedded DRAM technology for SOC applications [37, 38, 42–46]. In DT capacitor structures, thin dielectric exists to provide a high-capacitance element. For smart power applications, standard CMOS technology is extended to integrate high-voltage power devices. Trench structures exist in power electronics to provide sidewall vertical MOSFETs, and other novel structures. With the high-voltage CMOS applications, there is motivation to provide latchup robustness at levels significantly above the native power supply voltage of advanced CMOS. In automotive applications, there are states of mis-installation of electronics leading to reverse pin conditions (e.g. negative instead of positive voltage), to mis-installation of batteries. In addition, there are environments with high inductive load dumps, and switching conditions that lead to high transient states. Hence, there is motivation to provide high latchup immunity. For space applications, it is desirable to provide technology that is immune to SEL events.

Table 7.3 Deep trench depth vs. guard ring metrics (F, and $1/F$, and $1 - 1/F$, and guard ring efficiency).

Experimental split trench depth (μm)	Actual trench depth (μm)	F	$1/F$	$1 - 1/F$	Guard ring efficiency 100% $(1 - 1/F)$ (%)
1	1.6	4.64	0.215 5	0.784 5	78.45
2	2.46	6.29	0.159 0	0.841 0	84.10
3	3.38	8.71	0.114 8	0.885 2	88.52
4	4.14	10.6	0.094	0.906 0	90.60
6	6.25	17.1	0.058 48	0.941 5	94.15

Hence, process solutions that can provide a high trigger voltage, V_{TR}, or high holding voltage, V_H, are desirable (Table 7.3).

Deep trench isolation can be utilized in multiple ways to provide improved CMOS latchup robustness:

- Stand-alone DT guard rings structures can be used to isolate injection sources from SEL in conjunction with other independent guard ring structures.
- DT guard rings can be utilized on the perimeter of n-well regions, such as the n-well containing p-channel MOSFETs.

At the extracted value of 100 mA, the metric F shows a linear increase with trench depth. Evaluation of the DT depth in the form of the F factor shows that the expression becomes linear as a function of the DT depth. In linear form, we can assume a form $F = A L_{DT} + B$, where A and B are a constant, where F is a function of trench depth, leading to the fitting parameters [23]:

$$F = 2.7087 L_{DT} - 0.19$$

Inverting the factor, we can introduce the $1/F$ factor, which is more physically associated with the probability of escape and the guard ring efficiency. Assuming a functional and power relationship [23]:

$$\frac{1}{F} = \frac{A}{(L_{DT})_B},$$

where L_{DT} is the DT depth (in μm), and A and B are constants, the data fits for $A = 0.3575$ and $B = 0.9568$ (e.g. unity). From this definition, it can be shown that the guard ring efficiency can be expressed in a power form:

$$\Psi_{GRE} = 1 - \frac{1}{F} = 1 - \frac{A}{(L_{DT})_B}$$

In this form, it can be observed that as the factor $1/F$ approaches zero, the guard ring efficiency approaches unity (e.g. or in percentage form, 100%) (Table 7.4).

Hence the usage of DT independent guard rings lowers the probability of electron escape and improves the latchup margin to the "external latchup" problem observed in integrated electronics.

These results demonstrate the usage of DT structures as independent guard rings in a p− substrate wafer significantly improves the problem of "external latchup" where an

Table 7.4 Trench depth study and the "bipolar" and guard ring efficiency methodologies.

Experimental split trench depth (μm)	Actual trench depth (μm)	Forward bipolar current gain, β_F	Reverse bipolar current gain, β_R	F	$1/F$
1	1.6	0.205	0.244	4.64	0.215 5
2	2.46	0.191	0.323	6.29	0.159 0
3	3.38	1.36	0.381	8.71	0.114 8
4	4.14	1.20	0.424	10.6	0.094
6	6.25	0.050	0.513	17.1	0.058 48

injection source leads to triggering of latchup in adjacent circuitry. Electrons injected from the forward biasing of an n- or n-well region in a p− substrate can initiate latchup in an adjacent structure. From this experimental work, the deeper the trench depth, the probability of electron escape outside of the trench guard ring and collected by an adjacent well structure decreases.

7.11.1 Deep Trench (DT) within PNPN Structure

Trench structures can be used to improve the latchup robustness of a technology by reducing the lateral bipolar current gain of parasitic elements (Figure 7.6). To evaluate the DT structure, the standard latchup structure was modified to allow placement of the DT structure within the PNPN structure. The DT structure was placed centered on the perimeter of the n-well region which contains the p+ diffusion. The p+/n+ spacings were not altered to allow the placement of the DT structure [42–46] (Figure 7.26).

The lateral β_{PNP} can be reduced by placement of a trench region on the sidewall of the n-well region. Forward biasing of the p-diffusion in the n-well leads to minority

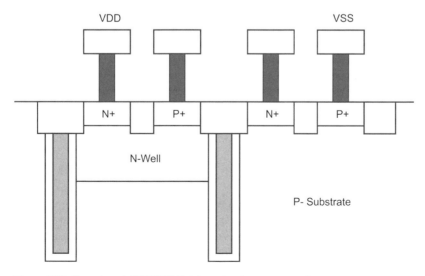

Figure 7.26 Deep trench (DT) PNPN latchup structure cross section.

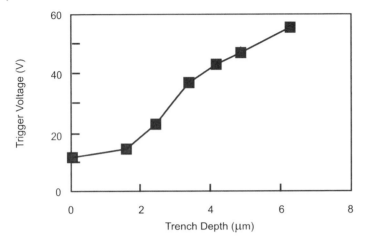

Figure 7.27 CMOS latchup trigger voltage, V_{TR}, results of the deep trench (DT) PNPN latchup structure as a function of deep trench depth.

carrier holes in the n-well region that diffuse to the collector region. The minority carrier holes flow from the low-doped well region to the heavily-doped retrograde well region. The built-in electric field established by the MeV retrograde well implant creates a drift component opposite to the flow of the holes to the collector region. Holes will flow laterally out of the well region in the case where no trench structure is present. Holes that flow from the p-diffusion to the p− substrate contribute to the forward PNP bipolar transport factor, α_{fPNP}. Placement of the trench region also reduces the lateral parasitic NPN formed between an n-diffusion outside of the n-well and the n-well itself. Minority carrier electrons which are transported from the n+ diffusion region (emitter) to the n-well (collector) contribute to the forward lateral NPN bipolar transport factor α_{NPN}.

The latchup trigger voltage is shown as a function of the DT depth (Figure 7.27). In the case of no DT, the latchup trigger voltage is approximately 10 V. When DT is below 2 μm depth, the trench depth has not penetrated through the retrograde n-well region. As the DT structure extends below the retrograde n-well region, V_{TR} increases by 2X. As the DT depth increases, V_{TR} continues to increase. With a 6-μm DT structure, the trigger condition increases 5X. For a 2.5 V technology, this establishes a latchup trigger condition over 25X the V_{DD} power supply value. For applications, such as 10 V V_{DD} conditions, this provides a 4X margin above V_{DD}. For automotive or space applications, this provides an increased assurance of high voltage spikes initiating latchup.

Holding voltages also increase with DT structures. Experimental results show that the V_H is approximately 2.5 V without a DT structure. As the DT depth is increased, V_H increases above the 2.5 V V_{DD} levels. As the DT depth increases greater than 3 μm, V_H increases over 10 V.

This provides a 4X margin to a 2.5 V V_{DD} power supply application. To fall into this state, a voltage impulse must exceed 55 V, and then be able to sustain 20 V condition to remain in this latched state. The trench structure provides a latchup robustness of 7X V_{DD} for this technology generation. Addressing the trigger voltage condition, we can further explore sensitivity in its design space. Trigger and holding voltage conditions form a loci of points in its design space, which satisfy the latchup stability criterion.

There are four parameters that strongly influence the latch conditions from the design perspective. These parameters include the NPN, and the PNP bipolar current gain transistors, as well as the n-well contact to p+ diffusion space, and the substrate contact to n-well space. The bipolar current gain terms in the plane can be modulated by varying the p+/n+ space between the two physical emitters. Well and substrate bypass resistors can be used in parallel with the emitter-base junctions in order to evaluate the stability in resistance space. In a two-dimensional (2D) plot of log (R_{sx}) and log (R_{well}), the loci of points for the holding and the trigger voltage can be plotted to demonstrate the values of the resistance of the well and the substrate that provide the latchup state for a given p+/n+ spacing. This two-dimensional log (R) space can be viewed as a three dimensional (3D) space by adding the DT parameter as an additional axis in the design space. In our experimental work, it was found the results were a weak function in log (R_{well}) space and a very strong function of the log (R_{sx}) space. For easier viewing an understanding, we will focus on the two dimensions of DT- and log (R_{sx}) space cuts in the 3D design space.

The utilization of the DT structure for CMOS peripheral circuits such as I/O output driver networks to isolate the p-channel MOSFET pull-up, the n-channel MOSFET pull-down, resistor elements and the electrostatic discharge (ESD) elements can provide a considerable latchup robustness improvement. The standalone guard ring structures as discussed in the first section demonstrates the improvement in the probability of escape for the trench guard rings. The usage of DT guard rings around n-channel networks improves the latchup robustness for negative undershoot phenomenon. The usage of DT guard rings abutting the n-well structure of the p-channel pull-up transistor demonstrates latchup robustness improvement from positive overshoot events.

7.11.2   Deep Trench Structure and Sub-Collector

The DT structure can also be integrated with a sub-collector implant [47]. A first advantage of the sub-collector is the lowering of the well shunt resistance. Lowering of the well resistance modulates the denominator of the generalized transport term for the PNP transistor, improving the stability of the PNPN network. Hence, the combination of the DT and the sub-collector leads to variation of three variables in the generalized tetrode relationship, leading to improved stability. A second advantage is that when using the DT structure integrated with the sub-collector, there are no additional cost or density implication concerns since the DT structure limits the out-diffusion of the sub-collector structure. Figure 7.28a shows the CMOS latchup structure with a DT and sub-collector implant.

An interesting issue is the turn-on voltage, V_{ON}. For a 1-μm DT depth, with or without sub-collector implants, there was no improvement in the turn-on mechanism. As DT depth increased (with sub-collector), the turn-on voltage, V_{ON}, increased as the trench penetrated through the sub-collector region. This effect was first discovered experimentally by A. Watson and S. Voldman [23, 46–48]. When the DT depth exceeded the sub-collector depth, V_{ON} remained constant; this experimental discovery shows that the pre-trigger voltage condition, the "turn-on" mechanism, can be eliminated when the trench penetrates through the sub-collector.

To evaluate the nature of the turn-on mechanism, the latchup structure was biased as an NPN transistor between the n-well/sub-collector structure, the substrate and the n-diffusion. From the $I_C - V_{CE}$ NPN characteristics, the avalanche breakdown

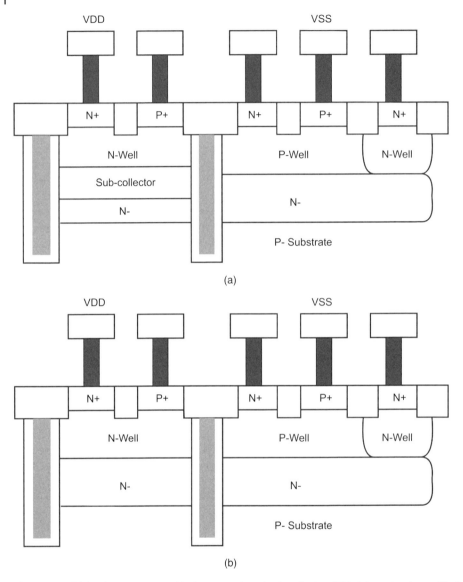

Figure 7.28 (a) Latchup structure with deep trench and sub-collector; (b) Latchup structure with deep trench and sub-collector.

voltage was extracted. As the DT depth increased, the reverse β_{NPN} decreased, and V_{AV} increased. From these experimental results, it is evident that the turn-on voltage condition tracks with this condition [26].

The trigger voltage condition, $V_{TR,}$ is a function of the deep trench depth. A key experimental discovery was the improvement of V_{TR} from 50 to 100 V. Hence, the combined effect of the trench structure and sub-collector allows for a significant latchup robustness improvement by an increasing in the trigger voltage of the technology providing immunity to SEL [47].

DT structures has the advantage of reducing the lateral bipolar current gain of both PNP and NPN transistor. The addition of the sub-collector leads to the elimination of the sidewall trench mechanism, lower β_{PNP} and lower internal well shunt resistance. Lowering of the well resistance modulates the denominator of the generalized transport term for the PNP transistor improving the stability of the PNPN network. Hence, the combination of the DT and the sub-collector leads to variation of three variables in the generalized tetrode relationship, leading to an improved stability. A second advantage is using the DT structure integrated with the sub-collector, there are no additional cost or density implication since the DT structure limits the sub-collector out-diffusion.

7.11.3 Deep Trench Structure and Merged Triple Well

As practiced today, circuit designers prefer to remap dual-well structures into triple-well implementations without a change in the on-chip design, ground-rules, or p+/n+ spacing rules. As a result, "triple well" is being practiced not as isolated regions but "merged triple well" where the n-well and associated isolating buried layers are integrated. With the introduction of merged triple well in advanced CMOS technologies, the vertical bipolar transistor is enhanced well above dual-well CMOS structures. In advanced BiCMOS technologies, for merged triple well CMOS, there are many options [52]:

- Merged triple well with deep trench
- Merged triple well, with deep trench with triple well implant only under p-well region
- Merged triple well, with deep trench with triple well implant under both n-well and p-well
- Merged triple well, with deep trench with triple well implant under both n-well and p-well, with additional BiCMOS implants (e.g. sub-collector)

This experimental work is completed in a high resistivity 50 Ω-cm p-substrate wafer in a 0.13-μm BiCMOS technology to highlight the enhancement of the NPN bipolar current gain; these results will be followed by the first measurements demonstrating integration of DT structures, and sub-collector implants in a triple well environment [52].

In order to reduce the lateral β_{NPN}, DT structures can be used to separate the triple-well buried layer from the n-well region (Figure 7.28b). Because of scattering phenomenon influencing MOSFET threshold implants, the spacing between the implanted buried layer edge and the n-channel MOSFET and the p-channel MOSFET must be large to avoid p-channel MOSFET and n-channel MOSFET threshold modulation. Hence, in some technologies, adequate spacing must be established between the edge of the buried layer implant, the n-channel MOSFET, and the p-channel MOSFET; this provides an opportunity to place deep trench structures between the p- and n-channel MOSFET in merged triple-well technology.

To avoid the issue of the buried-layer implant edges, some circuit designers would like to convert the technology from dual well CMOS to triple well using a blanket n− buried layer under all structures [33]. Using a "blanket" buried layer (e.g. unmasked), this can impact devices, and circuit elements. Additionally, using a continuous n− buried layer under the p+/n+ spacing allows for avoidance of mask edge scattering phenomenon and provides improved latchup.

In this structure, the role of the n− buried layer under the n-well provides latchup advantages:

- Low resistance shunt for the PNP transistor.
- Lower PNP bipolar gain due to increase base width and larger Gummel number (e.g. total implant dose integrated over the base region). The placement of the n− layer under the n-well prevents scattering mask phenomenon and improves the latchup robustness of the structure.

7.12 Layout Solutions

Radiation generate both EHP and also generates "fixed positive charge" in dielectrics. Fixed positive charge generated in dielectric isolation can introduce a surface potential problem for edge regions near a MOSFET channel. Electrons are mobile and, once generated, will quickly migrate to the nearest cathode. Positive charge from holes, however, in silicon dioxide cannot traverse the Si−SiO$_2$ interface. As such, hole charge becomes trapped in SiO$_2$ regions and will continue to accumulate until electrons are sourced to the fixed positive charge to force recombination. This is a concern for total dose testing of not only gate dielectrics but also isolation regions (e.g. LOCOS isolation, and STI). Solutions to avoid isolation concerns are as follows:

- Polysilicon bound structures without isolation edges
- Parasitic isolation device (PID)

7.12.1 Polysilicon Bound Structures

In MOSFETs that have isolation on the edge that shift, the device characteristics can introduce a layout that does not have isolation regions near the active part of the diode, or MOSFET. These "enclosed devices" can have a source or drain structure that does not have a LOCOS or STI edge. This also has advantages for a number of reliability issues:

- Leakage from silicide penetration at junction – silicide – isolation triple points
- ESD degradation

7.12.2 Parasitic Isolation Device (PID)

Gamma dose radiation introduces a fixed charge in the isolation regions. This can introduce flat band shifts in the MOSFET structure. The flat band shift is directly proportional to the square of the oxide thickness for any constant electric field. Thinning of the oxide will reduce the stored charge and minimize the flat band voltage shift. CMOS scaling offers improvements of this mechanism from both gate oxide scaling (to thinner oxide thicknesses) and the reduction in the operating voltage, which lowers the electric field (E). However, there is no improvement in the FOXs from MOSFET scaling.

An alternative solution to minimize threshold shifts induced by edge parasitic leakage is to introduce a PID. In this approach, a special extended silicon region is formed at each edge of the NMOS transistor active area (Figure 7.29a). This extension is referred to as the PID. This extension allows for a separate region that is doped uniquely to

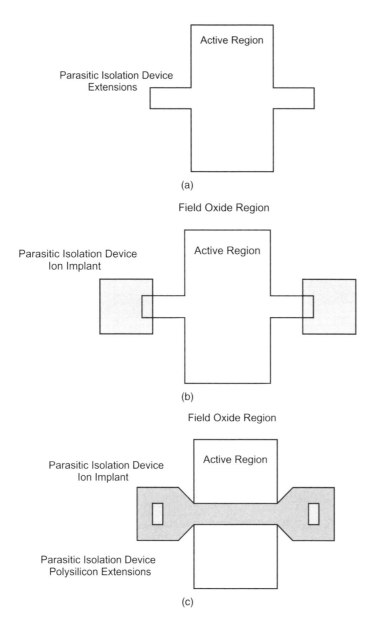

Figure 7.29 (a) MOSFET active region with parasitic isolation device (PID); (b) MOSFET active region with PID ion implant; (c) MOSFET with PID and polysilicon gate structure.

form a high-value field transistor threshold. Figure 7.29b shows the structure with the PID ion implant. The implanted area is then covered by a polysilicon gate structure (Figure 7.29c). The PID later will be covered by the poly gate electrode and as such becomes part of the self-aligned transistor channel. The PID is only needed by the NMOS transistor to avoid the offset of channel V_t by oxide charge arising from radiation exposure. Experimental results show MOSFETs with PID transistors show no degraded I-V characteristics up to gamma radiation doses over 1 Mrad. This result is particularly important to analog or mix signal circuits and power supply/controller circuits which are very sensitive to small radiation induced-shifts in the MOSFET subthreshold region. Note that without a PID, the threshold voltage shift of >10% is reached at 100 krads, given that most analog circuits, current mirrors, or voltage comparators would fail at <5% of V_t shifts this is a significant result.

7.13 Summary and Closing Comments

Chapter 7 focused on semiconductor process features for improving CMOS latchup. Chapter 7 also noted that although these features today are integrated in BiCMOS technology, all of these features can be integrated into mainstream CMOS technology.

Chapter 8 discusses single-event circuit solutions used for CMOS DRAMs and CMOS SRAMs, Bipolar SRAMS and other applications. SEU solutions shown will include resistor decoupling, to redundancy methods.

Problems

7.1 Draw a latchup PNPN test structure with a sub-collector implant where the sub-collector is under the n-well region. Draw a second latchup PNPN test structure with a sub-collector under both the n-well and p-well, where the sub-collector does not isolate the p-well from the p- substrate. Draw a third latchup PNPN test structure where an n-well encloses the p-well (isolates the p-well from the p- substrate).

7.2 As in Problem 1, create a circuit schematic of the PNPN structure for each of the three cases. Label the resistance terms and identify the resistances in the test structure cross section and in the circuit schematic.

7.3 A trench isolation is formed after the STI isolation, leading to a "dual-depth" isolation structure. What is the needed depth for maintaining latchup robustness?

7.4 Draw a PNPN test structure with shallow trench isolation, and the "trench isolation" (TI) structure. Assume three cases: (1) the trench isolation is shallower than the n-well region; (2) the trench isolation is equal to the n-well depth; and (3) the trench isolation is deeper than the n-well depth. Explain the lateral and vertical parasitic transistors in the three cases.

7.5 Derive a parasitic transistor model for trench isolation, where (1) the trench isolation is shallower than the n-well region; (2) the trench isolation is equal to the n-well depth; and (3) the trench isolation is deeper than the n-well depth.

7.6 Draw a PNPN test structure with shallow trench isolation, and the "deep isolation" (DT) structure, where the DT is polysilicon-filled, and borders the n-well. Assume the wafer is a p−/p++ substrate wafer. Assume the following cases: (1) the deep trench is shallower than the n-well depth; (2) the deep trench is equal to the n-well depth; (3) the deep trench is deeper than the n-well depth, but shallower than the p−/p++ interface; and (4) the deep trench is deeper than the n-well depth, but abuts the p−/p++ interface, and (5) the deep trench is deeper than the n-well, and extends beyond the p−/p++ interface. Explain the lateral and vertical parasitic transistors in the three cases.

7.7 For Problem 6, create a lateral bipolar model for all the above cases.

7.8 A series of deep trench rings are placed around a n+ injection source. Show the cases of the minority carrier transport for the following structure where the listed guard rings are from in sequence:
a. P+ diffusion, n-well
b. P+ diffusion, n-well, deep trench
c. P+ diffusion, n-well, deep trench, n-well
d. P+ diffusion, n-well, deep trench, n-well, deep trench

7.9 Explain the voltage conditions of the polysilicon region in a deep trench structure used as a border on the edge of an n-well region in a PNPN. What is the potential of the polysilicon?

7.10 Show the test structure for integration of deep trench, sub-collector, and triple well implant under both the n-well and p-well regions. Describe the lateral and vertical transport for the parasitic PNP and NPN structures.

References

1 Colinge, J.P. (1997). *Silicon-on-Insulator Technology*. Springer-Verlag.
2 Izumi, K., Omura, Y., and Sakai, T. (1983). Simox technology and its application to CMOS LSIS. *Journal of Electronic Materials* 12 (5): 845–861.
3 Tong, Q.Y. (2001). Wafer bonding and layer transfer for microsystems: an overview. In: *Semiconductor Wafer Bonding: Science, Technology, and Applications*, 1–16.
4 Aspar, B., Lagahe, C., Moriceau, H. et al. (2001). The smart-cut process: status and developments. In: *Semiconductor Wafer Bonding: Science, Technology and Applications*, 48–59.
5 Manasevit, H.M. and Simpson, W.J. (1964). Single-crystal silicon on a sapphire substrate. *Journal of Applied Physics* 35: 1349.
6 Aleksov, A., Li, X., Govindaraju, N. et al. (2005). Silicon-on-diamond: an advanced silicon-on-insulator technology. *Diamond and Related Materials* 14 (3–7): 308–313.

7 Jurczak, M. et al. (2000). *Silicon-on-nothing (SON)—an innovative process for advanced* CMOS. *IEEE Transactions of Electron Devices* 47: 2179–2187.

8 Oldham, T.R., Lelis, A.J., Boesch, H.E. et al. (1987). Post irradiation effects in field-oxide isolation structures. *Transactions of the IEEE Nuclear Science*, NS-34: 1184.

9 Davari, B., Koburger, C., Furukawa, T. et al. (1988). A variable size shallow trench isolation (STI) technology with diffused sidewall doping for submicron CMOS. In: *Proceedings of the International Electron Device Meeting (IEDM) Digest*, 92–93.

10 Shaneyfelt, M.R., Dodd, P.E., Draper, B.L., and Flores, R.S. (1998). *Challenges in Hardening Technologies Using Shallow-Trench Isolation*. Sandia Laboratories.

11 Nishioka, Y., Itoga, T., Ohyu, K. et al. (1990). Radiation effects on fluorinated field oxides and associated devices. *Transactions of the IEEE Nuclear Science* 37 (6) NS-34): 2026–2032.

12 Custode, F.Z., Poksheva, J.G. Radiation hardened field oxides for NMOS and CMOS-bulk and process for forming. U.S. Patent No., 4,994,407, issued February 19, 1991.

13 Borland, J.O. (1998) Method for CMOS latchup improvement by MeV BILLI (Buried Implanted Layer for Lateral Isolation) plus buried layer improvement. U.S. Patent No.5,821,589, issued October 13, 1998.

14 Morris, W. (2009). Buried guard ring structures and fabrication methods. U.S. Patent No. 7,629,654, issued December 8, 2009.

15 Morris, W. (2010). Buried guard ring and radiation hardened isolation structures and fabrication methods. U.S. Patent No. 7,804,138, issued September 28, 2010.

16 Morris, W. (2012). Methods for operating and fabricating a semiconductor device having a buried guard ring structure. U.S. Patent No. 8,093,145, issued January 10, 2012.

17 Morris, W. (2012). Fabrication methods for radiation hardened isolation structures. U.S. Patent No. 8,252,642, issued August 28, 2012.

18 Morris, W. (2012). Radiation hardened isolation structures and fabrication techniques. U.S. Patent No. 8,278,719, issued October 2, 2012.

19 Morris, W. (2013). Method for radiation hardening a semiconductor device. U.S. Patent No. 8,497,195, issued July 30, 2013.

20 Morris, W. (2014). Method and structure for radiation hardening a semiconductor device. U.S. Patent No. 8,729,640, issued May 20, 2014.

21 Morris, W. (2003). CMOS latchup. In: *Proceedings of the International Reliability Physics Symposium (IRPS)*, 86–92.

22 Chiu, H.K., Chen, F.C., and Tao, H.J. (2001). Top corner rounding for shallow trench isolation. U.S. Patent No. 6,265,317, issued July 24, 2001.

23 Voldman, S. (2007). *Latchup*. Wiley.

24 Boselli, G., Reddy, V., and Duvvury, C. (2004). Latch-up in 65 nm CMOS technology: a scaling perspective. In: *Proceeding of the International Reliability Physics Symposium (IRPS)*, 137–144.

25 Voldman, S., Gebreselasie, E., Liu, X. et al. (2005). The influence of high resistivity substrates on CMOS latchup robustness. In: *Proceedings of the Electrical Overstress/Electrostatic Discharge (EOS/ESD) Symposium*, 90–99.

26 Voldman, S. (2005). A review of CMOS latchup and electrostatic discharge (ESD) in bipolar complimentary MOSFET (BiCMOS) silicon germanium technologies: part II-latchup. *Microelectronics and Reliability* 45: 437–455.

27 Voldman, S. (2005). Latchup and the domino effect. In: *Proceedings of the International Reliability Physics Symposium (IRPS)*, 145–156.

28 Voldman, S. (2005). Cable discharge event and CMOS latchup. In: *Proceedings of the Taiwan Electrostatic Discharge Conference (T-ESDC)*, 23–28.

29 Cottrell, P., Warley, S., Voldman, S. et al. (1988). N-well design for trench DRAM arrays. In: *International Electron Device Meeting (IEDM) Technical Digest*, 584–587. IEEE.

30 Voldman, S., Marceau, M., Baker, A. et al. (1992). Retrograde well and epitaxial thickness optimization for shallow- and deep-trench collar merged isolation and node trench (MINT) SPT cell and CMOS logic technology. In: *International Electron Device Meeting (IEDM) Technical Digest*, 811–815. IEEE.

31 Voldman, S. (1994). *Optimization of MeV retrograde wells for advanced logic and microprocessor/Power PC and electrostatic discharge*. Smart and Economic Device and Process Designs for ULSI Using MeV Implant Technology: Semicon West GENUS Seminar, San Francisco.

32 Voldman, S. (1995). MeV implants boost device design. *IEEE Circuits and Devices* 11 (6): 8–16.

33 Voldman, S. (2004). *ESD and Latchup in Advanced Technologies, Tutorial*, in Tutorial Notes of the International Reliability Physics Symposium (IRPS).

34 Voldman, S. (2004) *Latchup Physics and Design, Tutorial*, in ESD Tutorials of the Electrical Overstress/Electrostatic Discharge (EOS/ESD) Symposium.

35 Voldman, S. (2005). *CMOS Latchup, Tutorial*, in Tutorial Notes of the International Reliability Physics Symposium (IRPS).

36 Voldman, S. and Gebreselasie, E. (2005). The influence of implanted sub-collector on CMOS latchup. In: *Proceedings of the Electrical Overstress/Electrostatic Discharge (EOS/ESD) Symposium*, 108–117.

37 Rung, R.D., Momose, H., and Nagabuko, Y. (1982). Deep trench isolated CMOS devices. In: *International Electron Device Meeting (IEDM) Technical Digest*, 237–240. IEEE.

38 Yamaguchi, T., Morimoto, S., Kawamoto, G.H. et al. (1983). High-speed latchup-free 0.5-µm channel CMOS using self-aligned $TiSi_2$ and deep trench isolation technologies. In: *International Electron Device Meeting (IEDM) Technical Digest*, 522–525. IEEE.

39 M. Bohr (1996). Isolation structure formation for semiconductor circuit fabrication. U.S. Patent No. 5,536,675, issued July 16, 1996.

40 Voldman, S., Gebreselasie, E.G., Lanzerotti, L.W. et al. (2005). The influence of silicon dioxide-filled trench isolation (TI) structure and implanted sub-collector on latchup robustness. In: *Proceedings of the International Reliability Physics Symposium (IRPS)*, 112–120.

41 Voldman, S., Gebreselasie, E.G., and Watson, A. (2005). Comparison of CMOS latchup with trench isolation (TI) and polysilicon-filled deep trench (DT) isolation structures for CMOS and BiCMOS technology. In: *Proceedings of the Taiwan Electrostatic Discharge Conference (T-ESDC)*, 23–28.

42 Voldman, S. (2003). The effect of deep trench isolation, trench isolation, and sub-collector on the electrostatic discharge (ESD) robustness of radio frequency (RF) ESD STI-bound p+/n-well diodes in a BiCMOS silicon germanium

technology. In: *Proceedings of the Electrical Overstress/Electrostatic Discharge (EOS/ESD) Symposium*, 214–223.

43 Watson, A., Voldman, S., and Larsen, T. (2003). Deep trench guard ring structures and evaluation of the probability of minority carrier escape for ESD and latchup in advanced BiCMOS SiGe technology. In: *Proceedings of the Taiwan Electrostatic Discharge Conference*, 97–103. Hsinchu City, Taiwan: National Chiao-Tung University.

44 Voldman, S., Perez, C.N., and Watson, A. (2005). Guard rings: theory, experimental quantification, and design. In: *Proceedings of the Electrical Overstress/Electrostatic Discharge (EOS/ESD) Symposium*, 100–107.

45 Voldman, S., Perez, C.N., and Watson, A. (2006). Guard rings: structures, design methodology, integration, experimental results and analysis for RF CMOS and RF mixed signal BiCMOS silicon germanium technology. *Journal of Electrostatics* 64: 730–743.

46 Voldman, S. and Watson, A. (2004). The influence of deep trench and substrate resistance on the latchup robustness in a BiCMOS silicon germanium technology. In: *Proceedings of the International Reliability Physics Symposium (IRPS)*, April 25–27, 135–142.

47 Watson, A. and Voldman, S. (2004). The effect of deep trench and sub-collector on the latchup robustness in BiCMOS silicon germanium technology. In: *Proceedings of the Bipolar Circuit Technology Meeting (BCTM)*, 172–175.

48 Voldman, S. and Watson, A. (2004). The influence of polysilicon-filled deep trench and sub-collector implants on latchup robustness in RF CMOS and BiCMOS SiGe technology. In: *Proceedings of the Taiwan Electrostatic Discharge Conference (T-ESDC)*, 15–19.

49 Voldman, S. (2004). *ESD and Latchup in Advanced Technologies*. Tutorial Notes of the International Reliability Physics Symposium (IRPS).

50 Voldman, S. (2004). *Latchup Physics and Design*. ESD Tutorials of the Electrical Overstress/Electrostatic Discharge (EOS/ESD) Symposium.

51 Voldman, S. (2006). The influence of a novel contacted poly-silicon filled deep trench (DT) biased structure and its voltage state on CMOS latchup. In: *Proceedings of the International Reliability Physics Symposium (IRPS)*, 151–158.

52 Voldman, S., Gebreselasie, E.G., Zierak, M. et al. (2005). Latchup in merged triple well structure. In: *Proceedings of the International Reliability Physics Symposium (IRPS)*, 129–136.

53 Voldman, S. and Gebreselasie, E.G. (2006). The influence of merged triple well, deep trench and subcollector on CMOS latchup. In: *Proceedings of the Taiwan Electrostatic Discharge Conference (T-ESDC), 2006*, 49–52.

54 Lin, H.Y. and Ting, C.H. (1989). Improvements of CMOS latchup using a high energy buried layer. *Nuclear Instrumentation and Methods in Physics Review* B38/39: 960–964.

55 Leong, K.C., Liu, P.C., Morris, W., and Rubin, L. (1998). Superior latchup resistance of high dose energy implanted p+ buried layers. In: *Proceedings of the XII International Conference on Ion Implantation Technology, Kyoto, Japan*, 99–108.

56 Kuroi, T., Komori, T.S., Miyatake, H., and Tsukamoto, K. (1990). Self gettering and proximity gettering for buried layer formation by MeV ion implantation. In: *International Electron Device Meeting (IEDM) Technical Digest*, 261–264. IEEE.

57 Bourdelle, K.K., Chen, Y., Ashton, R. et al. (2000). Epi-replacement in CMOS technology by high dose, high energy boron implantation into Cz substrates. In: *Proceedings of the International Conference on Ion Implantation*, 312–315.

58 Voldman, S., Lanzerotti, L., Morris, W., and Rubin, L. (2004). The influence of heavily doped buried layer implants on electrostatic discharge (ESD), latchup, and a silicon germanium heterojunction bipolar transistor in a BiCMOS SiGe technology. In: *Proceeding of the International Reliability Physics Symposium (IRPS)*, 143–151.

8

Single-Event Upset Circuit Solutions

8.1 Introduction

Single-event upsets can be minimized with semiconductor process solutions, as well as circuit solutions. Single-event upset solutions are dependent on the semiconductor technology type, circuits, and system. Figure 8.1 highlights the type of circuits that are prone to single-event upsets.

8.2 CMOS DRAM SEU Circuit Solutions

Single-event upsets occur in dynamic read access memory (DRAM) [1]. The rate of failure of the single-event upsets is also referred to as the soft error rate (SER). A single-event upset (SEU) from an ionized particle can occur from alpha particles, muons, protons, neutrons, and heavy ions [1–24]. Figure 8.2 highlights the factors that influence the SER in a DRAM. The SER is a function of the following:

- DRAM cell size
- DRAM cell stored charge
- DRAM technology type
- DRAM technology generation

Figure 8.3 shows an example of a CMOS DRAM cell. The DRAM cell stores charge in a trench capacitor used in a 4-Mb DRAM technology, known as a substrate plate trench (SPT) DRAM cell [4–6]. The advantage of a trench capacitor cell is that a large amount of charge can be stored in the DRAM cell, but with a small amount of semiconductor chip area. The SPT cell is provided on a p− epitaxial layer on a p++ substrate. The trench capacitor depth is 5.5 μm deep, extending into the p++ substrate. The p++ substrate provides a high capacitance between the substrate and the polysilicon-filled trench. A thin dielectric is formed between the polysilicon-filled region and the p++ substrate. The high capacitance of the DRAM cell increases the stored charge of the DRAM cell, increasing the critical charge (Qcrit) of the DRAM cell.

The SPT DRAM cell has a p-channel MOSFET transfer device contained within a retrograde n-well. The retrograde n-well limits the "funnel effect" from radiation events, such as alpha particles. Hence the collected charge from an ionizing event is limited.

Integrated Circuit Design for Radiation Environments, First Edition.
Stephen J. Gaul, Nicolaas van Vonno, Steven H. Voldman and Wesley H. Morris.
© 2020 John Wiley & Sons Ltd. Published 2020 by John Wiley & Sons Ltd.

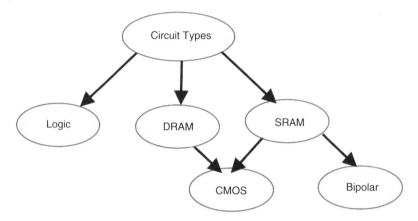

Figure 8.1 Types of circuits sensitive to single-event upsets (SEU).

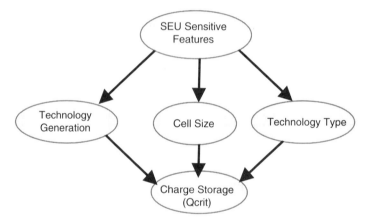

Figure 8.2 CMOS DRAM features that influence SEU.

8.2.1 CMOS DRAM Redundancy

Single-event upsets in a DRAM can be addressed using circuits that provide redundancy. Using circuit redundancy, the state of the cell can be evaluated to verify whether a soft error occurred [9].

8.2.2 CMOS DRAM with SRAM Error Correction

A technique to address single-event upset failures in CMOS DRAMs is to have static random-access memory (SRAM) cells serve as a means of error correction. In a single CMOS DRAM semiconductor chip, an array of DRAM bits can be sensitive to single-event upsets. Using an array of SRAM cells that are insensitive to single-event upsets, the failures within the DRAM bits can be verified to determine whether there was a single-event upset. The SRAM array serves as error correction to provide no net failures. Figures 8.4 and 8.5 shows an example of a CMOS DRAM that utilized an array of SRAM cells. This was utilized in a 16-Mb DRAM that used a deep trench capacitor cell.

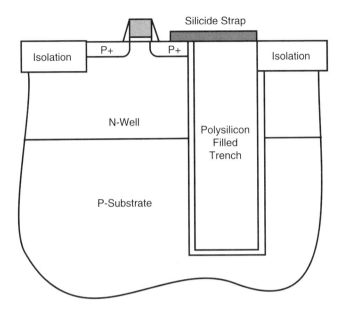

Figure 8.3 CMOS DRAM substrate plate trench (SPT) capacitor cell.

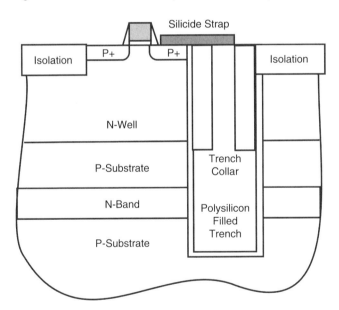

Figure 8.4 CMOS DRAM – 16 Mb DRAM trench cell.

Figure 8.4 shows an example of the 16-Mb merged isolation and node DRAM cell [7, 8]. This cell merges the isolation and trench regions to provide a dense array of DRAM cells. The merged isolation node trench (MINT) cell is provided on a p– epitaxial layer on a p++ substrate. A trench "collar" is formed near the top surface of the trench capacitor to eliminate trench gated-diode effects on the trench sidewall. An n-buried layer separates the p- region from the p++ substrate region. The trench capacitor depth is

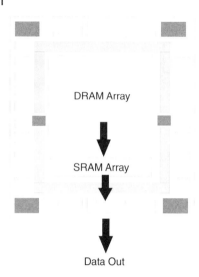

Figure 8.5 CMOS DRAM with SRAM array – 16 Mb DRAM.

DRAM Array

SRAM Array

Data Out

12 μm deep, extending into the p++ substrate. The p++ substrate provides a high capacitance between the substrate and the polysilicon-filled trench. A thin dielectric is formed between the polysilicon-filled region and the p++ substrate. The high capacitance of the DRAM cell increases the stored charge of the DRAM cell, increasing the critical charge (Qcrit) of the DRAM cell. With the thin dielectric and deep trench structure, the Qcrit of the cell is maintained as the cell size is scaled.

The MINT DRAM cell has a p-channel MOSFET transfer device contained within a retrograde n-well. The retrograde n-well and the n-buried layer limits the "funnel effect" from radiation events, such as alpha particles. Hence the collected charge from an ionizing event is limited.

Figure 8.5 shows the architecture of the DRAM array and the SRAM error correction code (ECC) array. Experimental results of the effectiveness of the on-chip ECC are discussed in publications [10, 11]. Soft errors that occur in the DRAM cells can be eliminated with the SRAM array.

8.3 CMOS SRAM SEU Circuit Solution

Single-event upsets in SRAM can be addressed by integration of the circuit solution in the SRAM cell itself (Figure 8.6). The technique for elimination of SRAM soft errors is as follows (Figure 8.6):

- CMOS SRAM latch decoupling [12–14]
- CMOS SRAM cell circuit redundancy [15]

8.3.1 CMOS SRAM Four-Device Cell

Single-event upsets in SRAM can occur in CMOS four-device (4-D) SRAM cell. Figure 8.7 shows an example of the CMOS SRAM 4-D cell. Any charge transfer that

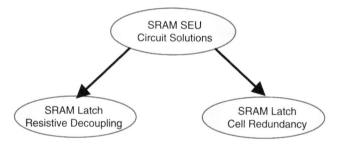

Figure 8.6 CMOS SRAM SEU circuit solution.

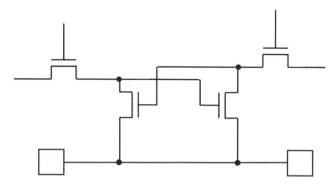

Figure 8.7 CMOS 4-D SRAM cell.

influences the differential voltage across the latch can disturb the SRAM cell. Four-device cells are vulnerable due to the small stored charge in this cell design [24].

8.3.2 CMOS SRAM Six-Device Cell

Single-event upsets in SRAM can occur in CMOS six-device (6-D) SRAM cell [15]. Figure 8.8 shows an example of the CMOS SRAM 6-D cell circuit topology. Any charge transfer that influences the differential voltage across the latch can disturb the SRAM

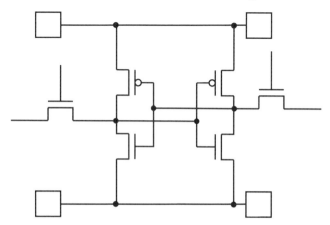

Figure 8.8 CMOS SRAM 6-D cell circuit topology.

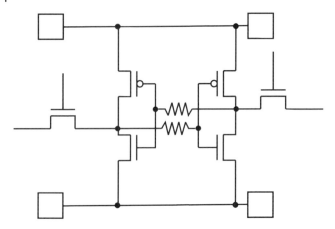

Figure 8.9 CMOS 6-D SRAM cell with cross-coupling resistor elements.

cell. An ionizing particle generates electron–hole pairs (EHPs) that can transfer charge to the p-channel MOSFET drain or n-channel MOSFET drain. Charge transfer to the MOSFET drains can lead to disruption of the differential voltage across the SRAM latch, leading to a soft error [15, 16].

A circuit topology that reduces the single-event upset is to introduce decoupling between the two sides of the latch during a transient event [12–14]. Figure 8.9 shows an example of the CMOS SRAM 6-D cell circuit topology with a resistive decoupling between the two sides of the latch. During a single-event, charge transfer that influences the differential voltage across the latch can disturb the SRAM cell. An ionizing particle generates EHP that can transfer charge to the p-channel MOSFET drain or n-channel MOSFET drain. In a standard SRAM cell, charge transfer to the MOSFET drains can lead to disruption of the differential voltage across the SRAM latch leading to a soft error. But, if the switching of the SRAM cell is such that the resistive decoupling avoids flipping the state of the SRAM cell, then no soft error occurs. This method has two disadvantages:

- Reduction of the circuit performance of the SRAM cell
- Increase in cell size for the cross-coupling resistor elements

8.3.3 CMOS SRAM 12-Device Cell

Single-event upsets in SRAM can be reduced by integration of the redundancy into the SRAM cell itself [15]. Figure 8.9 shows an example of the CMOS SRAM 13-D cell circuit topology that builds the redundancy into the CMOS SRAM 6-D cell. This method has two disadvantages:

1) Reduction of the circuit performance of the SRAM cell
2) Increase in cell size for the additional six transistor elements (Figure 8.10)

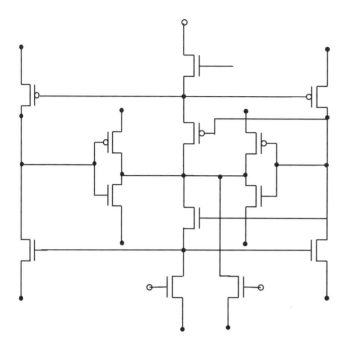

Figure 8.10 CMOS 13-D SRAM cell.

8.4 Bipolar SRAM

Before the innovation of CMOS SRAMs, high-speed memory used bipolar technology for SRAMs [17–22]. In the 1970s and 1980s, bipolar memory utilized bipolar technology for the memory cells. Figure 8.11 provides examples of the bipolar memory cell types. The bipolar SRAMs used were the following:

- Dual emitter npn transistor pull-down with resistor loads
- Dual emitter npn transistor pull-down with resistor load and Schottky clamps
- Dual emitter npn transistor pull-down with pnp transistor loads

Figure 8.11 Bipolar SRAM cell types.

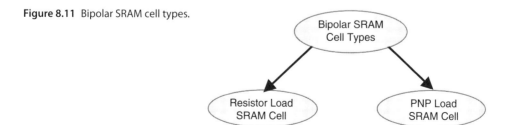

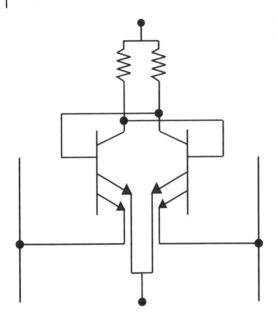

Figure 8.12 Bipolar SRAM cell with resistor loads.

8.4.1 Bipolar SRAM Cell with Resistor Loads

Figure 8.12 contains the circuit schematic for bipolar SRAM cell with resistor load elements. The bipolar SRAMs utilized resistor elements as a source of current to maintain the latch state. The npn transistors pull-down used a dual emitter structure for switching.

8.4.2 Bipolar SRAM Cell with Resistor Loads and Schottky Clamps

Figure 8.13 contains the circuit schematic for bipolar SRAM cell with resistor load elements and Schottky clamps. The bipolar SRAMs utilized resistor elements as a source

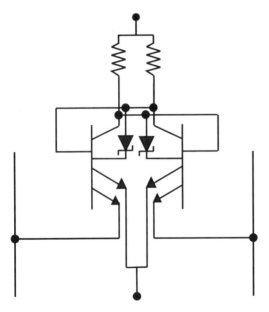

Figure 8.13 Bipolar SRAM cell with dual emitter, resistor load, and Schottky clamp elements.

Figure 8.14 Bipolar SRAM cell with PNP load.

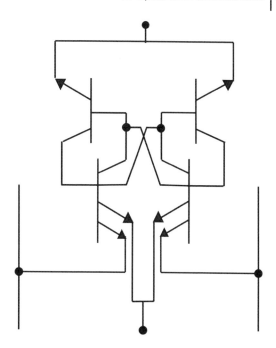

of current to maintain the latch state. The npn transistors pull-down used a dual emitter structure for switching. The Schottky clamps were placed parallel to the cross-coupling elements to prevent the bipolar transistor from undergoing bipolar transistor saturation. By avoiding bipolar transistor saturation, the stored charge in the base of the transistor was reduced, allowing for a high performance response of the bipolar SRAM latch circuit.

8.4.3 Bipolar SRAM Cell with PNP Transistors

Figure 8.14 contains the circuit schematic for bipolar SRAM cell with pnp loads. The bipolar SRAMs utilized the pnp elements as a source of current to maintain the latch state. The npn transistors pull-down used a dual emitter structure for switching. The pnp loads also formed a pnpn structure that allowed for faster switching characteristics.

8.5 Bipolar SRAM Circuit Solutions

To lower the SER of the bipolar SRAM cell, the critical charge to flip the latch was increased [17–22]. To improve the SER of the bipolar SRAM cell with the resistor loads, the Schottky clamps were removed to allow a higher stored charge in the SRAM cell. This was compensated by increasing the power used in the cell to maintain the performance of the SRAM cell without increasing the cell size.

For reduced SEU sensitivity, the Schottky clamps can be removed. Without the Schottky clamps, the transistors can intentionally be driven into saturation. The critical charge of the cell can be increased due to the "diffusion capacitance," which is proportional to the saturation current. Using this technique, a bipolar SRAM cell critical charge can be increased from 100 to 300 fC without an increase in the cell size; this is achieved by increasing the chip power. Ironically, the performance of the chip can also be improved.

8.6   SEU in CMOS Logic Circuitry

Single-event upsets can also occur in logic circuitry. Latch networks used in logic circuits are also sensitive to single-event upsets. As technology is scaled to smaller dimensions, there will be an increase in the SER of logic circuitry.

8.7   Summary and Closing Comments

This chapter discussed SEU circuit solutions in DRAMs and SRAMs. Examples of DRAM solutions using trench capacitor were discussed to increase the Qcrit of the DRAM cell. A second solution is usage of SRAM for error correction on a DRAM chip, leading to SEU immunity. For SRAM solutions, the usage of resistors in the SRAM latch networks was shown. Additionally, redundancy can be provided within the SRAM cell.

Chapter 9 discusses latchup circuit solutions that are used to provide latchup immunity. These will include local solutions within the chip, to global solutions on the power rails or power supplies.

Problems

8.1   Show the electron–hole pair (EHP) transport in a Emitter Coupled Logic (ECL) bipolar junction transistor SRAM cells. Which ones influence the cell? Which are common-mode collection versus differential charge collection?

8.2   In bipolar SRAMs, a Schottky diode is placed between the cross-couple region of the latch to avoid the bipolar transistors to undergo bipolar saturation. Explain why this is critical to bipolar transistor latch performance.

8.3   In bipolar SRAMs, for soft error rate (SER) improvement, the Schottky clamp diodes were removed to allow the bipolar junction transistors to enter a bipolar saturation. The charge stored in the bipolar base allows for an increase in the critical charge, where the diffusion capacitance adds to the critical charge of the SRAM cell. Derive the diffusion capacitance and the critical charge in the bipolar SRAM as a function of current.

8.4   In bipolar SRAM cells, there are ECL and merged transistor logic (MTL) configurations. Show both SRAM cells, and show where the charge is stored in both cases that influences the critical charge. Discuss the advantages and disadvantages of ECL versus MTL from a single-event substrate perspective.

8.5   For a CMOS 6-D SRAM cell, show the electron–hole pair transport that influence the critical charge. Which collection cases do not influence the state of the SRAM latch?

8.6   Adding a resistor element between the two sides of an SRAM latch can lead to prevention of a single-event from flipping the latch and cause a single-event upset (SEU) event. Show from the switching characteristics how this can be utilized.

References

1 May, T.C. and Woods, M.H. (1979). Alpha-particle-induced soft errors in dynamic memories. *IEEE Transaction of Electron Devices* 26: 2.

2 Ziegler, J.F. and Lanford, W.A. (1979). The effect of cosmic rays on computer memories. *Science* 206: 776.

3 Sai-Halasz, G.A., Wordeman, M.R., and Dennard, R.H. (1982). Alpha particle induced soft error rate in VLSI circuits. *IEEE Journal of Solid State Circuits* 17 (2): 355–361.

4 Lu, N.C., Ning, T.H. and Terman, L.M. (1987) Dynamic RAM cell with MOS trench capacitor in CMOS, U.S. Patent 4,688,063, issued August 18, 1987.

5 Lu, N.C. (1985). The SPT cell – A new substrate plate trench cell for DRAMs. In: *Proceedings of the International Electron Device Meeting (IEDM)*, December 1985, 771–772.

6 Lu, N.C., Cottrell, P.E., Craig, W.J. et al. (1986). A Substrate-Plate-Trench (SPT) capacitor memory cell for dynamic RAMs. *IEEE Journal of Solid State Circuits* SC-21 (5): 627–634.

7 Kenney, D.M. (1989) Semiconductor trench capacitor cell with merged isolation and node trench construction, U.S. Patent No. 4,801,988, issued January 31, 1989.

8 Kenney, D. (1988). A 16-Mb merged isolation and node trench SPT cell (MINT). In: *Symposium of VLSI Technology Technical Digest*, May 1998, 25–27.

9 Barth, J.E., Drake, C.E, Fifield, J.A. et al. (1992) Dynamic RAM with on-chip ECC and optimized bit and word redundancy, U.S. Patent No. 5,134,616, issued July 28th, 1992.

10 Zoutendyk, J. Single-event upset (SEU) in a DRAM with on-chip error correction code. *IEEE Transaction of Nuclear Science* 1987, NS-34 (6): 1310–1316.

11 NASA (1989). Internal correction of errors in a DRAM. In: *NASA Technical Briefs*, 30–31. NASA Jet Propulsion Laboratory (JPL).

12 Diehl, S.E. Error analysis and prevention of cosmic ion-induced soft error in CMOS RAMs. *IEEE Transaction on Nuclear Science* 1982, NS-29: 2032–2039.

13 Diehl, S.E., Vinson, J., Shafer, B.D., and Mnich, T.M. (1983). Considerations of single-event immune VLSI logic. *IEEE Transaction on Nuclear Science* NS-30: 4501–4507.

14 Diehl, S.E. (1984). A new class of single-event soft errors. *IEEE Transaction on Nuclear Science* NS-31 (6): 1145–1148.

15 Dooley, J.G., (1994) SEU-immune latch for gate array, standard cell, and other ASIC applications, U.S. Patent No. 5,311,070, issued May 10th, 1994.

16 Voldman, S., Corson, P., Patrick, L. et al. (1987). CMOS SRAM alpha particle modelling and experimental results. In: *Proceedings of the International Electron Device Meeting (IEDM)*, vol. 29, 344–347.

17 Sai-Halasz, G.A. (1983). Cosmic ray induced soft error rate in VLSI circuits. *IEEE Electron Device Letters (EDL)* 4 (6): 172–174.

18 Sai-Halasz, G.A. and Tang, D.D. (1983). Soft error rates in static bipolar RAMs. *Proceedings of the International Electron Device Meeting (IEDM)* 29: 344–347.

19 Voldman, S., Patrick, L., and Wong, D. (1985). Alpha particle effects on bipolar ECL static array. In: *Proceedings of the International Solid State Circuits Conference (ISSCC)*, 262–263.

20 Voldman, S. and Patrick, L. (1984). Alpha particle induced single-event upsets in bipolar static ECL cells. *IEEE Transaction on Nuclear Science* 31 (6): 1196–1200.

21 Idei, Y., Hanna, N., Nambu, H., and Sakurai, Y. (1991). Soft error rate characteristics in bipolar memory cells with small critical charge. *IEEE Transaction on Nuclear Science* NS-38 (6): 2465–2471.

22 McCall, D., Kroesen, P., Jones, F. et al. (1991). Alpha particle immune ECL latches. *Proceedings of the Bipolar Circuits and Technology Meeting (BCTM)*: 124–129.

23 Ziegler, J.F. and Puchner, H. (2004). *SER-History, Trends, and Challenges*. Cypress Semiconductor Company.

24 Chappell, B., Schuster, S.E., and Sai-Halasz, G.A. (1985). Stability of SER analysis of static RAM cells. *IEEE Transaction of Electron Devices* 32 (2): 463–470.

9

Latchup Circuit Solutions

9.1 Introduction

Single-event latchup (SEL) can occur within a circuit, or between circuits. Latchup is highly influenced by the types of circuits, and how these circuits interact during design integration and synthesis. In this chapter, we will focus on the types of circuits that cause latchup, the integration of these networks, and the interactions that can occur [1–20]. We will close the discussion by providing examples of circuit solutions, novel circuits, active clamps, and active guard rings to address latchup.

System-level latchup can be addressed by introduction of latchup prevention specific semiconductor components. Today, specific control chips can be integrated into systems to avoid latchup. For example, spacecraft system electronics have incorporated techniques into systems to provide SEL immunity to events. The latchup protection circuitry is integrated on the same physical package as the latchup sensitive device. These latchup prevention networks can set a "threshold" based on a known sensitivity level, provide a means of detection of particle event, provide a current limitation of the device that is undergoing latchup, can shut down the system, and then reset to its original state. These are techniques used today to address latchup to avoid system-level latchup concerns.

9.2 Power Supply Concepts

Chip design architecture also plays a role in latchup prevention. In latchup, the semiconductor chip cannot latchup given that (i) the bipolar transistors do not become forward active; (ii) the circuit does not reach the knee voltage (e.g. trigger voltage); or (iii) the load line of the circuit cannot support the holding voltage or the holding current condition. In this discussion, it is inherent that the power supply voltage and the relationship of the circuit to the power supply influences all of the above conditions. The inter-relation between the circuit subfunction and the power rails can influence latchup.

9.2.1 Power Supply Current Limit – Series Resistor

The objectives of a latchup circuit design concept are to avoid failure of semiconductor chips and to provide a circuit element in the power grid that limits the current flow to the circuitry. A first method is to place a resistor element between the power supply

Integrated Circuit Design for Radiation Environments, First Edition.
Stephen J. Gaul, Nicolaas van Vonno, Steven H. Voldman and Wesley H. Morris.
© 2020 John Wiley & Sons Ltd. Published 2020 by John Wiley & Sons Ltd.

rail and the entire circuit set. This method is not a very elegant solution but is suitable for semiconductor chips that do not draw significant current into the functional circuit block. As the chip is in functional operation, the series introduces a small voltage drop during semiconductor chip operation. To avoid destruction of the system, the semiconductor chip, and the internal metal power bus of the semiconductor chip, resistance can be added to introduce an *IR* drop at high current levels. The size of the resistance can be estimated by (i) allowable current through the chip without thermal failure; and (ii) an allowed current, which is less than the holding current (Figure 9.1).

9.2.2 Power Supply Current Limit – Current Source

A second solution is the usage of a current source within a semiconductor chip (Figure 9.2). Using a current source, the amount of allowed current through the chip can be maintained as a circuit within the network attempts to undergo latchup. The current source can be chosen so that the allowed current magnitude is below I_H of the circuits within the subfunction or chip. Examples of the current source can be a bipolar or CMOS current mirror.

To avoid system-level latchup issues, integration of "on-chip" or "off-chip" latchup-prevention networks can provide this current-limiting function. The role of the current-limiting device is to sense the "event" according to a preset "threshold" level. The current-limiting network detects an increase in the current during a SEL event above the preset threshold level. When the current exceeds this threshold level, a "shutdown" is initiated of the chip that is undergoing latchup. The shutdown remains in a shutdown mode for a preset time interval. After a preset time, the chip supply voltage returns to its initial state. In electrostatic discharge (ESD) events, ESD circuitry is initiated by external ESD event itself (e.g. RC-triggered ESD power clamps); analogously, internal single events that can lead to SEL can be used to sense the current

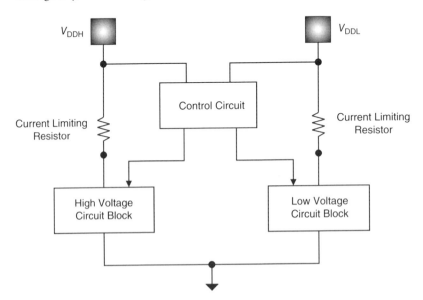

Figure 9.1 Power supply current-limiting resistor element.

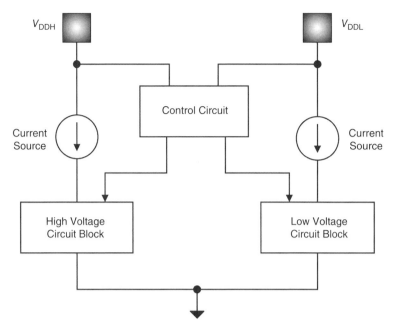

V_{DDH}

V_{DDL}

Control Circuit

Current Source

Current Source

High Voltage Circuit Block

Low Voltage Circuit Block

Figure 9.2 Power supply current-limiting current source.

increase on the power rails, and latchup prevention networks can be initiated to shut down the event, and then restore the chip to its initial state. A key difference between ESD event-prevention networks and latchup-prevention networks, is "powered" versus "unpowered" conditions. A latchup circuit design practice is as follows:

- Introduce a latchup-prevention network to provide current limitation based on a preset magnitude.
- The current-limit preset magnitude is set based on known system-level transients, space environment, or known device sensitivity.
- Provide a current-limiting network to the device.
- Identify detection of system-level or single events from detection of current-level increase.
- Provide a means of shutdown of the chip undergoing latchup.
- Set a preset time for shutdown during the latchup event.
- Reset the chip power after recovery from the transient event to the original power state.

9.2.3   Power Supply Solutions – Voltage Regulator

Another method of controlling the voltage levels within a chip is the usage of a voltage regulator. Figure 9.3 shows an example of a regulator network. In many applications, the voltage regulator is placed between the external power rail and an internal power rail. Voltage regulators control the power supply internal to the chip from external noise and perturbations from outside sources.

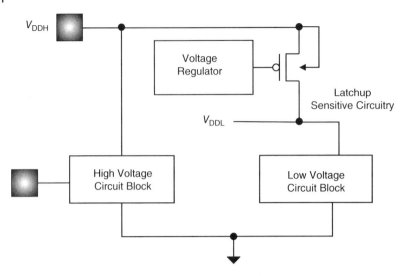

Figure 9.3 Power supply current-limiting voltage regulator.

A CMOS latchup circuit design practice is as follows:

- A current-limiting element or network is placed between the power supply and the circuit functional block wherein the element is chosen to guarantee to limit the power in the circuit.
- A current-limiting element or network is placed between the power supply and the circuit functional block wherein the element is chosen to guarantee to limit the current in the circuit such that it less than the holding current.
- A regulator is placed between the external power rail and the internal power rail so to avoid overvoltage of the core circuitry.

9.2.4 Latchup Circuit Solutions – Power Supply Decoupling

Latchup is a concern in both internal circuits and peripheral circuitry. Latchup may also occur as the result of interaction of the ESD device, the I/O off-chip driver, and adjacent circuitry initiated in the substrate from overshoot and undershoot phenomenon. These may be generated by CMOS off-chip driver (OCD) circuitry, receiver networks, and ESD devices. In order for latchup to occur, the anode of the pnpn structure must be electrically connected to a power supply source (such as the V_{DD} power supply) and the cathode of the pnpn structure must be electrically connected to a low potential (such as the V_{SS} power supply). Hence, a latchup design discipline is the decoupling of the parasitic pnpn structure from the power supply voltages; this is achieved by the decoupling of the CMOS inverter circuits from the power supplies during overshoot and undershoot phenomena [1].

In CMOS I/O circuitry, undershoot and overshoot may lead to injection in the substrate. Hence, both a p-channel MOSFET and an n-channel MOSFET may lead to substrate injection. Simultaneous switching of circuitry, where overshoot or undershoot injection occurs, leads to injection into the substrate which leads to both noise injection and latchup conditions. An interesting case study is the design methodology of the interaction of "activated" and "unactivated" elements in a gate-array environment. In a gate array environment, a "sea of gates" philosophy allows customization and personalization of circuit elements at a metallization level where the silicon regions are predefined. Unused n-diffusion regions are grounded (e.g. V_{SS} power supply), and unused p-diffusion regions are connected to the high power supply (e.g. V_{DD} power supply). As the substrate voltage potential rises relative to the n-diffusion, all the n-diffusion elements of the gate arrays will tend toward the forward active state. As the substrate voltage potential lowers, the unused p-diffusion elements, the n-well and the substrate may activate the vertical pnp. This may occur as a result of minority carrier injection in wells and substrate regions.

In gate array design, latchup can occur from external latchup injection. An additional issue is the propagation of the CMOS latchup process. As an initial source injects electrons into the substrate, a first circuit element may latchup. The latchup of a first circuit leads to the turn-on of a pnp parasitic element, leading to more injection into the substrate. Hence, a circuit solution of detachment of the rails from a vulnerable pnpn parasitic structure is a viable solution to "truncate the latchup propagation" through the semiconductor chip and array region [1]. In this case, the methodology may be addressed by certain functional blocks instead of spatial dependence. Hence the methodology of detachment and connection to the latchup control networks may be according to the circuit type as well as physical localness (placement) to the injection source.

A latchup circuit design practice to eliminate latchup is the electrical de-coupling of the power supplies during transient events. A latchup decoupling network consists of a latchup control isolation network electrically coupled to the substrate. In an ASIC environment, the latchup control isolation network purpose is to electrically isolates the "sea of gates" (e.g. CMOS logic circuits) from the power rail. Figure 9.4 shows a latchup

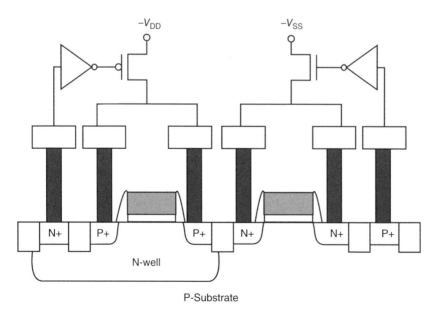

Figure 9.4 CMOS latchup power supply decoupling network.

control isolation network integrated into a sea of gates environment. The circuit consists of the following [1]:

- A p-channel MOSFET array whose source is in series with a p-channel MOSFET "switch" between the p-channel MOSFET array and the V_{DD} power supply.
- A n-channel MOSFET array whose source is in series with a n-channel MOSFET "switch" between the n-channel MOSFET array and the V_{SS} power supply.
- The gate of the p-channel MOSFET "switch" is electrically connected to the output of an inverter circuit, where the inverter circuit input is electrically connected to the n-well contact.
- The gate of the n-channel MOSFET "switch" is electrically connected to output of an inverter circuit, where the inverter circuit input is electrically connected to the p-well (e.g. or p- substrate) contact.

In this circuit, perturbations to the well or substrate voltage potential are sensed by well and substrate contacts. This perturbation signal is then sensed on inverter circuit. When a perturbation signal is undesirable, leading to a latchup event, the "switches" isolate the p-channel (or n-channel) transistor from its respective power supply rails.

Another implementation of this concept can utilize an *active clamp network* [1]. Instead of inverter networks, active clamp networks generally have the advantage of being able to respond at excursions outside of the normal voltage operational regime (Figure 9.5). While diode-based implementations respond to excursions greater than or equal to $V_{DD} + V_{be}$, (e.g. V_{be} being the forward bias voltage of a MOSFET junction), active clamp networks may respond to excursions greater than or equal to the V_{DD} power supply voltage. Additionally, active clamp networks may respond to voltage excursions below V_{SS}, instead of below $V_{SS} - V_{be}$ (e.g. as in diode-based scheme). An active clamp network is designed utilizing a reference control network. In this case, the active clamp network is electrically coupled to the n-well, and a second active clamp is electrically connected to the p-well substrate contacts. As will be discussed in this chapter, an active clamp is turned on for excursions outside of normal voltage range (e.g. undershoot where $V^* < V_{SS}$ and overshoot where $V^* > V_{DD}$).

In an active clamp network, a reference control network is used where the reference potential is set to V_{Tn}, and a second reference control network is used whose reference potential is $V_{DD} - V_{Tp}$; these references may be established using a MOSFET whose gate is coupled to its own drain connection.

The network is established where a second NFET has its gate connected to reference V_{Tn}. The network is established where a second PFET has its gate connected to a reference $V_{DD} - V_{Tp}$; these control elements sense the V_{DD} and V_{SS} substrate local potential. When the local substrate connection potential, or the local well connection potential extends outside of the normal voltage range, these elements "turn on." The turn-on of these elements leads to the turn-off of a transistor element between the gate array diffusions and corresponding power supplies [1] (Figure 9.5).

A latchup discipline circuit solution is as follows:

- A latchup decoupling network is established that senses perturbations outside of the normal operational range.
- The latchup network decouples the parasitic pnpn element from at least one power supply (e.g. V_{DD} or V_{SS}), preventing CMOS latchup.
- The power supply decoupling networks can utilize inverter logic, or active clamp networks that are electrically connected to a local well or substrate contact.

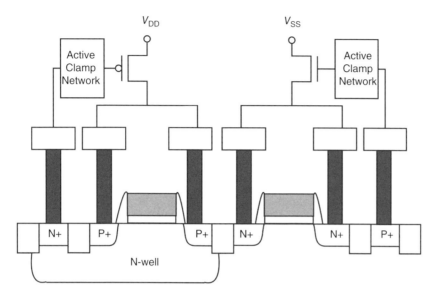

V_{DD} V_{SS}

Figure 9.5 Power supply decoupling network utilizing active clamp networks.

9.3 Overshoot and Undershoot Clamp Networks

Latchup events can be minimized with the utilization of a class of circuits that are used to minimize electrical overshoot and undershoot [2–11]. Given undershoot and overshoot can be minimized, forward biasing as well as substrate injection can be reduced.

Latchup occurs as a result of circuits that have a first node serving as an anode electrically connected to a high power supply (e.g. V_{DD}), and a second circuit whose cathode is electrically connected to a low power supply (e.g. V_{SS}). A p-channel MOSFET (or any p-diffusion) electrically connected to V_{DD} serves as an anode of the pnpn, and a n-channel MOSFET (or any n-diffusion) electrically connected to V_{SS} serves as the cathode of the pnpn. Hence, a key concept of latchup is:

- Circuits that have p-channel MOSFETs electrically connected to V_{DD}, and an n-channel electrically connected to V_{SS} are prone to latchup from overshoot events on V_{DD} and undershoot events on V_{SS}.

Where this may seem obvious, the compliment case is less obvious. If these circuits are latchup prone, then for the inverse – are they latchup immune? or do they assist in lowering the latchup concern? There are circuits whose p-channel MOSFETs (or p-diffusions are electrically connected to V_{SS}), and whose n-channel MOSFETs (or n-diffusion) is electrically connected to V_{DD}. Examples of circuits that have n-channel MOSFETs (or n-diffusions) electrically connected to V_{DD}, are V_{DD}-to-V_{SS} ESD RC-triggered n-channel MOSFET power clamps, HSTL OCD (e.g. with a n-channel MOSFET pull-up transistor), and an active clamp network; these networks prevent overshoot phenomenon either through clamping, parasitic bipolar turn-on. Hence, there are networks that provide "anti-overshoot" and "anti-undershoot" characteristics that assist in preventing latchup. Hence a key concept of CMOS latchup is:

- Circuits that have p-channel MOSFETs electrically connected to V_{SS}, and/or an n-channel electrically connected to V_{DD} are not prone to latchup from overshoot events on V_{DD} and undershoot events on V_{SS}, and provide "anti-latchup" characteristics.

Therefore, there are circuit families or classes that can assist in the latchup prevention. In the following sections, we will learn about some interesting concepts that provide "anti-overshoot" and "anti-undershoot" characteristics.

9.3.1 Passive Clamp Networks

Latchup events can be minimized with the utilization of a class of circuits that are used to minimize electrical overshoot and undershoot. This class of circuits are called *passive clamp networks*. "Passive" clamping circuits are unable to effectively meet these opposing requirements for high-performance applications. Passive clamping circuits have been used in both bipolar and CMOS high-performance environments for 2.0–0.5 μm technology generations. These concepts were first utilized using bipolar transistors in high-performance environments. In CMOS technology, passive clamp networks have historically taken advantage of the parasitic bipolar transistors for overshoot and undershoot, directing the current between the MOSFET source and drain. The advantage of passive clamps is the current is directed to the power rails instead of the injection into the substrate. Without active clamping, undershoot current is directed into the substrate, which can lead to injection.

Figure 9.6 is an example of a passive clamp network. For example, n-channel and p-channel MOSFET parasitic bipolar transistors were connected between the input and the power rails such that the MOSFET devices remained "off" and the MOSFET parasitic bipolar devices would turn "on" during overshoot and undershoot conditions. By connecting the n-channel MOSFET to V_{DD} power rail and the p-channel MOSFET to V_{SS} power rail, and disable the MOSFET gate electrode, the overshoot and undershoot of the signal pad would activate the parasitic bipolar transistors. In these cases, the current will flow to the power grid V_{DD} and V_{SS} for overshoot and undershoot phenomenon. For negative undershoot, the MOSFET parasitic npn bipolar transistor must become

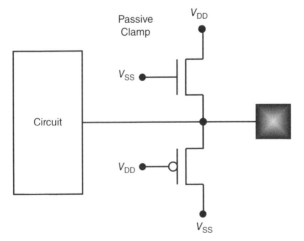

Figure 9.6 Passive clamp network.

forward active, at a $V_{BE} = 0.7$ V, and discharge to the V_{DD} power rail. For positive overshoot, the MOSFET pnp bipolar transistor must become forward active, at a $V_{BE} = 0.7$ V, and discharge the current to the V_{SS} power rail.

In this practice, there is an interesting feature to these circuits. Note that this "circuit feature" of "sinking" a negative undershoot to the V_{DD} power supply, and a positive undershoot to the V_{SS} power supply, is the same "utility" that a guard ring provides. An n-well guard ring serves as a "sink" for negative undershoot to the V_{DD} power supply. A p+ substrate guard ring serves as a "sink" for a positive overshoot to the V_{SS} power supply. Hence, in essence, these passive MOSFET networks serve as a well-controlled scalable "pseudo guard ring" circuit. But the advantage is that these are scalable, self-enclosed designs, can be modeled, and do not inject to the chip substrate. ESD networks, such as diodes, were also used as passive undershoot and overshoot clamps at 0.7 V noise levels. ESD diodes and MOSFET parasitic bipolar devices have served a useful role when the power supply voltage V_{DD} was above 5 V.

Distinctions between the ESD diode networks and passive clamp networks are as follows:

- ESD diode networks inject current into the semiconductor chip substrate V_{SS} for negative polarity events.
- Passive clamps inject current into the V_{DD} power rails for negative polarity events.
- Passive clamps inject current into the V_{SS} power rails for positive polarity events.
- ESD networks are designed to "sink" high current levels well above the functional application (e.g. 1–10 A current levels).
- Passive clamp networks are design to "sink" overshoot and undershoot current levels (e.g. 10–100 mA current levels).

Passive clamps have a functional limitation as the power supply voltage is scaled. A "clamping figure-of-merit" is the ratio of the clamping voltage, ΔV, and power supply voltage, V_{DD}, expressed as $\Delta V / V_{DD}$.

$$FOM = \frac{\Delta V}{V_{DD}}.$$

In order to damp out ringing and noise, a constant ratio must be maintained as the native power supply or peripheral supply voltage is scaled. A CMOS latchup circuit design practice is as follows:

- Passive clamps are used to limit overshoot and undershoot on signal nodes.
- Passive clamp networks "sink" current to the power supply rails.
- Passive clamp networks can "sink" undershoot current to the V_{DD} power supply, serving as a pseudo guard ring.
- Passive clamp networks can "sink" overshoot current to the V_{SS} power supply, serving as a pseudo guard ring.

9.3.2 Active Clamp Networks

Latchup events can be minimized with the utilization of active clamp networks [2–11]. Active clamp circuits are key to minimize electrical overshoot and undershoot, minimize reflected signals, and achieve performance objectives and reliability requirements in high-performance circuits; these challenges include more significant challenge impedance matching conditions, MOSFET gate dielectric reliability, MOSFET off-current levels, low power, latchup, and ESD protection. Active termination elements are of interest in the transfer of signals between receivers and transmitters [2–11]. Active termination elements exists in peripheral circuits; in the high-performance peripheral I/O environment, there are the receivers, the OCDs, the active clamp elements and ESD networks. L. DeClue, and H. Muller [2], E.Davidson and R.Lane [3], G.Slaughter [4], J. Kosson [5], C. Petersen [6], A. Furman [7], E. Honningford [8], S. Voldman and D. Hui [9, 10], and B. Marshak, D. Hui, S.Voldman, and R.R. Williams [11] addressed the issues of voltage clamping, overshoot and undershoot, reflections, and performance impacts.

In an ideal system, the input voltage switches instantaneously between the high-voltage state, a "digital 1," and the low-voltage state, "digital 0," never exceeding the power supply and ground states, and zero transition time. In a real environment, the signal has a finite transition time, overshoots and undershoots the power supply and ground voltages, oscillates, and undergoes ringing. The ideal "clamp" circuit eliminates ringing and noise such that the signal at the input remains at or near the two desired voltage states and switches between those states in the minimum time. The ideal clamp must drain or supply current instantaneously to/from the network at the input to the circuit being clamped whenever the voltage at the input exceeds or falls below the desired voltage state. In order to achieve this, the clamping must have a low dynamic impedance (e.g. resistance) and a low reflection coefficient in the vicinity of the upper and lower voltage corresponding to the two digital logic states. On the other hand, in order to maximize switching speed between the two logic states, the impedance of the clamping circuit and the reflection coefficient should be very high during switching for a brief time when the input voltage is between the upper and lower digital voltages.

From a latchup circuit design practice, active clamps have significant benefits. Active clamp networks improve latchup tolerance of peripheral circuits by shunting the overshoot and undershoot signal to the power supply rails without forward biasing of the n- and p-channel MOSFET devices, and resistors in the peripheral circuits. Active clamp circuits also prevent electrical overstress (EOS) on the MOSFET gate dielectric in receiver networks and driver networks; this is very important with the rapid scaling of the MOSFET gate dielectric to avoid the EOS.

Figure 9.7 shows an example of a CMOS active clamp circuit. In this circuit, an n-channel MOSFET source is connected to the pad node, the drain is connected to

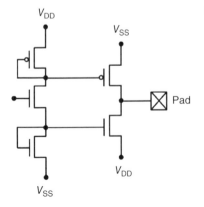

Figure 9.7 CMOS active clamp network.

V_{DD} power supply rail and a p-channel MOSFET source is connected to the pad node and MOSFET drain is connected to the V_{SS} power supply rail. The n-channel MOSFET gate is biased using a reference control element at the n-channel MOSFET threshold voltage, V_{TN}. The p-channel MOSFET gate is biased using a second reference control element at the voltage $V_{DD} - V_{TP}$, where V_{TP} is the p-channel MOSFET threshold voltage. These clamping transistors are serially connected between the input pad and the power rails. The reference voltage is set so that when there is noise, ringing or undershoot event below the ground potential, the n-channel clamp element turns on. In this case, the input node is rapidly damped to the ground potential. Similarly, the p-channel reference circuit is set such that whenever there is an excursion above the V_{DD} power supply, the p-channel MOSFET turns on and rapidly reduces the voltage to the power supply voltage. The center reference transistor controls current flow through the reference elements [9, 10].

Active clamps are "pseudo-zero V_T" networks in that they turn on at the power supply voltage rails. An "ideal" active clamp would have no "on resistance" providing an "ideal switch" characteristics, which is "off" for voltages between $V = V_{SS}$ to V_{DD}, and "on" for all voltage excursions. Active clamp networks would also turn on prior to any ESD circuit network. For example, a dual-diode ESD networks turn-on occurs at $V = -V_{BE}$ for undershoot, and $V = V_{DD} + V_{BE}$ for overshoots, whereas the active clamp network turns on at $V = 0$ and $V = V_{DD}$.

Figure 9.8 shows the low current I-V characteristic of a clamp network for a 1.8 V power supply chip as a function of the input voltage, V_{in}. When the input pad voltage is less than the V_{SS} potential, the n-channel clamp element is on and the current is flowing out of the clamping circuit. When the input voltage V_{in} is between V_{SS} and V_{DD}, no current is flowing through the active clamp network. When the input pad voltage V_{in} exceeds the power supply voltage V_{DD}, the p-channel MOSFET clamp element is on and the current is flowing in the opposite direction from the input pad through the p-channel MOSFET to the V_{SS} power rail. Using a MOSFET control switch serially connected between the two reference MOSFET circuits, the active clamp can be turned off by controlling the current through these two MOSFET control references [9, 10]. Note in Figure 10.24 that the magnitude of the current level for these elements are tens of

Figure 9.8 I-V characteristics of an active clamp network.

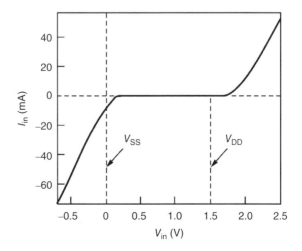

milliamperes (e.g. on the order of overshoot and latchup events but well below the level of ESD events).

A figure of merit of a clamp circuit is the reflection coefficient. The reflection coefficient is defined as the difference of the terminator (receiver) load, Z_1, and the characteristic impedance of the transmission line, Z_0, over the sum of Z_1 and Z_0,

$$\Gamma = \frac{Z_1 - Z_0}{Z_1 + Z_0}$$

For a perfect matched system, the reflection coefficient would be zero for V_{in} below the substrate ground potential, unity for V_{in} equal to the substrate potential and the native power supply voltage V_{DD}, and zero for greater than the native power supply voltage V_{DD}. For a perfectly matched active clamp network, the reflection coefficient should be -1 for $V_{in} < V_{SS}$, unity for $V_{SS} < V_{in} < V_{DD}$, and -1 for $V_{in} > V_{DD}$. Figure 9.9 shows the reflection coefficient of the circuit as a function of the input voltage. For the condition that $V_{in} < V_{SS}$, the reflection coefficient of the circuit is near ideal (e.g. $\Gamma = -1$). This corresponds to a low impedance state and allows current to flow rapidly from the clamp circuit (e.g. p-channel MOSFET) to the input node to prevent undershoot. For $V_S < V_{in} < V_{DD}$, the reflection coefficient is near unity [9, 10].

A latchup circuit design practice is as follows:

- Active clamp networks can be utilized to reduce undershoot and overshoot on signal pads at the power supply rails prior to a forward biasing of any other circuit element.
- Active clamp networks can be used to sink overshoot and undershoot voltage excursions outside of the V_{SS} to V_{DD} functional voltage range.
- Active clamp networks can be utilized to sink overshoot and undershoot signal currents to the power supply rails (e.g. 10–100 mA levels).

9.3.3 Dynamic Threshold Triple Well Passive and Active Clamp Networks

A third class of clamp networks that can be utilized in triple well CMOS technology is the dynamic threshold triple well clamp networks. Triple-well CMOS technology has an advantage over twin-well bulk CMOS in that the n-channel MOSFET substrate is not bound to the substrate potential. As a result, the triple-well CMOS can construct

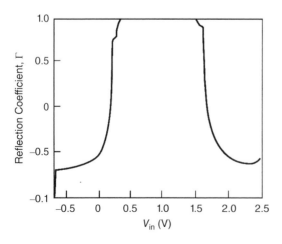

Figure 9.9 Reflection coefficient of an active clamp network.

symmetric circuits and apply dynamic threshold techniques. As a result, active clamp networks can use complimentary methods for the n-channel and p-channel transistor to address positive and negative voltage and current excursions [10].

Dynamic threshold voltage MOSFETs provide a higher transconductance (g_{msat}), a higher drain saturation current (I_{dsat}), a dynamic threshold voltage that can be used as a lower voltage triggering device, a high I_{on}/I_{off} ratio, and scalable properties to sub-0.7 V power supply voltages. Figure 9.10 shows an example of a dynamic threshold active clamp. This network will limit voltage excursions in triple-well environments.

Figure 9.11 shows a modified version of the bulk CMOS active clamp network. In this case, the bodies of the network are electrically coupled to the input. In the case of triple-well technology, this leads to the current injection into the triple-well structures and not the chip substrate.

A latchup circuit design discipline incorporates the following:

- Anti-overshoot and anti-undershoot networks can minimize system level transients, and reflections.
- Anti-overshoot and anti-undershoot prevent the forward bias of semiconductor devices used in semiconductor peripheral circuits.
- Active clamp networks provide a means of discharging the transient currents to the power rails and avoids substrate and well minority carrier injection.

Figure 9.10 Dynamic threshold triple well body- and gate-coupled ESD network.

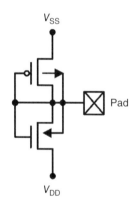

Figure 9.11 Dynamic threshold triple well active clamp network.

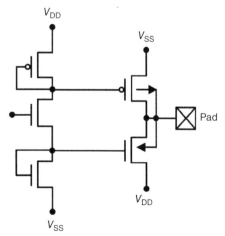

9.4 Passive and Active Guard Rings

Latchup and noise are initiated in the substrate from overshoot and undershoot phenomenon. These can be generated by CMOS OCD circuitry, receiver networks, and ESD devices. Unfortunately, parasitic pnpns are often unrecognized or anticipated.

Today, system-on-chip (SOC) solutions have been used for solving the mixed signal and radio frequency (RF) requirements. SOC applications have a wide range of power supply conditions, number of independent power domains, and circuit performance objectives. Different power domains are established between digital, analog, and RF functional blocks on an integrated chip. The integration of different circuits and system functions into a common chip has also resulted in solutions for ensuring that noise from one portion or circuit of the chip does not affect a different circuit within the chip [12–20]. With the chip integration issues, the need for better guard rings and alternative guard ring solutions have had increased interest [18–31]. In the mid-1990s, there was a focus in semiconductor chip design to achieve both noise isolation and ESD protection. This was achieved by establishing isolated power rails, which were then electrically reconnected with ESD solutions. Since 2000, there has been an increased focus on guard ring solutions that achieve the following objectives:

- Achieve noise isolation, latchup robustness, and ESD results.
- Do not inject current back into the power grid.

With the growth of interaction between digital, analog, and RF domains, guard ring concepts have increased in importance. In addition, with the growth of smart power technology, solutions are needed for avoidance of interaction of the high-voltage CMOS (HVCMOS) chip sectors and the low-voltage sectors of a CMOS chip [18–38]. In smart power technology, the concern of latchup between the lateral-diffused MOS (LDMOS) power devices and low-voltage CMOS has been a focal point for development of new guard ring concepts from 1990 to present day [18–38]. New guard ring solution for smart power and BiCMOS applications were proposed by the following teams: Bafluer et al. [20, 22]; Peppiettte [21]; Winkler and Herzl [24]; Gonnard and Charitat [25]; Zhu, Parthasarathy, Khemka, and Bose [26]; Gonnard et al. [27]; Schenkel et al. [28]; Parthasarathy et al. [29, 30]; Laine et al. [31]; Khemka et al. [32]; Horn [33]; Voldman et al. [34]; Gupta, Kosier, and Beckman [35]; Stella, Favilla, and Croce [36]; Singh and Voldman [37, 38]; and Zhu et al. [39]. In the next sections, examples of both the passive and active guard rings circuits and structures will be shown.

9.4.1 Passive Guard Ring Circuits and Structures

Passive guard ring structures can be used in a SOC application by utilization of process features. Passive guard rings are valuable between digital, analog, and RF domains to avoid latchup. In a BiCMOS technology with trench technology, passive guard ring elements are as follows:

- N-well rings
- Sub-collector rings
- N-well ring and deep trench (DT) ring
- Plurality of N-well and DT rings
- Plurality of deep trench rings and high dose buried layer (HDBL)

With the integration of trench structures and implants, p−/p++ substrates, and HDBL can be used to influence the vertical and lateral transport between domains.

In smart power applications, technologies integrate bipolar, CMOS, and DMOS devices (e.g. referred to as BCD technology). In BCD smart power applications that integrate high-voltage DMOS with low-voltage CMOS, passive guard ring solutions are utilized between the different power domains. These solutions are as follows:

- DeMOS implant guard rings
- LDMOS n-well and n-body guard rings
- LDMOS n-buried layer (e.g. n-tub) guard rings
- p− Epitaxy/p++ substrate wafers
- Heavily doped buried layer implants below the LDMOS n-tub regions

9.4.2 Active Guard Ring Circuits and Structures

One of the problems with diffused junction passive guard rings is that in order for the metallurgical junction-based guard rings to improve CMOS latchup, current enters the semiconductor chip power grid. The guard ring improves the latchup tolerance; however, the overshoot noise that can initiate the latchup is injected into the V_{SS} ground rail (through the vertical pnp), and possibly spread to other circuits. Solutions for improving latchup tolerance have been used, but these circuits introduce noise into the power rails (e.g. V_{DD} or V_{SS}). In a digital, analog, and RF chip application, a circuit solution that improves latchup tolerance and at the same time limiting the amount of the current injection or noise introduced into the power rails is valuable to achieve both CMOS latchup objectives as well as noise objectives.

In HVCMOS technology, a concern is the interaction of the LDMOS transistors and the adjacent low voltage CMOS circuitry [18, 19]. In HVCMOS, inductive "load dumps" initiate injection of minority carriers in the chip substrate. As a result of the physical size of the HVCMOS LDMOS devices, as well as the magnitude of the current injection, it is critical not to disturb the other chip functions on the smart power chips.

Different "active" guard ring circuit concepts have been introduced for latchup improvement [20–31, 35–39]. In "active" guard rings, the objective is to not only collect minority carriers but to actively compensate the effect. The latchup circuit design discipline includes the following concepts:

- Electrically collecting minority carriers at a metallurgical junction, and whose junction is electrically connected to the chip substrate to alter the substrate potential.
- Electrically collecting minority carriers at a metallurgical junction, and whose junction is electrically connected to the chip substrate, to alter the substrate potential, with the objective of reduced forward bias of the injection structure.
- Electrically collecting minority carriers at a metallurgical junction, and whose junction is electrically connected to the chip substrate, to alter the substrate potential, with the objective of introduction of a lateral electrical field assist to reduce the lateral bipolar current gain.
- Electrically sensing the substrate potential drop, and inverting the polarity of the potential drop using inverting amplifier networks.

Figure 9.12 is an example of an active guard ring. Typically, in a passive guard ring concept, a p+ substrate contact is electrically connected to a V_{SS} power rail, and an

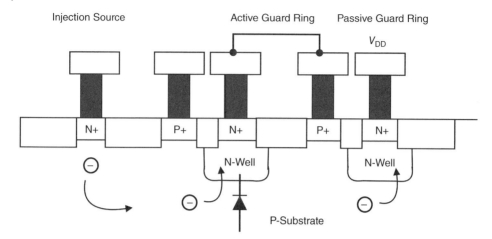

Figure 9.12 Active guard ring with n-well junction electrically connected to a p+ substrate contact to establish lateral opposing electric field.

n-well ring is electrically connected to a V_{DD} power rail. But in an active guard ring, an n-well region is not electrically connected to a power rail. In an active guard ring, the n-well structure collects the minority carrier electrons in its metallurgical junction formed with the p− substrate region. The n-well ring is electrically connected to a "soft grounded" p+ substrate contact. When the minority carrier electron traverse the metallurgical junction, it reduces the electrical potential of the n-well region (e.g. noted in the figure is ΔV). By electrically connecting the n-well to the p+ substrate contact, the electrical potential of the substrate is also lowered by the same potential magnitude. In this case, the electrical potential of the region is lowered. The lowering of the substrate potential can be utilized as two means. First, given the p+ diffusion is near a forward biased structure (e.g. an injecting structure), the reduction of the potential can lower the forward bias state, turning off the injection process. Second, given another p+ substrate contact, a lateral electric field can be established that inhibits the flow of minority carriers. Given a parasitic npn bipolar transistor is formed between the injection source, and a collecting victim circuit, if the lateral electrical field opposes the current transport, the lateral npn bipolar current gain is reduced. The derivation of "electric field assist" is the transport solution, where in this case, the electric field reduces the lateral bipolar current gain. In this methodology, the placement of the p+ region can be on the injection side, or collection side of the n-well region. Gupta, Kosier, and Beckman [35] added an additional p++ substrate diffusion inside of the n-well ring/p+ substrate contact to establish a well-defined lateral electric field where a p+ region is flanking both sides of the n-well ring, with the outer p+ substrate contact electrically connected to the n-well ring. In a second implementation [35], two p+ diffusions are placed within the n-well ring, forming three p+ substrate rings, and a single n-well ring. In this case, the two "floating" p+ diffusions flanking the n-well are electrically connected to the n-well, and the third p+ diffusion is shorted to a power rail V_{SS}.

Various implementations of guard rings are utilized where a plurality of p+ substrate contacts and n-wells are integrated, mixing both the active and passive concepts, where some of the wells are "floating" and some electrically connected to the power supplies.

In these implementations, a plurality of trench structures can also be added to reduce the lateral bipolar current gain. In all cases, as the number of additional guard rings are increased, and the effective base width, the bipolar current gain decreases. In the work of Gupta et al., it was shown that without a guard ring, the lateral bipolar current gain decreased from 2×10^2 to 1×10^{-1} as the base width increased from 20 to $500\,\mu m$; with a unbiased guard ring, a bipolar current gain of 5×10^{-3} was achieved at a spacing of $20\,\mu m$ [35]. The measurement of Gupta et al. demonstrates that greater than three orders of magnitude of improvement was achieved with the active guard ring concept.

The concept of the active guard ring can also be integrated with passive guard ring solutions. Figure 9.13 shows the structure of Figure 9.12 with an additional n-well region. Electrons that bypass the active guard rings can be collected by the second n-well structure.

In order to improve the guard ring efficiency, additional solutions can be added to prevent the migration of injected carriers. Gupta, Kosier, and Beckmann, as well as Stella, Favilla, and Croce, added additional implant structures to the active guard ring concept to enhance the guard ring depth, as well as restricting the minority carriers vertical transport [35, 36]. Figure 9.14 shows a guard ring concept that includes the n-well, and an n-doped buried layer underneath the n-well to deepen the guard ring surface area. The n-doped buried layer can be a sub-collector, a HDBL implant, or an LDMOS n-tub or deep n-well implant. In addition, a p−/p++ substrate wafer is used to restrict the vertical trajectory of the minority carrier. Hence, the following concepts are introduced in this implementation:

- N-buried layer under the n-well can provide an increase in the guard ring collection area.
- An n-well/n++ buried layer region electrically connected to a p++ substrate contact can modulate the substrate potential and induce an opposing lateral electric field.

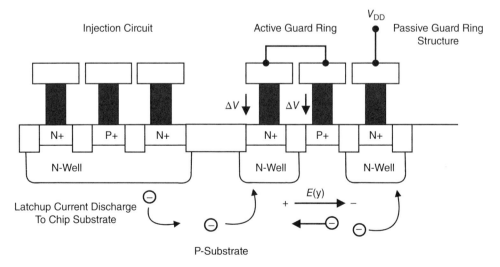

Figure 9.13 Active guard ring with n-well junction electrically connected to a p+ substrate contact to establish lateral opposing electric field, and passive n-well guard ring.

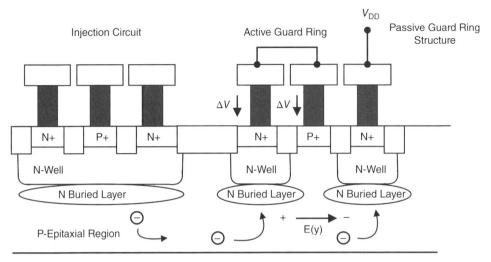

Figure 9.14 Active guard ring with n-well, and a n+ buried layer, and a passive guard ring on a p−/p++ substrate wafer.

- A low-high (L-H) p−/p++ step junction can induce a vertical electric field to (i) decrease transport deep under the guard ring structure, and (ii) increase the current density in the region between the p− epitaxial region.

The first concept increases the guard ring efficiency with the deeper guard ring structure. The second concept provides a means of lowering the lateral transport with the introduction of the lateral electrical field. Third, the p−/p++ step junction reflecting boundary redirects the minority carrier toward the guard ring structure for collection. Fourth, the combination of the p−/p++ step junction and the guard ring region, restricts the carrier flow in the epitaxial region. At high currents, this lowers the parasitic bipolar current gain as well. Stella, Favilla, and Croce noted that in this structure, due to the p−/p++ reflecting barrier, and collection by the active guard ring, the characteristic decay length, L_d, can be defined as [36]

$$L_d = \frac{1}{\sqrt{\left(\frac{1}{L_d}\right)^2 + \left(\frac{\pi}{2T_p}\right)^2}}$$

where L_n is the diffusion length, and T_p is the epitaxial thickness between the n-well guard ring and the p−/p++ interface.

Another approach is the active compensation through inversion of the signal to nullify its effect. A circuit solution to address both noise and CMOS latchup is the usage of *active guard rings*. Winkler and Herzl introduced this concept to eliminate noise in mixed signal chips [24]. An active guard ring circuit is a circuit that senses an undershoot phenomena in the chip substrate. The concept of the active guard ring is to sense the negative change in the substrate potential; this negative signal is fed as an input into an inverting amplifier. The output of the inverting amplifier is then electrically connected to the substrate. In this fashion, the negative polarity undershoot voltage is inverted and

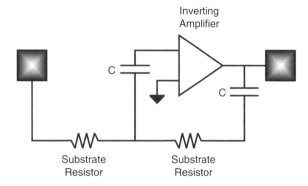

Figure 9.15 Active guard ring noise and latchup suppression network.

amplified, whose signal is then re-injected into the same electrical region to nullify the undershoot voltage transient state [24, 37, 38].

Figure 9.15 shows the p+ noise suppression element is electrically connected to the input of a differential amplifier. Two resistors are formed across the differential amplifier between the input and output of an amplifier, where the resistors use the silicon substrate itself. In addition, an active noise suppression circuit is also added with an input connected to p-region, and an output connected to substrate contact. In the differential amplifier, one input is electrically connected to the p+ noise suppression element and the second input is electrically connected to the ground [37, 38].

Figure 9.16 shows an example of a capacitance-coupled implementation with ESD protection [37, 38]. In this case, the differential input is coupled to the p+ guard ring through a capacitor. An RC network is formed on the differential input node. In addition, the differential amplifier output is electrically connected to the resistor feedback network through a capacitor on the output node, forming an RC network on the output node.

In the presence of multiple "victim" circuits, there is a plurality of potential circuits that may be influenced by a single injection source. In the case of multiple collection regions, the parasitic bipolar npn transistor formed between the injector source and the victim circuit is a function of the victim circuit "collector" area, and relative distance to the injection source (e.g. effective base width). Each circuit will see a different level of current injection; hence, the level of active compensation will be different for each circuit. Compensation-based active guard ring structures can increase in sophistication with additional analog function circuits to provide the circuit compensation

Figure 9.16 Active guard ring noise and latchup suppression network with RC network on input and output of differential amplifier.

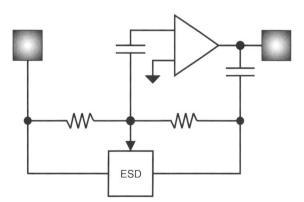

techniques [33]. Additionally, of interest is the case of large injection structures near large collecting structures. In this case, when the space between the injector and the collector is small, the distribution of the currents over the areas must address the spatial distances between points in the injector and the collector.

In the compensating active guard ring concept, the compensation current can be evaluated with the evaluation of segments of the collecting structures. Compensating active guard rings are of interest for the case of large injecting sources and large collecting regions where the spacing between the space between the two regions are small. There are many examples of these situations. A first example is the case of Huh et al., as discussed in this chapter, where the injecting source was large I/O MOSFET OCD, and the collecting structure was the n-well of decoupling capacitor network. In smart power, the injection from a DMOS injection transistor to a second adjacent collecting structure at close proximity can be estimated as a one-dimensional transport, which has an exponential decay characteristic spatially. Horn expressed the injection in the form as follows [33]:

$$I_s(x) = \iint_{A_C} k_1 \exp\{-k_2'x\}$$

where the I_S is the current collected at the "collector," collector area, A_C, and k_1 and k_2 are coefficients. The first coefficient k_1 is proportional to the injection current and diffusion properties, and k_2 is proportional to the inverse of the diffusion length in the x-direction [33].

This development was generalized in two dimensions, within the plane of the wafer. In the x- and y- dimensions, the equation is defined as

$$I_s(x,y) = \iint_{A_C} k_1 \exp\{-k_2x - k_3y\}dxdy$$

$$I_s(x,y) = \iint_{A_C} k_1e - (k_2x + k_3y)dxdy$$

Given the collection region is a rectangle in the x- and y-dimension, the derivation can be expressed as the integration over the x-direction and the y-direction as follows:

$$I_s = \int_{y_1}^{y_2} \int_{x_1}^{x_2} k_1e - (k_2x + k_3y)dxdy$$

Placing the derivation in another form, there exists an $x = x_n$, and $y = y_n$ where

$$\int_{y_1}^{y_2} \int_{x_1}^{x_2} f(x,y)dxdy = \frac{\int_{x_1}^{x_2} f(x,y_n)dx \int_{y_1}^{y_2} f(x_n,y)dy}{f(x_n,y_n)}$$

Horn defined the following terms [33], where the first term is the current along the dimension $y = y_2$, by taking the integral of the collection over the x-dimension:

$$I_A = \int_{x_1}^{x_2} f(x,y_2)dx$$

and the second current is the integral along $x = x_2$, integrating over the y-dimension:

$$I_B = \int_{y_1}^{y_2} f(x_2,y)dy$$

and the current at the point $x = x_2$, and $y = y_2$.

$$I_C = f(x_2, y_2)$$

In this form, the minority current collected can be expressed as these terms:

$$I_s = \int_{y_1}^{y_2} \int_{x_1}^{x_2} k_1 e - (k_2 x + k_3 y) dx dy = \frac{I_A I_B}{I_C}$$

Hence, an active compensation current can be evaluated from the terms I_A, I_B, and I_C. As shown [33], this method can be modified by using a strip in the x-direction from x_3 to x_4 (e.g. $x_3 < x < x_4$) and a second strip in the y-direction, y_3 to y_4 (e.g. $y_3 < y < y_4$). It can be show that

$$\int_{y_1}^{y_2} \int_{x_1}^{x_2} f(x, y) dx dy = \frac{\left\{ \int_{y_3}^{y_4} dy \int_{x_1}^{x_2} f(x, y_n) dx \right\} \left\{ \int_{x_3}^{x_4} dx \int_{y_1}^{y_2} f(x_n, y) dy \right\}}{\int_{y_3}^{y_4} \int_{x_3}^{x_4} f(x, y) dx dy}$$

where we redefine the three currents for the compensation current for the active guard ring according to the following relationships:

$$I_A = \int_{y_3}^{y_4} \int_{x_1}^{x_2} f(x, y) dx dy,$$

$$I_B = \int_{y_1}^{y_2} \int_{x_3}^{x_4} f(x, y) dx dy$$

and

$$I_C = \int_{y_3}^{y_4} \int_{x_3}^{x_4} f(x, y) dx dy$$

where

$$\int_{y_1}^{y_2} \int_{x_1}^{x_2} f(x, y) dx dy = \frac{I_A I_B}{I_C}$$

Figure 9.17 shows an example of a compensation-based active guard ring concept that introduces a multiplier-divider network. Each victim circuit is compensated based on its injection current. In the figure, the three currents I_A, I_B, and I_C are shown.

A latchup circuit design practice is as follows:

- Passive guard rings are used to provide a means of collecting minority carrier current to the power supply rails.
- Passive guard rings can be used to lower the parasitic bipolar transistor current gain formed in the parasitic transistors between an injection source, and a collection region.
- Active guard rings provide a means of collecting minority carriers and not injecting the current into the power supply rail, and reestablish the substrate potential.
- Active guard rings provides a means of introducing lateral electric fields to minimize transport between an injection source and a collecting region.
- Compensating active guard rings can process the injection signal and provide a means to neutralize the injection to reestablish the substrate potential.

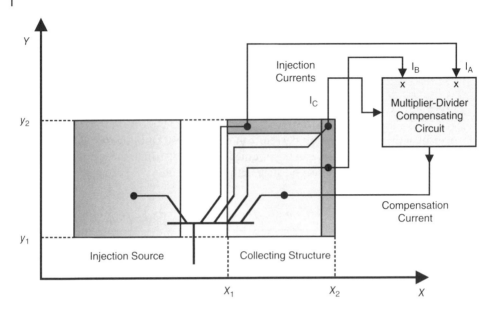

Figure 9.17 Compensation based active guard ring with analog multiplier-divider network.

9.5 Triple-Well Noise and Latchup Suppression Structures

Other methods to avoid noise and latchup interaction in circuits exist in triple-well technology and BiCMOS technology. Physical structures such as triple-well implants, and HDBLs can improve the latchup immunity of digital devices while providing isolation structures that provide noise isolation for both the digital and analog devices. A HDBL or sub-collector implants can be used to isolate physical chip sectors. These can be integrated with arrays of deep trench structures; an array of deep isolation trenches provides increased isolation between devices where needed.

9.6 System-Level Latchup Issues

Latchup can occur within systems leading to system-level failure. Latchup issues can occur in laptops, servers, personal computers, cell phones, handheld electronics, or power systems. System-level latchup problems can occur as a result of the following interactions:

- Improper power-sequencing of system level connections
- System-level noise
- Transient systems
- Cable discharge events (CDE)
- Board-to-chip capacitive coupling leading to overshoot and undershoot events on the semiconductor chip
- Semiconductor chips with low latchup immunity
- Improper assembly leading to reverse pin voltages (e.g. reverse polarity on electrical pin)

- Improper installation of battery (e.g. reverse polarity of battery)
- Poor grounding of ground supply
- Inductive load dumps
- Board-level Inductance mismatch in a multiple component system
- Bond wire (wirebond) mis-match in a multiple component parallel system
- Bond wire (wirebond) bond breakage
- Nonuniform thermal shunting of power

Solutions for system level CMOS latchup can include the following:

- Board-level diode anti-overshoot clamps
- Board-level diode anti-undershoot clamps
- Board-level active clamp networks
- Board-level spark gaps
- Voltage suppression cable elements (e.g. cables whose resistance increases at high current)
- Cabling with discharge connectors (e.g. BNC connectors with shorting means)
- System-level "touch pad" to discharge cables, and personnel
- Board-level transient voltage suppression devices (e.g. polymer voltage suppression shunt elements)
- Sockets that prevent battery reversal
- Sockets that prevent negative polarity reversal
- Latchup robust semiconductor products
- Latchup-proof switches (e.g. SOI technology) [40]
- Fault-protected switches [40]
- Channel protector clamps [40]

9.7 Summary and Closing Comments

Chapter 9 focused on CMOS latchup and circuits, from circuit problems to circuit solutions. Today, there are hundreds of patent inventions in the area of circuits for latchup. In this chapter, the objective was to provide a small sample of the circuit art that exists today.

Chapter 10 will focus on issues in the future in advanced technologies and the technology implications. The chapter will address radiation effects in nanotechnologies and advanced packaging concepts.

Problems

9.1 In an integrated circuit, subfunctions necessary to construct a microprocessor design include I/O books, on-chip cache SRAM memory, core logic, PLL, and DLL circuits. In addition, decoupling capacitors are used in between chip sectors. List the possible layout combinations that can occur.

9.2 In an integrated circuit, subfunctions necessary to construct a microprocessor design include I/O books, on-chip cache SRAM memory, core logic, PLL, and

DLL circuits. In addition, decoupling capacitors are used in between chip sectors. List the possible power rail to power rail combinations that can occur that may lead to a CMOS latchup concern.

9.3 In a mixed-signal chip, there are digital, analog, and RF chip sectors. In the analog chip sector, the circuits may contain bipolar, CMOS, or BiCMOS circuits. In the RF chip sector, npn bipolar transistors exist. What are the possible interactions between the chip sectors? What are the possible pnpn structures that can be formed?

9.4 In "complimentary bipolar," there are both pnp transistors, and npn transistors in a common process. What are possible parasitics when the pnp and npn transistor interact that can lead to latchup? Given complimentary BiCMOS, what are the possible parasitics formed between the pnp, the npn and the CMOS logic circuitry?

9.5 In triple-well technology, what are the potential chip function interactions? Is triple-well better or worse than dual-well CMOS for subfunction-to-subfunction latchup interactions?

9.6 Given a chip where there is a large digital chip sector, a smaller analog sector, and very small RF sector, assume each is on a separate power supply. Assume the digital and analog sectors have the same V_{DD} voltage levels but different power rails, and assume that the RF sector is at a higher V_{CC} voltage. Assume the capacitance of each chip sector is proportional to its chip area. How would you sequence the semiconductor chip? Develop a circuit using circuit comparators and switches for the three power rail system. Provide a solution extending the patent of Lin, Yu, and Peng [41].

9.7 Using a p-channel MOSFET well bias network, show the states of the input relative to the V_{DD} power supply. Using a p-channel MOSFET well bias network, show each state when used between two power supply rails.

9.8 An insulated gate bipolar transistor (IGBT) device is constructed on a p+ substrate. The structure consist of an enclosed n+ MOSFET drain, within a MOSFET polysilicon gate structure. The n+ doped MOSFET drain is in a p-body region. The p-body region is contained within a n− region serving as the MOSFET source. The MOSFET gate structure overlaps the p− body and n− region providing a MOSFET lateral channel within the p-body. The n-source region is on a p+ substrate region. A vertical bipolar pnp is formed between the p+ source and the p-body region, and the n-region serves as the vertical bipolar pnp base region. The n-channel MOSFET serves as a switch and base current drive to the vertical pnp structure. Assume a load in series with the vertical pnp transistor. Draw the vertical cross section of the device and the circuit schematic. Show the existence of the vertical parasitic npn transistor and the parasitic pnpn device.

9.9 Assume N parallel IGBT devices are electrically connected in parallel through wire bonds. Show how a wire-bond breakage can lead to nonuniform current distribution within a system. Assume N IGBT devices in parallel, electrically connected to an inductor. Assume that the inductance of each inductor is not identical. Show how this leads to nonuniform current distribution, and latchup (Hint: $DV = (L_i - L_j) \, dI/dt$).

References

1 Voldman, S. (2006) A method, apparatus, and circuit for latchup suppression in a gate-array ASIC environment. U.S. Patent No. 7,102,867, issued September 5th, 2006.

2 DeClue, L. and Muller, H. (1976) Active termination network for clamping a line signal, U.S. Patent No. 3,937,988, issued February 10th, 1976.

3 Davidson, E. and Lane, R. (1997) Low power transmission line terminator, U.S. Patent No. 4,015,147, issued March 29th, 1977.

4 Slaughter, G. (1990) Non-reflecting transmission line termination, U.S. Patent No. 4,943,739, issued July 24th, 1990.

5 Kosson, J. (1990) Voltage clamping with high current capability. U.S. Patent No. 4,958,093, issued September 18th, 1990.

6 Petersen, C. (1992) High speed anti-undershoot and anti-overshoot circuit. U.S. Patent No. 5,103,118, issued April 7th, 1992.

7 Furman, A. (1993) Zero power transmission line terminator. U.S. Patent No. 5,227,677, issued July 13th, 1993.

8 Honningford, E. (1996) CMOS input voltage clamp, U.S. Patent No. 5,528,190, issued June 18th, 1996.

9 Voldman, S. and Hui, D. (2000) Switchable active clamp network. U.S. Patent No. 6,075,399, issued June 13th, 2000.

10 Voldman, S. (2000) Silicon-on-insulator body- and dual gate-coupled diode for electrostatic discharge (ESD) applications. U.S. Patent No. 6,034,397, issued March 7th, 2000.

11 Mashak, B.W., Williams, R.R, Voldman, S., and Hui, D. (2001) Active clamp network for multiple voltages. U.S. Patent No. 6,229,372, issued May 8th, 2001.

12 Li, T., Tsai, C.H., Rosenbaum, E., and Kang, S.M. (1999). Substrate modeling and lumped substrate resistance extraction for CMOS ESD/Latchup circuit simulation. In: *Proceedings of the 36th ACM/IEEE Conference on Design Automation, June 1999*, 549–554. IEEE.

13 Owens, B., Birrer, B., Adluri, P. et al. (2003). Strategies for simulation, measurement and suppression of digital noise in mixed-signal circuits. In: *Proceedings of the IEEE 2003 Custom Integrated Circuits Conference (CICC), Sept. 2003*, 361–364. IEEE.

14 Koukab, A., Banerjee, K., and Declercq, M. (2002). Analysis and optimization of substrate noise coupling in single-chip RF transceiver design. In: *Proceedings of the 2002 IEEE/ACM International Conference on Computer-Aided Design, Nov. 2002*, 309–316. IEEE.

15 Chan, H.H.Y. and Zilac, Z. (2001). A practical substrate modeling algorithm with active guard-band macro-model for mixed-signal substrate coupling verification. In: *Proceedings of the 8th IEEE International Conference on Electronics, Circuits, and Systems. Sept. 2001*, 1455–1460. IEEE.

16 Nagata, M., Nagai, J., and Morie, T. (2000). Quantitative characterization of the substrate noise for physical design guides in digital circuits. In: *Proceedings of the IEEE Custom Integrated Circuits Conference (CICC). May 2000*, 95–98. IEEE.

17 Li, T. and Kang, S.M. (1998). Layout extraction and verification methodology for CMOS I/O circuits. In: *Proceeding of the Design Automation Conference (DAC), June 1998*, 291–296. IEEE.

18 Murari, B. (1978). Power integrated circuits: problems, tradeoffs, and solutions. *IEEE Journal on Solid State Circuits* SS-13 (3): 307–319.

19 Chow, T.P.D., Patttanayak, D., Baliga, B.J. et al. (1991). Interaction between monolithic junction-isolated lateral insulated-gate bipolar transistors. *IEEE Transactions on Electron Devices* ED-38 (2): 310–315.

20 Bafluer, M., Buxo, J., Vidal, M.P. et al. (1993). Application of a floating well concept to a latchup-free low-cost smart power high-side switch technology. *IEEE Transactions on Electron Devices* ED-40 (7): 1340–1342.

21 Peppiette, R. (1994). A new protection technique for ground recirculation parasitics in monolithic power IC's. *Sanken Technical Report* 26 (1, 1994): 91–97.

22 Bafluer, M., Vidal, M.P., Buxo, J. et al. (1995). Cost-effective smart power CMOS/DMOS technology: design methodology for latchup immunity. *Analog Integrated Circuits and Signal Processing* 8 (3): 219–231.

23 Chan, W.W.T., Sin, J.K.O., and Wong, S.S. (1998). A novel crosstalk isolation structure for bulk CMOS power IC's. *IEEE Transactions on Electron Devices* ED-45 (7): 1580–1586.

24 Winkler, W. and Herzl, F. (2000). Active substrate noise suppression in mixed-signal circuits using on-chip driven guard rings. In: *Proceedings of the IEEE 2000 Custom Integrated Circuits Conference (CICC), May 2000*, 356–360. IEEE.

25 Gonnard, O. and Charitat, G. (2000). Substrate current protection in smart power IC's. In: *Proceedings of the International Symposium on Power Semiconductor Devices (ISPSD), 2000*, 169–172. IEEE.

26 Zhu, R., Parthasarathy, V., Khemka, V., and Bose, A. (2001). Implementation of high-side, high-voltage RESURF LDMOS in a sub-half micron smart power technology. In: *Proceedings of the International Symposium on Power Semiconductor Devices (ISPSD), 2001*, 403–406. IEEE.

27 Gonnard, O., Charitat, G., Lance, P. et al. (2001). Multi-ring active analogic protection (MAAP) for minority carrier injection suppression in smart power IC's. In: *Proceedings of the International Symposium on Power Semiconductor Devices (ISPSD), 2001*, 351–354. IEEE.

28 Schenkel, M.P., Pfaffli, P., Wilkening, W. et al. (2001). Transient minority carrier collection from substrate in smart power design. In: *Proceedings of the European Solid State Device Research Conference (ESSDERC), 2001*, 411–414. IEEE.

29 Parthasarathy, V., Zhu, R., Khemka, V. et al. (2002). A 0.25-μm CMOS based 70V smart power technology with deep trench for high-voltage isolation. In: *International Electron Device Meeting (IEDM) Technical Digest, 2000*, 459–462. IEEE.

30 Parthasarathy, V., Khemka, V., Zhu, R. et al. (2004). A multi-trench analog + logic protection (M-TRAP) for substrate crosstalk prevention in a 0.25-μm smart power platform with 100V high-side capability. In: *Proceedings of the International Symposium on Power Semiconductor Devices (ISPSD), 2004*, 427–430. IEEE.

31 Laine, J.P., Gonnard, O., Charitat, G. et al. (2003). Active pull-down protection for full substrate current isolation in smart power IC's. In: *Proceedings of the International Symposium on Power Semiconductor Devices (ISPSD), 2003*, 273–276. IEEE.

32 Khemka, V., Parthasarathy, V., Zhu, R. et al. (2003). Trade-off between high-side capability and substrate minority carrier injection in deep sub-micron smart power technologies. In: *Proceedings of the International Symposium on Power Semiconductor Devices (ISPSD), 2003*, 241–244. IEEE.

33 Horn, W. (2003) *On the Reverse-Current Problem in Integrated Smart Power Circuits.* Ph.D. Thesis, Technical University of Graz, Austria, April 2003.

34 Voldman S., Johnson R.A., Lanzerotti L.D., and St Onge, S.A. (2003) Deep trench-buried layer array and integrated device structures for noise isolation and latch up immunity. U.S. Patent No. 6,600,199, issued July 19th , 2003.

35 Gupta, S., Kosier, S.L. and Beckman, J.C. (2004) Guard ring structure for reducing crosstalk and lath-up in integrated circuits. U.S. Patent No. 6,747,294, issued June 8th 2004.

36 Stella, R., Favilla, S., and Croce, G. (2004). Novel achievements in the understanding and suppression of parasitic minority carrier currents in p-epitaxial/p++ substrate smart power technologies. In: *Proceedings of the International Symposium on Power Semiconductor Devices (ISPSD), 2004*, 423–426.

37 Singh, R. and Voldman, S., (2004) Method and apparatus for providing ESD protection and/or noise suppression in an integrated circuit. U.S. Patent No. 6,826,025, issued November 30th, 2004.

38 Singh, R. and Voldman, S. (2006) Method and apparatus for providing noise suppression in an integrated circuit. U.S. Patent No. 7,020,857, issued March 28th, 2006.

39 Zhur, R., Khemka, V., Bose, A., and Roggenbauer, T. (2006). Substrate majority carrier induced LDMOS failure and its prevention in advanced smart power technologies. In: *Proceedings of the International Reliability Physics Symposium (IRPS), 2006*, 356–359.

40 Redmond, C. (2001). Winning the battle against latchup in CMOS analog switches. *Analog Devices: Analogue Dialogue* 35 (5): 1–20.

41 Lin, S.T., Yu, T.L., and Peng, Y.C. (2002) Latch-up protection circuit for integrated circuits biased with multiple power supplies and its method. U.S. Patent No. 6,473,282, issued October 29th, 2002.

42 Hsu, C.T., Lin, L.C., Tseng, J.C. et al. (2006). A pin latchup failure and the latch-up trigger current induced npn snapback effect in a high-voltage IC product. In: *Proceedings of the Taiwan Electrostatic Discharge Conference (T-ESDC)*, 53–57. IEEE.

43 Salcedo-Suner, J.R., Cline, R., Duvvury, C. et al. (2003). A new I/O signal latch-up phenomenon in voltage tolerant ESD protection circuits. In: *Proceedings of the International Reliability Physics Symposium (IRPS)*, 85–91. IEEE.

44 Huh, Y., Min, K., Bendix, P. et al. (2005). Chip level layout and bias considerations for preventing neighboring I/O cell interaction-induced latchup and inter-power supply latchup in advanced CMOS technologies. In: *Proceedings of the Electrical Overstress/Electrostatic Discharge (EOS/ESD) Symposium, 2005*, 100–107. IEEE.

45 Huh, Y., Min, K., Bendix, P. et al. (2005). Inter-circuit and inter-power supply CMOS latchup. In: *Proceedings of the Taiwan Electrostatic Discharge Conference (T-ESDC)*, 32–36. IEEE.

46 Weger, A., Voldman, S., Stellari, F. et al. (2003). A transmission line pulse (TLP) pico-second imaging circuit analysis (PICA) methodology for evaluation of ESD and latchup. In: *Proceedings of the International Reliability Physics Symposium (IRPS)*, 99–104. IEEE.

47 Brennan, C.J., Chatty, K., Sloan, J. et al. (2005). Design automation to suppress cable discharge events (CDE) induced latchup in 90 nm ASICs. In: *Proceedings of the Electrical Overstress/Electrostatic Discharge (EOS/ESD) Symposium*, 126–130. IEEE.

48 Voldman, S. (2004). *ESD: Physics and Devices*. Chichester: Wiley.

49 Voldman, S. (2005). *ESD: Circuits and Devices*. Chichester: Wiley.

50 Voldman, S. (2006). *ESD: RF Technology and Circuits*. Chichester: Wiley.

51 Voldman, S. (2002) BiCMOS ESD circuit with sub-collector/trench-isolated body MOSFET for mixed signal analog/digital RF applications. U.S. Patent No. 6,455,902, issued September 24th, 2002.

52 Pequignot J. Sloan, J.H., Stout, D.W. and Voldman, S. (2005) Electrostatic discharge protection networks for triple well semiconductor devices. U.S. Patent No. 6,891,207, issued May 10th 2005.

53 Lundberg, J. (1990) Low voltage CMOS output buffer. U.S. Patent No. 4,963,766 issued October 16th, 1990.

54 Adams, R.D., Flaker, R.C., Gray, K.S. and Kalter, H.L. (1988) CMOS off-chip driver circuits, U.S. Patent No. 4,782,250, Nov. 1st, 1988.

55 Adams, R.D., Flaker, R.C., Gray, K.S., and Kalter, H.L. (1988). An 11ns 8K x 18 CMOS static RAM. In: *Proceedings of the International Solid State Circuits Conference (ISSCC)*, 242–243. IEEE.

56 Austin, J.S., Piro, R.A., and Stout, D.W. (1992) CMOS off chip driver circuit, U.S. Patent No. 5,151,619, issued September 29th, 1992.

57 Hoffman J., Jallice D., Puri, Y., and Richards, R. (1992) CMOS off chip driver for fault tolerant cold sparing, U.S. Patent No. 5,117,129, issued May 26th, 1992.

58 Dobberpuhl, D.W. (1992) Floating-well CMOS output driver. U.S. Patent No. 5,160,855, issued Nov. 3rd, 1992.

59 Voldman, S. (1994). ESD protection in a mixed voltage interface and multi-rail disconnected power grid environment in 0.5- and 0.25-μm channel length CMOS technologies. In: *Proceedings of the Electrical Overstress/Electrostatic Discharge (EOS/ESD) Symposium, 1994*, 125–134. IEEE.

60 Voldman, S. (1999) Electrostatic discharge protection circuits for mixed voltage interface and multi-rail disconnected power grid applications. U.S. Patent No. 5,945,713, issued August 1, 1999.

61 Voldman, S. (1997) Power sequence-independent electrostatic discharge protection circuits. U.S. Patent No. 5,610,791, issued March 11th, 1997.

10

Emerging Effects and Future Technology

10.1 Introduction

In the future, as technologies and applications change, the issue of radiation effects will change. In this chapter, the text will close on discussions associated with radiation effects, technology and system scaling, nano-technology issues, technology lifetime and reliability, future terrestrial and space issues, and 2.5-D and 3-D components and systems [1–5].

10.2 Radiation Effects in Advanced Technologies

Radiation effects in advanced technologies will change with the device types, circuits, and systems evolve. With future technology, the scale of the devices and circuits will decrease as the radiation particles and issues remain the same. Future devices will change in geometry, orientation, and size. This will influence the circuitry sensitivity and failure rates. As the circuits decrease in physical size, particles that did not impact the circuits in the past may lead to failure in future devices. Particles such as alpha particles, muons, and pions may begin to influence circuits that were in the past immune to the events [1, 2].

Space environments contain high-energy cosmic radiation, which is destructive to microelectronics. Terrestrial environments, however, are shielded by Earth's atmosphere, and radiation at sea level is minimal. Commercial off-the-shelf (COTS) devices are widely used in terrestrial electronic systems (computers, communication networks, mobile devices, and electronic control systems). COTS have no reason to be hardened to radiation since Earth's atmosphere (depth) shields it from cosmic radiation. However, neutrons from the cosmic background are still present at a flux near $14 \, \text{n/h cm}^2$ at sea level.

At commercial flight levels, the neutron flux increases by less than three orders of magnitude. The probability of a random strike by a neutron to upset a memory bit (single-event upset [SEU] event) is very low, so radiation isn't highlighted in device specifications. However, dynamic read access memories (DRAMs) suppliers first investigated the effect of terrestrial neutron upsets caused by packaging materials emitting trace amount of alpha particles with spallation releasing both neutrons and protons in the 1980s. As DRAMs bit density increased, an increase in soft bit errors were noted and the use of error correction code (ECC) was implemented to correct these soft bit errors.

Integrated Circuit Design for Radiation Environments, First Edition.
Stephen J. Gaul, Nicolaas van Vonno, Steven H. Voldman and Wesley H. Morris.
© 2020 John Wiley & Sons Ltd. Published 2020 by John Wiley & Sons Ltd.

10.2.1 Moore's Law, Scaling, and Radiation Effects

Moore's law (scaling of the CMOS dimensions) increased chip bit/logic density and led to scaling the supply voltage, reducing power consumption. The negative impact is that technology scaling made CMOS devices more sensitive to radiation at decreasing energies. The reliability trend of CMOS scaling was that as feature size reduced, sensitivity to radiation increased. Scaling of voltage, which is good for power reduction and reliability, led to reduction of the bit stray capacitance; this reduced the magnitude of critical charge (Q_{crit}) needed to upset the single memory bit state or a flip a logic state. As a result, the DRAM terrestrial SEU rates went up.

Technology evolution will continue to scale, with the desire for higher performance at a lower cost. The trend in the past follows the Moore's law scaling relationship. With the desire to scale the devices, the circuit elements will be smaller in the future. Smaller devices lead to less charge collection within a given circuit. Additionally, there will be an increase in multiple bit failures, circuit-to-circuit interactions. With system-on-chip (SOC) applications, there will be a higher rate of interaction of events between multiple functions. This will include digital-to-analog failures and core-to-core SEUs. Single-event latchup (SEL) will change with the increase in "latchup domino" effect, where multiple circuits are influenced by a single particle [3, 4]. As the proximity of circuits are increased, there is a higher likelihood of latchup propagating through multiple circuits. Additionally, there may be a higher level of interaction between SEU and SEL. In the future, as technologies and applications change, the issue of radiation effects will change.

In future technologies, SEL will continue to be an issue, but of a lower magnitude and probability. However, the likelihood of single-event transients (SETs) will increase in importance as power supplies are reduced in the core of semiconductor chips.

10.2.2 Technology Lifetime and Reliability

In future technologies, there will be new issues associated with technology lifetime and reliability. There are new questions about technology in space environments:

- *Mission life*. How long do you want a mission to last in space?
- *Radiation hardening time length requirement*. How rad hard does a system need to be to survive? Days, weeks, or months?
- *New missions*. What will the radiation hardening requirement be required for missions to the moon, Mars, and other distant missions?
- *Radiation hardening requirements for extreme environment missions*. Are there places in space where we will not be able to fulfill the radiation requirements, limiting our ability to go to those locations?
- *Radiation hardening requirements for nuclear radiation accidents*. Will new requirements be required for nuclear accidents?
- *Commercialization*. Will the increase in access to space (commercialization) cause a throw-away mentality?
- *New space entrants*. Will there be learning curves for new entrants to the space age that create more junk?
- *Impact of space junk*. Is there a point where there will be too much junk, limiting further usage of some orbits? Making things dangerous for humans? Is it a problem already today?

10.2.2.1 New Missions

Today, the mission to Jupiter is considered a mission that cannot fly as the satellites must be at least 1 Mrad to survive two to three years. Additionally, the mission to Venus is also limited by both radiation and high temperatures.

10.2.2.2 Throwaway Mentality

Will the increase in access to space lead to a throwaway mentality? There seems to be a mentality, but there is not many choices. At this time, small satellites are being designed for "short lifetime roads" where the launch cost is the main cost overhead.

10.2.2.3 New Space Entrants

Will there be new space entrants and more junk? Already, things are in less severe environments, and it is very dense.

10.2.3 Terrestrial Issues

In the future, radiation effects in terrestrial environments will change with scaling. With the scaling of technology, circuits that were not sensitive to some single events will become sensitive. Particles such as muons, pions, and other particles that once did not influence circuits will begin to have visible failure rates [1, 2]. In the past, this occurred with alpha particles in both DRAM and static random access memory (SRAM) products in both bipolar and CMOS technology. In the future, with scaling, even more particles will have visible SEU failure rates. Another particle is that atmospheric thermal neutrons will have a higher failure rate in the future. Will there be a background upset level that will be "unacceptable" for operation?

In the future, dense circuit applications that were not designed in radiation hardened technology may be required. Will we eventually be designing radiation hardened circuitry by virtue of the process technology node? Additionally, with the increase of server farms, system failure will be more visible, where in the past with a single server it was not observed.

10.2.4 Space Mission Issues

With scaling of technology in the future, there are many questions on space missions that are unanswered today. A question on mission lifetime will be a greater issue in the future. There will be a tradeoff of radiation hardness requirements, mission lifetime, and cost. There will be missions that are not suitable due to heavy ion events, with a high probability of failure of the components and systems in future systems. Even today, the survivability of technology from radiation events are limiting the lifetime of some missions.

10.2.5 Server Farms

With the increase of server farms, system failure will be more visible, where in the past with a single server, it was not observed. As the number of servers in a given application or function increase, there will be a visible failure rate. Will components in servers be chosen based on their SEU failure rates? Will servers require a higher level of ECC than in the past? Will there be a performance bottleneck with the addition of ECC?

The same effects are true in terrestrial applications with a lower rate of accumulation. Hence, power, performance, and duty cycle can be achieved with multiple scrub engines in SRAM used in the I- and D-cache.

10.2.6 Automotive

Today, and in the future, electronics will increase significantly as the industry move toward "smart cars." Today, the electronics will include the electronic system, power train, infotainment, networks, chassis, safety and control, to comfort and control. The power train will include engine control for engine, fuel injection, knock control, and transmission. The chassis electronics will include the steering, brakes, traction control, and suspension. The safety control may include airbags, tire pressure monitoring systems, collision warning, parking assistant, back monitor, and night vision. The comfort control will include power doors, power windows, climate control, seat controls, and mirror and wiper controls. The electronic system will include the alternator, battery, starter, lighting, diagnostics, to in-car data bus. Analog and digital circuits will be important for these applications.

10.2.7 Internet of Things (IoT)

In the future, the internet of things (IoT) and new applications will change. Semiconductor foundries predict that IoT as a giant market that will hit greater than USD 1 trillion per year in 2025. And, as the AI and IoT connectivity goes global, foundries sees society in the future as everything is smart (e.g. home, phone, car, factory, grid, multiple internets, and future applications). The increase of handheld applications will require new evaluation of modeling and simulation. For example, wearable electronics may require new modeling and simulation methods due to the radiation shielding of a human on the electronics by combining the person and the device being evaluated. New questions will arise in the failure rate of handheld devices, laptops, and portable electronics.

In IoT, reliability is a main concern as IoT markets will take CMOS technology outdoors more than before. Historically, applications remained inside but with IoT there will be an increase of usage of electronics in outside environments. On Earth, there will be a greater concern of radiation effects from terrestrial neutrons. Analog will be important for IoT applications in the future.

10.2.8 More than Moore

In the semiconductor industry, there is a desire to extend the Moore's law plot into the future. The desire was to have it to extend beyond what was anticipated based on scaling theory. Today, we are interested in "more than Moore" for the shrinking of devices.

10.3 Radiation Effects in Semiconductor Nanostructures

In future nanostructures, radiation influences will change as a result of the new technologies, device size, geometrical changes, as well as the density of structures. These devices will include the following:

- Scaled planar MOSFETs
- Bulk FinFETs
- Silicon on insulator (SOI) FinFETs [5]
- 3-D circuits

10.3.1 Planar MOSFETs in Sub-25 nm

With technology scaling, bulk planar MOSFET channel length and width will be reduced. Some advanced technologies remain as planar MOSFETs, prior to migration to a FinFET layout. In the future, the proximity to adjacent planar transistors will be decreased. This will lead to multiple-bit upsets (MBUs). Injection into multiple circuit elements can occur in the same circuit or multiple adjacent circuits.

CMOS circuits are directly controlled by signals, clocks, and voltage. However, the parasitic silicon-controlled rectifier (SCR) circuit network is not directly controlled. And if a signal overvoltage (timing), radiation particle strike, or increase in temperature causes any diode to forward bias, the SCR parasitic is immediately activated and minority carriers (hole current) is injection by the pnp emitter (anode) to source current to the VSS collector (cathode) terminal that causes a ΔV shift in the p-well (*local potential*) which is forward biased n^+/p^- diode; at that point the parasitic network fully triggers into latchup. Depending on supply voltage, the latchup saturation may or may not reach the saturation stage and destructive failure. However, a transient trigger event *always causes* digital data upsets and logic failure, that can be fatal to the system.

Particles striking the bulk CMOS devices have measurably differential transient currents based on individual diffusion layouts to well diffusions and the SEU striking angle is varied from orthogonal (0 angle) to high angle strikes. As can be seen in simulation, the currents generated in the bulk devices increases as the angle increases, making the SEU upset more disruptive and longer lasting. So one mitigation would be to reduce this effect as much as possible.

Low-angle incident particles can traverse many device elements leading to different failure events. However, all commercial CMOS devices are manufactured with bulk silicon in which electrical isolation depends on reverse-biased diodes and as such parasitic SCR can cause electrical shorting between VSS and VDD terminals if the parasitic SCR is triggered, which then causes disruptive electrical transient upsets of core logic by forward biasing diodes and the catastrophic destruction of I/O devices that operate at higher voltage. A key issue in sub-25 nm technology is that the I/O circuitry is still at a higher power supply voltage, making them vulnerable to SEL.

SEL is a persistent reliability risk for commercial bulk sub-25 nm CMOS devices operating in radiation environments and can't be eliminated by the circuit design, but can be mitigated by circuit latchup layout rules. However, commercial circuits are laid out at minimum n^+/p^+ spacing and are sensitive to latchup and microtransient upset even at particle cross sections <15 MeV-cm^2.

Shielding is not effective for single-event effects (SEE) unless high Z material is used for shielding materials, and even then, with limited success. Any CMOS device that can latchup when struck by a radiation particle is generally considered "unsuitable" or "at risk" to mission lifetime reliability in low Earth orbits (LEOs) due to solar flares. For high LEOs or orbits outside of LEO, the risk of SEL is a serious reliability risk, and as such, mitigation at the system level is required.

10.3.2 Bulk FinFET

In foundries, semiconductor technologies have migrated to bulk FinFET devices (Figure 10.1) FinFETs consist of a plurality of Fins electrically connected into a single MOSFET circuit element. With FinFET devices, the incident radiation particle will traverse many of the fins in a given device. With the new vertical geometry, the distribution of the charge in the device will be significantly different from the planar MOSFET. In a bulk FinFET, electron–hole pair (EHP) within the fin and the substrate have the possibility to reach the sensitive junctions.

In sub-25 nm technology, latchup is still a concern in bulk FinFETs. The application of these structures in space environments will still be vulnerable to CMOS SEL. Transient upsets, micro-latchup, and SEL will be issues. Although the core power supply voltage is decreasing, the I/O voltage remains higher. Destructive latchup in the core circuitry will not occur in sub-40 nm technology, but transient effects and micro-latchup effects will be worse.

For FinFETS, the core supply voltage is scaled to 1 V (or less), which eliminates the risk for destructive latchup, since the parasitic pnp bipolar current drive strength cannot drive the parasitic pnpn parasitic network into the saturation level. However, the parasitic SCR can still trigger into a micro-latch event (nondestructive transient latch) and will be destructive to the chip digital (data) state. This type of upset interrupts circuit functionality and persists for several nanoseconds, affected subcircuits that remain off-line until the parasitic diodes recover and supply voltage is restated.

However, although the core voltage is decreasing, higher supply voltages (e.g. 1.8 V or 3.3 V) are still required in I/O interface circuits and some analog/power management/radio frequency (RF) circuits in sub-25 nm CMOS technologies. As FinFETs are scaled, the shallow trench isolation (STI) depth is reduced in the core as well as the

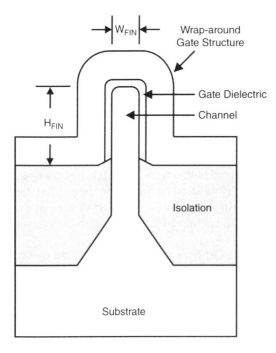

Figure 10.1 Bulk CMOS FinFET structure.

Table 10.1 npn bipolar current gain comparison of planar vs FinFET.

Collector current (mA)	Bipolar current gain npn planar	Bipolar current gain npn bulk FinFET
0.1	0.12	0.55
0.2	0.10	0.35
0.8	0.08	0.25

Table 10.2 pnp bipolar current gain comparison of planar vs. FinFET.

Collector current (mA)	Bipolar current gain npn planar	Bipolar current gain npn bulk FinFET
1	2.0	4.5
4	0.8	1.8
6	0.1	0.8

I/O circuitry. The reduction of the STI depth leads to an increase in the lateral bipolar current gain of the parasitic devices in all circuitry.

A study of latchup in the I/O interface circuits in a bulk FinFET technology was first reported by Dai et al. [6]. The small fin structure seems to be impact latchup immunity due to the increased parasitic resistance and the reduced guard ring efficiency. In addition, a narrower fin pitch with a reduced STI depth can also impact the latchup sensitivity. Dai et al. [6] investigated the characteristics of the pnp and npn bipolar current gain in a bulk FinFET and planar technology. The test structure was fabricated in both bulk FinFET and planar technologies for comparison. A bulk FinFET technology with a target fin width and pitch of 10 and 45 nm was used. The fin height is 30 nm with S/D silicon epitaxial growth without merging fins. The advanced local interconnect (LI) process with a pitch of 110 nm was integrated in both FinFET and planar technologies. Tables 10.1 and 10.2 show a comparison of the bipolar current gain for the npn, and pnp, respectively.

In Dai et al. [6], the equivalent active region of the planar CMOS structure is ~4.5× wider due to the 10-nm fin width with 45-nm fin pitch; but the STI depth is ~3.6× deeper. From the test results, the parasitic vertical resistance is mainly influenced by the STI depth, not the fin-confined active region in the FinFET technology. Due to the smaller vertical resistance, the bipolar current gains of the FinFET bipolar junction transistors (BJTs) are higher than those in the planar ones. These results demonstrate that FinFET devices are more susceptible to latchup.

10.3.3 SOI FinFET

For SOI FinFETs, the EHP generation in the bulk substrate does not reach the sensitive nodes of the device. As a result, particles at high angle of incident do not lead to single-event failures. Particles at low angle of incidence can generate EHP in the fin structure above the buried oxide (BOX) region. Cosmic rays and heavy ions that initiate low-angle incident secondary tracks can still influence the devices. The design layout, fin-to-fin spacing, and physical dimensions can influence the collected charge [5].

SOI has been considered the only silicon solution for hardening CMOS circuits to radiation or enabling latchup immunity for fail-safe operation. However, using substrate engineering techniques, structures such the buried guard ring have proven capable of eliminating latchup even when exposed to both radiation and high-temperature environments for both I/O and core circuits. Such techniques can be added to the commercial process to enable hardening of the CMOS device at high and low voltages that can be manufactured using bulk Czochralski (CZ) silicon wafers, scaled to any generation and manufactured at commercial silicon foundries at low cost.

10.3.4 3-D Circuits

In advanced technologies, three-dimensional circuits (also known as 3-D circuits) provide layers of circuit elements within a given chip. Some three-dimensional circuits are distributed in a CUBE, or a stack. These 3-D circuits may be complex to evaluate from SEUs for testing and simulation. For testing, the angle of incidence and orientation may influence the SEU failure rate for these 3-D devices and circuits.

For simulation, SEU simulation will have to account for correlated strikes to different layers of a stacked die or multilevel circuit. The 3-D devices and circuits open new questions on how to design and layout 3-D structures to minimize or mitigate SEU events. The design of the stack must be addressed to avoid correlated effects. There are methods to avoid correlated events, which are observed through both 2-D and 3-D simulation. Most of the 3-D simulation are done with profile statements. Mitigation of various SEE behavior can be done with 2-D simulation.

10.4 Radiation Effects and Advanced Packaging

In future systems, radiation effects in advanced packages can lead to different failure rates, and interactions. Having the die stacked, or adjacent, can lead to different interaction and failure rates. Two directions that the industry has taken are as follows:

- 2.5-D package
- 3-D package

Recent projections from large semiconductor foundries is the push into wafer scale integration and chip-to-chip integration. Packaging will have increased utilization to "integrate more devices and transistors" into the package.

The new path to the future is in wafer level integration combined with vertical integration to make 3-D devices which will be "packaged" together. It is believed at this time that this will be the focus due to scaling to 5 nm technology will be difficult due to heating and voltage scaling.

Figure 10.2 2.5-D multi-chip assembly.

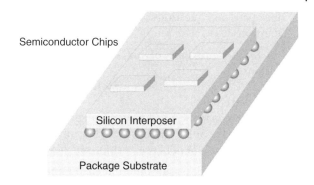

Semiconductor Chips

Silicon Interposer

Package Substrate

Figure 10.3 3-D multi-chip structure.

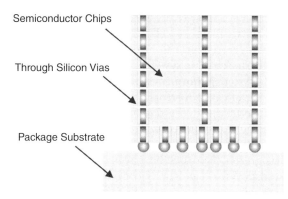

Semiconductor Chips

Through Silicon Vias

Package Substrate

10.4.1   Radiation Effects and 2.5-D Circuits and Technology

In 2.5-D designs, semiconductor chips are placed on top of another and interconnected with solder balls or wirebonds (Figure 10.2). For different ionized particles, the incident particle can penetrate through both elements in a 2.5-D assembly. The failure rate is a function of the charge track generation in the different parts of the chip. Consideration of placement of sensitive circuit of the first and second chip may be valuable to evaluate the implications. For example, sensitive memory may be in the top chip, and insensitive logic may be placed in the second lower chip.

Additionally, it may be advantageous to have multiple chips for some particle events, since the upper chip may help "shield" the lower chip. As a result, consideration of the issue of ionized radiation going through multiple chips should be evaluated.

10.4.2   Radiation Effects and 3-D Circuits and Technology

For 3-D technology, semiconductor chips are stacked on top of each other (Figure 10.3). Through silicon via (TSV) structures are used to interconnect the adjacent chips. In these 3-D systems, the memory chips can be stacked on top of the logic chip. The issue for advanced technology and radiation effects is that a single particle can penetrate multiple chips in the stack.

This opens new questions about design and architecture in advanced chip integration. Questions exist, such as:

- *Proximity.* Is proximity of each circuit function in the stack a more important consideration?
- *Correlated events.* Are there ways to avoid some correlated events in stacked die or multilevel circuits?
- *Angle of incidence.* Considering the angle of incidence, do we need to rethink the idea of a correlated SEE event in stacked or multilevel circuits?
- *Error correction.* Will new algorithms be needed for multistacked chips?

Proximity of each circuit function is an important consideration, where diffusion of minority carriers to adjacent diffusions or circuits. This can be addressed by the introduction of doping layers, moats, and guard rings. Doping layers, moats, and guard rings can collect EHPs, as well as lead to a higher recombination.

Error correction can also be introduced to reduce the impact of single events in multi-stacked chip systems. Error correction "scrubbing" can be introduced with ECC, but a system performance can be impacted. There is a SEU "over-time build-up," which is a bigger cause of errors, which leads one to the conclusion that the system must have ECC scrubbing in the chip or memory core, since it requires too much time and power to go outside of the core.

For large memory, the risk of a MBU is greater than unity if the memory over time has greater than 1000 bit errors (to an uncorrected word). This is known as "birthday statistics" for "big data." Hence, the only way to ensure MBU failure is to get less than 1000 bit errors prior to the ECC fixes the addresses.

Although there are many uncertainties in the future, what is clear is that radiation hardening and failure rates will be worse in future. With scaling and heating, the sensitivity will continue to increase. With heating, the trigger voltages for latchup will decrease as devices increase from ambient temperature to 85–125 °C, leading to higher latchup sensitivity.

10.4.3 More than Moore and 3-D Integration

As the technologies extend into the future, the industry refers to as "more than Moore." Power consumption reduction will drive future design of technologies and architectures, that require less energy hungry devices and interconnect systems. The electronic market will be able to face an exponential growth thanks to the availability and feasibility of autonomous and mobile systems necessary to societal needs.

The increasing complexity of high-volume fabricated systems will be possible if we aim at zero intrinsic variability, and generalize 3-D integration of hybrid, and heterogeneous technologies at the device, circuit, and system levels.

10.5 Ruggedized Capability

In the present and the future, "ruggedness" will be a metric of interest as electronics are used in different mobile environments. Today, two issues will be ruggedness for radiation and temperature. Additionally, systems that are waterproof will be of high value in future mobile systems. Hence, it is desirable to have "ruggedized silicon devices."

10.5.1   Ruggedized Capability for Radiation

Today, there is a need for rugged electronics for terrestrial and space applications. These include the following applications:

- Medical digital logic
- Medical analog devices
- Satellites in all Earth orbits (e.g. LEO, MEO, GEO, equatorial, and polar)
  So "ruggedized" then has a wide range of applications topics:
- Ruggedized to radiation (SEE and TID)
  - LEO (almost no radiation and *the* reason why people confuse this)
  - MEO (very bad TID, SEE)
  - GEO bad SEE, higher-energy particles latchup worse
  - LEO (easy, use commercial devices, works, short mission lifetimes make expensive)
  - LEO disadvantages (too low, short fly-by, low radio Horizon, so your need lots of satellites)

10.5.2   Ruggedized Capability for High Temperature

In the future, electronics will need to be both suitable for radiation environments and high-temperature operation. With high-temperature operation, it influences the semiconductor device, making them more vulnerable to radiation effects. High-temperature operation applications can include the following:

- IoT industrial
- Auto-grade 0 drive train
- Smart grids
- Oil and gas drills
- Electric motor controls
- Satellites
- Aviation jet engines and controls
- Defense

10.6   Radiation Models

In the future, new space radiation simulation tools will be updated to include new models and features. The Space Environment Information System (SPENVIS) is a new tool supporting new radiation environment models. For example, recent added features of SPENVIS are:

- Slot Region Radiation Environment Models (SRREMs)
- Solar Accumulated and Peak Proton and Heavy Ion Radiation Environment (SAPPHIRE) models
- Coupling SPENVIS-4 and SPENVIS-NG for running the latest versions of GRAS, MULASSIS, and IRENE

  Other upgrades include:

- IRENE (AE-9/AP-9) trapped proton and electron model version 1.5
- Geant4 based tools GRAS (4.0) and MULASSIS (1.26)

- AZUR 3G30 cells in MC-SCREAM
- Increased maximum upload size for the trajectory file

Note that the default parameters of some model input pages have been changed, which could produce different results. It is therefore strongly recommended to rerun existing model runs for single-event phenomena (SEP) fluences and fluxes and the downstream models that use their outputs.

10.7 A Nuclear World

As we enter a nuclear world, with both safety and accidents, different scenarios have occurred. In recent times, some of the intense radiation environments, such as Fukushima, have prevented measurement of the radiation level, since the electronics cannot survive during the measurement process. This has been true for Fukushima, as well as Chernobyl and Three Mile Island.

Even with radiation hardened electronics, the monitoring, repair, and cleanup are hampered by the survivability of the electronics. In the future, there will be a need for electronics that can survive these radiation environments for safe repair and cleanup. Electronics that can survive 4 Mrad will be required for these environments.

Another issue is a nuclear detonation. A High-Altitude Nuclear Detonation (HAND) can impact electronics in a high LEO, increasing the total ionizing dose (TID) in the LEO satellites. A LEO around the Earth is an orbit altitude of 1200 mi or less. Satellites placed in this orbit will experience rapid orbital decay and altitude loss due to atmospheric drag.

10.8 Summary and Closing Comments

Chapter 10 discusses issues that are anticipated in future nanostructures, and systems. In the chapter, there is a focus on both terrestrial and space applications of these future devices. In advanced technologies, there will be both dimensional, and orientation differences in comparison to older technology generations. This will influence modeling, simulation, and anticipated failure rates.

Problems

10.1 How will technology scaling have an influence on future space missions? Will we be able to have longer space missions with scaled technology?

10.2 At what point will technology scaling not be possible due to cosmic ray influence in terrestrial applications?

10.3 Provide cases of cosmic ray influence on terrestrial chip applications. What circuit types are sensitive today, and what will be sensitive in the future?

10.4 What are solutions for server farms to avoid SEU events? What are the limitations on performance?

10.5   How would one design a three-dimensional circuit to minimize charge collection during an SEU strike?

10.6   How would one design a stack of die to avoid collective effects?

10.7   Are there ways to avoid correlated events in a stack die or multilevel circuit?

10.8   What role will SEE angle of incidence play a role in stack or multilevel circuits?

10.9   What is the role of proximity of circuit function in a 2.5-D chip system?

10.10   Compare the response to a particle strike for a planar versus bulk FinFET. Show the electron–hole pair (EHP) generation and collection process.

10.11   Compare the response to a particle strike of a planar MOSFET in bulk CMOS versus a planar MOSFET on partially depleted silicon on insulator (PD-SOI). Show the electron–hole pair (EHP) generation and collection process.

10.12   Compare the response of a particle strike for a bulk FinFET versus an SOI FinFET. Show the electron–hole pair (EHP) and collection process.

References

1   Ziegler, J.F. and Lanford, W.A. (1979). The effect of cosmic rays on computer memories. *Science* 206: 776.
2   Ziegler, J.F. and Puchner, H. (2004). *SER-History, Trends, and Challenges*. Cypress Semiconductor Company.
3   Voldman, S. (2007). *Latchup*. Wiley.
4   Voldman, S. (2005). Latchup and the domino effect. In: *Proceedings of the International Reliability Physics Symposium (IRPS)*, 145–156. IEEE.
5   Artola, L., Hubert, G., and Schrimpf, R.D. (2013). Modeling of radiation-induced single event transients in SOI FinFETs. In: *Proceedings of the International Reliability Physics Symposium (IRPS)*, SE-1.14. IEEE.
6   Dai, C.-T., Chen, S.H., Linten, D. et al. (2017). Latchup in bulk FinFET technology. In: *IEEE Transaction of Electron Devices*, EL 1.1–EL 1.3. IEEE.

Index

Integrated Circuit Design for Radiation Environments, First Edition.
Stephen J. Gaul, Nicolaas van Vonno, Steven H. Voldman and Wesley H. Morris.
© 2020 John Wiley & Sons Ltd. Published 2020 by John Wiley & Sons Ltd.